Building Procurem[ent]

30107 005 402 042

Building Procurement

Roy Morledge
Adrian Smith
Dean T. Kashiwagi

Blackwell
Science

© 2006 by Roy Morledge, Adrian Smith and Dean T. Kashiwagi

Editorial offices:
Blackwell Publishing Ltd, 9600 Garsington Road, Oxford OX4 2DQ, UK
Tel: +44 (0)1865 776868
Blackwell Publishing Inc., 350 Main Street, Malden, MA 02148-5020, USA
Tel: +1 781 388 8250
Blackwell Publishing Asia Pty Ltd, 550 Swanston Street, Carlton, Victoria 3053, Australia
Tel: +61 (0)3 8359 1011

The right of the Author to be identified as the Author of this Work has been asserted in accordance with the Copyright, Designs and Patents Act 1988.

All rights reserved. No part of this publication may be reproduced, stored in a retrieval system, or transmitted, in any form or by any means, electronic, mechanical, photocopying, recording or otherwise, except as permitted by the UK Copyright, Designs and Patents Act 1988, without the prior permission of the publisher.

First published 2006 by Blackwell Publishing Ltd

ISBN-10: 0-632-06466-8
ISBN-13: 978-0-632-06466-3

Library of Congress Cataloging-in-Publication Data
Morledge, Roy.
Building procurement/Roy Morledge, Adrian Smith, Dean T. Kashiwagi. – 1st ed.
p. cm.
Includes bibliographical references and index.
ISBN-13: 978-0-632-06466-3 (pbk. : alk. paper)
ISBN-10: 0-632-06466-8 (pbk. : alk. paper)
1. Construction industry–Great Britain. 2. Construction industry–United States. 3. Construction industry–China. 4. Industrial procurement–Great Britain. 5. Industrial procurement–United States. 6. Industrial procurement–China. I. Smith, Adrian. II. Kashiwagi, Dean T. III. Title.

HD9715.G72M667 2006
658.15′242–dc22
2005028884

A catalogue record for this title is available from the British Library

Set in 10/12.5pt Minion
by Graphicraft Ltd, Hong Kong
Printed and bound in India
by Replika Press Pvt Ltd, Kundli

The publisher's policy is to use permanent paper from mills that operate a sustainable forestry policy, and which has been manufactured from pulp processed using acid-free and elementary chlorine-free practices. Furthermore, the publisher ensures that the text paper and cover board used have met acceptable environmental accreditation standards.

For further information on Blackwell Publishing, visit our website:
www.blackwellpublishing.com

Contents

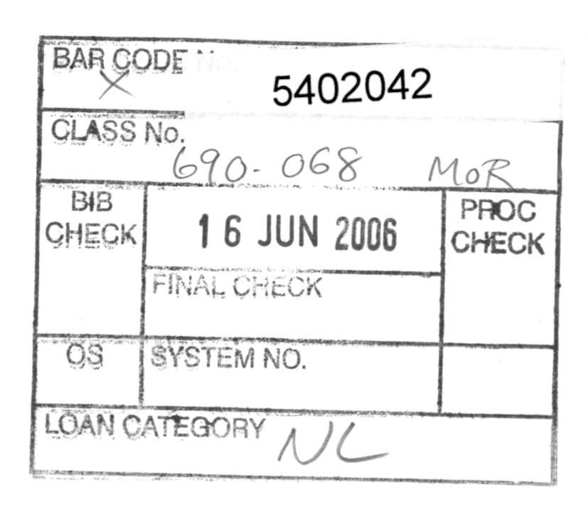

BAR CODE No. 5402042
CLASS No. 690.068 MOR
BIB CHECK
16 JUN 2006
PROC CHECK
FINAL CHECK
OS
SYSTEM NO.
LOAN CATEGORY NL

Foreword

The scale and complexity of construction projects is ever increasing and with advances in technology, possibilities in the design and construction of buildings become increasingly wide. As knowledge of what is technically feasible becomes increasingly known, customers expect an ever greater performance from their buildings. We have got used to this with cars and computers, where yesterday's ground-breaking model is taken for granted today and will be thrown away tomorrow. But, to ensure the successful delivery of buildings, the sophistication of the design needs to be matched by the processes that are used to procure the buildings. By bringing together in one volume the latest thinking on effective building procurement, this book makes a real contribution to ensuring that the construction industry can actually deliver the product that the customer wants and expects – what can be more important than that, after all?

Stephen Brown, *Head of Research, Royal Institution of Chartered Surveyors*

Preface

The procurement of construction work is complex, and a successful outcome frequently elusive. It is remarkably different from purchasing almost any other commodity associated with commercial activity, recreation, domestic activity or travel. Most products are the result of a manufacturing process based in an environment that is permanent and controlled, where the product can be designed, prototyped and tested prior to final production, where quality can be readily assured, and where potential purchasers can view and even try the product before making their final decision to buy.

Projects, and particularly construction projects, generally follow a process that has few of the above characteristics and which can provide none of the benefits.

Experience over the past few decades has shown that the successful completion of construction projects, both large and small, within budget, to time and to the satisfaction of customers has been by no means as common as it ought to be. Some of the issues have been comprehensively analysed over the past decade in the high-profile reports produced by Sir Michael Latham and Sir John Egan, and in enquiries held into the failure of recent major projects, of which that conducted by Lord Fraser into the development of the Scottish Parliament building in Edinburgh is perhaps the best known contemporary example. In every case the procurement phase has been identified as being crucial to eventual project success and a significant component in project failure.

This book therefore seeks to explore approaches to improving the successful procurement of buildings. The importance of understanding the need to establish realistic and measurable objectives, both for clients and contractors, linked to a carefully evaluated business case is particularly emphasised.

It is worthy of note that purchasers of construction projects are usually referred to as clients rather than customers. This suggests that they are purchasing services rather than products, which is largely the case, and hints at their possible legal position if the project does not meet that requirement of fitness for purpose which is inherent in the purchase of other products. The term 'client' also suggests the difference in approach that needs to be adopted for a satisfactory outcome to be achieved.

The management framework necessary for building procurement is largely temporary (i.e. it is assembled for a specific project and dissolves away once the project is complete) and the link between client and constructor is frequently fractured. Methods adopted in manufacturing industry to assure primary functions such as communication, co-ordination and collaboration between segregated participants are difficult to establish and maintain in the construction context. Consequently focus upon individual elements tends to prevail, often to the detriment of the overall project objectives. Whilst this issue is to some extent being addressed through the development of improved supply chain management practices among the larger contractors, there is plainly still a long way to go.

This book highlights the importance of those early project stages which are so often given insufficient emphasis or time prior to focus upon concept design. The initiation of the project and the need for the development of a justified business case are explored along with the need to identify measurable outcomes against which the completed project can be judged.The *value* to the client of the project output is discussed as one primary driver which is often forgotten as direct project issues such as time and cost dominate the construction phase. Upon completion these issues become historical fact and emphasis returns to the value or performance benefit derived from the completed project.

The UK is an example of an island nation which has steadfastly hung on to those traditional procurement practices with which it is most comfortable. Data regarding the relative use of different types of procurement strategy indicate a numerical dominance of these traditional approaches with little evident sensitivity to varying client needs, although the relatively few experienced clients often do adopt different approaches. Traditionalism is sustained largely through the conservatism of professional institutions, universities and colleges in terms of the nature of professional education. This tendency towards the traditional has been frequently criticised and adjudged to deliver less than satisfactory outcomes but still dominates the construction procurement landscape. Innovative approaches requiring increased collaboration and trust tend to be resisted where a culture rooted in price and strict conformance with the contract exists.

Tools and techniques suitable for adoption in the case of most projects are explored and evaluated in this book. Some considered to be innovative and beneficial have existed for many years, but their adoption has been limited largely to those projects where experienced clients understand the benefit and ask their consultants to provide the necessary services.

Construction clients differ widely in terms of their experience, purpose and size. Numerically, inexperienced clients dominate and on average their projects are relatively small and short-lived. On the other hand the minority of regular and experienced clients are by far the biggest spenders, and they tend to benefit from their experience and buying power. Nevertheless, whether the client is an inexperienced small to medium enterprise, a major corporate organisation or a public body, exposure to risk is inherent in the procurement of construction. Perhaps the inexperienced SME is likely to be proportionally most exposed, but managing that risk is a key part of the process and the issues surrounding client risk are explored in Chapter 11.

The supply-side of the industry is dominated by small, often specialist, firms usually acting as subcontractors to contractors who generally perform a role of co-ordination and contractual management. As such the skill base is separated from the client and design team and consequently the benefits associated with collaboration and innovation tend to be limited. Innovation in construction technologies tends to result from developments by suppliers or specialists and is only adopted slowly.

Collaborative approaches to procurement have gained increasing support from experienced and regular construction clients where they can see that continuous measurable value-based improvement has been achieved by adopting such practices in preference to price-led traditional practices. The benefits of these techniques are, however, perhaps less obvious to inexperienced clients or those with infrequent demand for construction, and collaborative arrangements such as partnering have therefore tended to gain less support in these areas. Chapter 7 discusses some procurement strategies where collaboration is enabled and Chapter 12 explores the culture underlying the management of collaborative business relationships such as partnering.

Internationally, procurement practices in Europe, China and the USA are reviewed to enable some comparison to be made, not just about the practices themselves, but also about the context and attitudes of clients. In Europe, although traditional differences are still evident in the way in which construction is procured in different countries (and there may well be potential benefits for clients in some of the adopted practices, particularly the post-completion protection required in some countries), the European Union progressively imposes an increasing level of regulation in the constant search to provide a Europe-wide open market with standardised procurement procedures.

Governments are commonly adopting an approach to procurement based upon long-term agreements of various kinds with project funding provided by private sector investors. Chapters 13 and 14 explore public private partnerships and private finance and outline the processes involved, the benefits and the inherent difficulties of complex schemes where traditionally disparate participants have to collborate if successful bids and outcomes are to be achieved.

The USA has few of these pressures, and construction procurement there has developed historically along rather different lines. These issues are explored in Chapter 16, and they are evaluated against business and management principles, thus giving an unusual perspective of procurement. This issue continues in Chapter 17, where we introduce the Performance Information Procurement System (PIPS) – an innovative and unique approach to procurement developed, tried and tested by Dean Kashiwagi and his team. A number of case studies where the approach has been applied are provided in the Appendix.

China represents a fascinating instance of an economy developing at a tremendous pace, where procurement influences from many parts of the world, largely focused on geographical areas with extensive traditions of western influence such as Hong Kong and Shanghai, collide with traditional Chinese values and a centrally controlled state construction industry.

Some key issues have emerged during the writing of this book. The importance of the preconstruction stage to the value or worth of the completed project is one such issue: here an increased emphasis upon project definition and the business case will pay off. A shift away from price-led team selection to a selection based predominantly upon proven ability and capacity is another. Not that price is unimportant, but often a small increase in price has relatively little impact on total cost, but has a huge potential impact upon team attitude and commitment and thus indirectly upon the *value* of the completed project to all of the stakeholders.

Our intention in writing this book is to provide students, researchers, practitioners and those involved in the provision of construction services with a wide-ranging appreciation of the issues involved in the procurement of building projects. The book covers an exceedingly broad canvas, and we are well aware that some areas are treated somewhat superficially. This is partly deliberate in the search for readability (this is after all not a research thesis) and partly imposed by pressures of space, but we trust that the references given will enable those seeking a deeper understanding of specific issues to explore the subject in greater depth.

Finally may we jointly express our grateful thanks to all of those colleagues and friends with whom we have debated issues and discussed techniques and points of principle over the years. Many of the outcomes from these discussions have found their way into this book in one form or another.

Particular thanks to the staff and students of:

The School of Architecture, Design and the Built Environment
Nottingham Trent University
City University of Hong Kong
The College of Estate Management, Reading
Del E. Webb School of Construction, Arizona State University
Knowles Management – a division of J.R. Knowles

Roy Morledge, *Nottingham Trent University*
Adrian Smith, *College of Estate Management*
Dean Kashiwagi, *Arizona State University*

1 Introduction

The worldwide construction industry embraces the sectors of building, civil engineering and the process plant industry. It includes projects of dramatically different types, size and complexity and requires extensive professional and trade skills. Groak (1994) suggested that the construction sector was more an agglomeration of projects than a discrete industry or a fixed constellation of firms. Winch (2002) suggested that construction is essentially a service industry. He argued that what is sold to the client is not a product but a capacity to produce.

However construction is described, it is an important contributor to the national economy, and without adequate construction capacity aspirations for economic growth cannot be achieved. Economic activity, self-evidently, is the primary driver for construction activity and the two are inextricably linked.

There is, however, an inherent dysfunction between demand and supply because of the extended time period between the initiation of the project procurement process and its eventual delivery. This period of time can encompass significant changes in economic activity and this may provoke changes to the initial rationale for the construction process.

There is constant demand for the construction sector to source the physical assets necessary to live and work in modern society. The building industry produces a diverse range of outputs, ranging from products verging on mass production in the construction of houses, through bespoke service facilities such as schools and hospitals, elements of production in the form of industrial premises to house the manufacturing operations of organisations both large and small, to minor repair and maintenance work. Civil engineering on the other hand provides for many of our transportation needs in the form of roads, tunnels and bridges, railways, docks and airports, and for our energy needs in the form of pipelines and powerlines, and for the essentials of civilised life in the form of water and sewage treatment facilities, distribution and disposal systems. Indeed the capital assets of a country consist predominantly of built environment assets, and in the UK housing, infrastructure and other buildings represent 76% (£3800 billion) of a total asset base worth around £5000 billion at 2005 prices.

There is therefore no doubt that construction forms a major aspect of the economy. For example, in the UK it currently contributes approximately 8% to GDP and provides direct employment for more than 5% of the working population.

At the time of writing, the UK market in construction indicates a continuing strong public sector demand to meet the policies of the Government, and a recovery of confidence in the private sector. Consistently low interest rates are enabling investment in real estate driven by a strong economy and a housing market which appears insatiable. This situation is set to continue, and the award of the 2012 Olympic Games to London will provide even stronger challenges to the construction industry well into the future.

There is, however, a massive shortage of traditional construction skills, particularly trades such as bricklayers, plasterers and carpenters. The industry is not attractive to younger people and particularly not to women who are grossly under-represented in both workforce and professional roles. The Construction Industry Training Board suggests that the current shortage is in excess of 300 000 crafts people and with the increasingly ageing existing workforce this skills-gap is growing wider each year.

Paradoxically, innovation in design is limited, with a tendency towards traditionalism driven to some extent by building regulations but also by the aversion of designers to take risk.

Despite current trends, history shows that over time construction demand is variable and is often affected by government policy changes. Also few clients are regular or significant purchasers from the industry and consequently most are inexperienced. It is likely that as many as 80% of clients are inexperienced, but by far the largest proportion of expenditure is by the more experienced clients. There is little, or no, cascading of knowledge and experience from the regular buyers to the inexperienced majority, and little understanding by inexperienced clients of the need for that knowledge. The result is a heavy dependence upon construction professionals who tend, in the main, to limit their exposure to risk by leaning heavily towards traditional practice.

Few construction companies employ skilled craftsmen, preferring to outsource rather than to retain and train. Most skilled tradesmen are self-employed or employed in small specialist companies because the returns are greater than employment by contractors. This scenario results in huge fragmentation and specialisation, with 95% of firms in the industry employing less than seven people and most taking work on a project by project basis. Small firms rarely feel able to afford to train new people and consequently there are very few young trades people in the industry. At the same time the role of traditional contractors has moved from one where they manage their own men to one where they coordinate the activities of other (small) companies and organisations.

This fragmentation of the industry means that it tends to be a diverse supply market from which clients may source their specific needs. There are subcontractors or specialists who occupy small, specialised niches in order to survive by avoiding direct competition with established market-leaders. There are also firms who are spanning these 'niches' in the provision of complete 'solutions' to large organisations and who act as 'integrators'. The end result is a very large number of specialists with which any client may do business in the delivery of his or her construction requirements.

A further consequence of fragmentation is that the industry takes little or no responsibility for seeking improvement in terms of either design or process. Attempts at developing innovative solutions are often strongly resisted by established professions or trades unless there is clear gain to *them* rather than the client customer. Even 10 years after Sir Michael Latham highlighted the difference between construction and other industries, investment in construction research and development is still significantly lower than other industries. Consequently if there is innovation it tends to be driven either by the client or by changes in technology in other industries which can be imported to construction through specialist suppliers.

The construction industry is thus unique in the way that it establishes projects to deliver one-off products. In this sense the industry is quite different from the manufacturing or retail sectors which have continuous demand and are thus able to construct and refine stable supply chains. Where these industries can focus on the improvement of the product or streamlining their supply chain management processes, the product of construction is bespoke in nature and the supply process is more an inconsistent network than a coordinated and carefully managed supply chain.

These problems often cause conceptual difficulties for clients coming to the industry with a value proposition associated with the delivery of a project. Clients who usually purchase defined and specified goods in the course of their normal business, and who concentrate on price and delivery, usually need considerable assistance to purchase undefined, unspecified projects where price and delivery are both vague at the outset.

The process has been described and analysed as having six primary steps (Egan 2002) commencing with a business case for the project rooted in the client's established need for the project. From this business case or value proposition the client will be able to identify a multiplier from the investment. Usually user-value is significant as a multiplier of spend.

Whilst the benefits from the construction process can be immense in business terms, the perceived performance of the construction industry appears historically poor. Many, perhaps most, projects are delivered late, over budget and with variable quality standards. Often these are unwelcome outcomes which frequently surprise the client organisation and cause significant problems. Although these deficiencies have been addressed by some sectors of the industry, the lamentable performance of the Holyrood Project for the Scottish Parliament (three years late and almost ten times the initial budget) leads us to believe that the improvements reported in some sectors are by no means universal.

Table 1.1 summarises the UK construction industry's performance over six years to 2004. This is based upon construction industry key performance indicators (KPIs) but is broadly in line with other surveys such as those carried out by client groupings (Construction Clients' Forum 1999, 2000).

At first sight this performance does not look very good. Some improvement has been identified in relation to client satisfaction of both the product and the service but there is little evidence of improvement in project cost predictability and only a little in respect of project time predictability. It is very difficult, however, to estimate the cost of a future design or when it can be delivered. Each project is usually unique even if only because it is set upon a unique site.

There is no consistently reliable industry-wide database of project costs and project times and there are so many variables. Consultants and constructors may be tempted

Table 1.1 Summary of UK construction industry performance 1999–2004 based upon industry key performance indicators (KPIs). (Source: *DTI Construction Statistics Annual* (DTI 2004).)

KPI	Measure	1999	2000	2001	2002	2003	2004
Client satisfaction – product	8/10 or better	72%	73%	72%	73%	78%	80%
Client satisfaction – service	8/10 or better	58%	63%	63%	65%	71%	74%
Defects	8/10 or better		65%	53%	58%	68%	68%
Safety	Accident incident rate	1354	1217	1318	1217	1097	1172
Predictability of construction cost	% on target or better	37%	45%	48%	50%	52%	49%
Predictability of project cost	% on target or better		50%	46%	48%	52%	50%
Predictability of construction time	% on target or better	34%	62%	59%	61%	59%	60%
Predictability of project time	% on target or better		28%	36%	42%	44%	44%

to be optimistic in order to seek to serve their client's best interest or to keep a marginal project alive. Ironically those clients who are disappointed in the performance of the construction industry often have extended design periods associated with their own non-construction products.

Nonetheless, it is the client who takes the initiative to start a construction project, and, therefore, construction is often a customer-driven one-off production process. Frequently the client will set targets based upon a set of expectations drawn from his or her own experience, which may well be unsuitable for a construction project. Inevitably construction professionals and/or constructors will strive to meet the client's demands, often taking a much too optimistic approach in the absence of a completed design to what can be achieved for the price or in the time allocated.

The start of the construction process is frequently matched by the client to strategic or business need for a constructed asset, e.g. a school, factory or office complex. After establishing a construction project organisation to provide the necessary expertise to finalise the design and specification, the client will undertake a tendering process to select a contractor. In most cases, this contractor will take care of the employment of specialists and subcontractors and the procurement of materials. When contracts are signed, and a sufficient amount of information is available, the physical execution of the construction project can start. This includes production of materials, manufacturing, engineering and assembly of elements, and final construction on site. After the successful completion of the project, there will be a hand-over, and use of the completed asset by the end user.

Construction is largely a site operation, confined to the specific location where the final assembly takes place, and it is important to note that construction takes place at the site of consumption as opposed to manufacturing industry where products are produced in bulk for subsequent consumption at a distance.

Construction project organisations, in contrast to manufacturing, tend therefore to be temporary. The participants involved with this temporary multi-organisation experience frequent changes of membership but are totally interdependent of each other, operating through a variety of contractual arrangements and specific procedures.

In summary, it can be clearly seen that the construction industry is vital to the economy but both demand and supply are fragmented, inconsistent and complex. Collaboration and innovation are the exception, not the rule, and most purchasers of construction are inexperienced. Nonetheless its products are valuable, useful and serve their purpose, and the key to their successful production is effective procurement. This book aims to explore those factors that are influential on procurement success.

References

Construction Clients' Forum (1999, 2000) *Survey of Clients' Satisfaction*. CCF.

DTI (2004) *DTI Construction Statistics Annual*. The Stationery Office.

Egan Sir J. (2002) *Accelerating Change*. Strategic Forum for Construction. Construction Industry Council.

Groak S. (1994) Is construction an industry? Notes towards a greater analytical emphasis on external linkages. *Construction Management and Economics*, **12** (3), 289–293.

Winch G. (2002) *Managing Construction Projects*. Blackwell Publishing.

2 Procurement strategy: a literature review

Introduction

The efficient procurement of construction work (defined here as the framework through which construction is brought about, acquired or obtained (Sharif and Morledge 1996)) through choice of the most appropriate procurement strategy has long been recognised as a major determinant of project success (Bennett and Grice 1990). Indeed Newcombe (1992) argued that the selection of the procurement path is much more than simply establishing a contractual relationship. Rather, building on the work of the Tavistock Institute (1965) and Cherns and Bryant (1984), he argues that the procurement strategy not only creates a unique set of social relationships but also 'forms a power structure within a coalition of competing or co-operating interest groups'. In terms of quantifiable project deliverables, Gordon (1994), in an American study, reported that it was possible to reduce project capital cost by an average of 5% through selection of the most appropriate procurement methodology.

In the modern world, client satisfaction is increasingly seen by all concerned with the development and construction process to be largely dependent upon the selection of the most appropriate procurement methodology, and failure to select an appropriate procurement approach is now well recognised as a primary cause of project failure (Masterman 1996). Hibberd and Djebarni (1996) reported that whilst 64% of the clients they surveyed were happy with the procurement methodologies they were using at the time, 89% stated that they had been previously dissatisfied. Smith and Wilkins (1996) reported that in a study of eleven publicly funded major hospital projects constructed in the UK, USA and Hong Kong, projects procured using a traditional approach consistently failed to achieve the client's objectives in terms of time or cost or both whereas projects procured using a design and build route consistently fared much better. They do, however, point out that this is not to say that the design and build route is generically better than the traditional path; simply that, at the time they were procured, the choice of design and build path was the result of a careful analysis of project characteristics and client expectations, whereas the traditional path was generally chosen 'because it's always been done that way'. This last point may well be a reflection of a lack of skill by those making the procurement decision.

These issues were well recognised by early researchers in the field, and authors such as Franks (1984) and Nahapiet and Nahapiet (1985) offered critiques of the major procurement methods in common use together with some advice as to how to go about making the most appropriate choice, based largely upon the client's perceptions of success in terms of time, capital cost and quality. Nonetheless as recently as 1997 Bowen et al. (1997) reported that their research led them to believe that relatively few construction industry professionals fully understood the differences between the various procurement systems, and would be unable to make sensible recommendations as to which system would be most appropriate for a specific project.

Early attempts to examine the strategic procurement selection process led a number of academics and practitioners to develop structured methodologies, tools and models of various types to aid this selection process. Among the models published in the UK, one of the earliest was *Thinking About Building* (Building EDC 1985), and the approach was further developed, and made progressively more complex, using multi-attribute decision analysis by Skitmore and Marsden (1988), Bennett and Grice (1990), Cook et al. (1993) (use of multilevel decision trees), Chan et al. (1994) (an adaptation of Skitmore and Marsden's model to suit the Australian construction industry), Zavadskas and Kallauskas (1996), Dell'Isola et al. (1998) and Love et al. (1998b). Ambrose and Tucker (2000) continue the theme with the development of yet another multi-attribute model, but this time based on a three-dimensional matrix. Al-Tabtabi (2002) presents a further variation on the theme with a model using an analytical hierarchy process built upon the work of earlier researchers including Saaty (1980) and Skibnewski and Chao (1992). The practical use of Al-Tabtabi's model is illustrated with a live case study showing successful use of the model in the procurement of the Kuwait University Expansion Program, a substantial project costing approximately 132 million Kuwait Dinars (approximately US$427.7 million) comprising more than 40 design and construction packages. Seydel and Olson (1990), in the USA, took a somewhat different approach presenting a model based on the use of fuzzy set theory.

Elsie, developed under a Royal Institution of Chartered Surveyors (RICS)-funded project based at Salford University (Brandon et al. 1988; Brandon 1990) and PASCON (Mohsini and Botros 1990) are two examples of attempts to automate the process through the development of computer-based expert systems, and, again in the USA, the Performance Information Procurement System (PIPS) (see Chapter 17) has been developed over a period of almost 20 years as an integrated and detailed structured methodology including selection of the initial procurement route, contractor selection, bidding and price analysis (Kashiwagi et al. 1996; Kashiwagi and Mayo 2001).

Newcombe (2000) adopts a somewhat different approach, and presents the development and field testing of a construction procurement simulator designed to aid the development of skills in analysing client needs and translating them into an appropriate procurement path.

More recent commentators have recognised that selection of the most appropriate strategy for large and/or complex projects, particularly those involving multiple stakeholders, is a difficult and complicated process dependent upon the interaction of many variables and incorporating a high degree of subjective and in many cases intuitive judgement. We should also note that the problems of complexity tend to be magnified greatly where the overall complexity of the project depends not only upon the technical complexity of the work but also upon the complexity of the context in which the project is to be developed and constructed. These issues came to a head in the UK in the early 1990s, when client concerns over the generally unsatisfactory performance of the construction industry in terms of its ability to deliver on time, within budget and to the expected quality and performance standards placed the role of the client in construction procurement firmly in the spotlight. Liu (1994) understood this well, suggesting that successful procurement approaches would be more likely to result from a deeper mutual understanding of the organisational and performance goals of the parties involved. She clearly pointed to the importance of understanding the cultural and organisational aspirations of all stakeholders in the procurement process, in particular those based upon a collaborative relationship.

In the light of the above, it is no surprise that procurement issues featured strongly in the government-sponsored 1994 report on the UK construction industry by Sir Michael Latham (Latham 1994). Latham's recommendation that the construction industry should move towards more collaborative methods of procurement built largely upon experience elsewhere, for example in the international oil exploration industry where alliancing as a concept had been in use for some years as a method of sharing risk and reward in high-risk activities, and in the USA where partnering had been used in various forms for some years. A growing insistence by major clients that, rather than being considered simply as passive customers of construction services, they should undertake a major role at the centre of the construction procurement process as a full partner added further impetus to the debate. A number of academics and researchers examined various facets of the client's role (see for example Wilkins and Smith 1995), and concluded that, particularly in the case of complex buildings, there was evidence that closer involvement of the client in the procurement process appeared to be a significant factor in project success. This enhanced client interest in the procurement process, coupled with demands for clearer guidance on procurement issues, led directly to the publication by the RICS of *The Procurement Guide* (1996).

A second government-sponsored review of the UK construction industry by Sir John Egan (Egan 1998) set targets for improvement in construction performance designed to improve customer satisfaction, and the implementation of Egan's recommendations was subsequently accepted as government policy for centrally funded public sector projects. The targets were also made implicit in the public sector 'best value' legislation introduced in 1997, which required all public bodies to show that they were achieving maximum value for money (defined in terms of whole life cost not just initial capital cost) in all of their projects including construction work. The message was further reinforced in a report by the National Audit Office, *Modernising Procurement* (1999), which amongst other things emphasised the need, particularly in the case of strategic projects, for close client involvement in allocating and managing risk and setting out effective mechanisms to incentivise superior project performance.

More recently, concerns over the environmental performance and social sustainability of the built environment have led some to question how these issues ought to be addressed in the construction procurement process.

The recent intense interest in the procurement of construction work has, predictably, led to a more detailed and extensive study of the construction procurement problem as an element of the wider discipline of construction management. In recent years therefore we have seen considerable attention being paid to understanding how the various elements of the procurement process work – individually, in combination with each other and in interaction with the rest of the construction management process. This process has been carried out largely by attempting to relate practical empirical experience of both successful and unsuccessful projects to established theoretical concepts in general management, industrial psychology and motivation. A number of major threads in procurement research have emerged from this work including, in no particular order, the following:

- Procurement strategy selection models, both manual and computer aided.
- Team relationships, supply chain management and communication and their effect on project performance.
- Defining in greater detail the role of the client in the procurement process.
- The interrelationship between the various components of the overall procurement approach (e.g. the overriding strategy, contractor and consultant selection processes, contractual framework and tender evaluation models) in enhancing project success (the general presumption seems to be that for the process to be deemed successful then the whole must be greater than the sum of the parts).
- Legal and contractual issues, including public sector procurement legislation, and their effect on project performance.
- Comparative analysis of international procurement practice and the impact of cultural differences.
- Risk allocation and reward.
- Bid evaluation techniques.
- The impact of environmental and sustainability issues upon the procurement process.

Procurement strategy selection models

It has already been shown that many attempts have been made to develop selection models designed to match a range of project performance indicators to the project characteristics with the aim of achieving improved overall 'project success'. The problem here, however, lies in the definition of 'project success'. Many construction professionals believe overall project success to be a comprehensive assessment arising out of a consensus of all key stakeholders. Others, however, appear to believe that this approach is much too complex, and that client satisfaction with the final outcome is perhaps the most important indicator of whether a project can be considered successful or not. The problem here is that many complex projects are commissioned by complex and multiheaded client bodies, who may find it difficult even among themselves to agree on a generally acceptable definition of overall project success.

If the key objectives can be isolated, then there is evidence that the selection of an appropriate procurement strategy which adequately matches the objectives of the key stakeholders is an important contributor to overall project success (Akintoye 1994; Naoum and Mustapha 1995).

Kumaraswamy and Dissanayaka (1996) reviewed existing attempts to model the procurement process, and suggested that the debate on what constitutes project success may be moved forward by considering not only the identified project success factors but also the linkages between them. Based on this presumption they go on to hypothesise how a theoretical client advisory model based upon weighted client objectives might be constructed.

Team relationships, supply chain management and communication and their effect on project performance

Teamwork and team performance

Teamworking has long been recognised as a key management area in construction (see for example Walker 1989; Bennett 1991), and in recent years researchers have become increasingly interested in the contribution good team performance can make to project success. The problems faced by construction project teams operating as temporary multi-organisations are well known, and have been addressed by a number of researchers including Shoesmith and Langford (1991), Mohsini and Davidson (1992) and Bowen and Edwards (1996).

Walker (1997) provides a useful review of the relevant literature concerning the effect of team performance on project timescale through a series of 64 major Australian projects (defined as projects in the range A$3–80 million). He concludes that:

> 'Team performance appears to be a complex balance of the management action, undertaken through planning, co-ordination and communication, to aid and inform decision making. Various team members employ different levels of task or people oriented management styles and different organisational responses in terms of rigidity or flexibility towards rules and regulations. These are undertaken in response to situational factors and different groups respond in different ways depending on their management maturity and/or willingness, as well as the perceived level of complexity of the tasks they face.'

He also concludes that inhibited team management will inhibit team performance, and he agrees with other researchers (e.g. Smith and Wilkins 1996) that, particularly for complex projects, non-traditional procurement methods are more likely to enhance construction time performance. Newcombe (1999) supports the idea that the chosen procurement path has considerable influence upon the project team's ability to learn, a contention for which he presents a detailed theoretical justification, whilst Moore and Dainty (2000) examine a series of issues relating to communication specifically within design and build teams. Kumaraswamy (1998), however, concludes that irrespective of the procurement system used, project success depends primarily upon the attitude, determination and capabilities of those involved.

Partnering and multicultural teams

Partnering as a concept for improving team performance has a significant history, and has been well described by, amongst others, Baden Hellard (1995) and Stephenson (1996). The idea for the concept originated many years ago. In the USA MacNeil (1974) foresaw collaboration as an improved contracting strategy for long-term work, although some see the origins of the approach as reaching much further back and seek to draw lessons from biblical history (Lynch 1989; Ogunlana 1999), and the long-running strategic alliance in the UK between Marks and Spencer and Bovis was extremely successful for both parties through the 1970s and into the 1980s.

In the USA Weston and Gibson (1993) reported average cost savings of 18.49% and time savings of 12.33% on a sample of 120 projects let by the Arizona Department of Transport, Larson (1995) reported significantly improved results in terms of controlling cost, technical performance and client satisfaction arising from the use of partnering on a sample of 280 projects, and Maloney (1997) reported that the use of partnering was increasing.

Uher (1999) presents a review of Australian experience, and concludes that from its introduction in 1991, despite its relatively short (and somewhat chequered) history, partnering is now firmly established as an integral part of the Australian construction industry. Lenard (1999), also commentating on contemporary Australian practice, reports that in a study of 17 construction projects in and around Sydney it was found that a symbiotic contractual and working relationship between the client and the contractor encouraged innovation and facilitated the development of a learning environment within the project team. Kenley et al. (2000) provide yet another view of Australian practice in the context of the *Construct Australia* report published by the Australian Procurement and Construction Council (APCC 1997).

In the UK partnering was strongly advocated by Sir Michael Latham in his 1994 review of the UK construction industry (Latham 1994), and has since been endorsed by the Government as the way public sector projects should be let whenever possible. The technique has since received significant academic approval and widespread endorsement in the UK by clients, contractors and researchers alike. See for example Bennett and Jayes (1995; 1998), Walter (1998), and CIB (1997).

There are, however, those who question the effectiveness of partnering as an improved method of procurement (see for example Cox and Thompson 1997; Smyth and Thompson 1999), whilst others maintain that it is merely seductive rhetoric encouraged by large corporations eager to flex their commercial muscle (Green 1999), and one would have to admit that in some supposed partnering relationships, although the parties speak of trust and commitment, the thinly veiled threat of differential commercial muscle appears to lurk only just beneath the surface. Miller et al. (2003) explore these issues from the point of view of small construction enterprises in the construction industry in South Wales. They conclude that although 'the degree of harmonisation between the contractor and small sub contracting firms is inextricably linked to client satisfaction', it is nonetheless 'evident that the partnering philosophy is not currently embodied within the industry in industrial South Wales despite continued attempts by policy makers to encourage construction firms to subscribe to such concepts'. They also suggest that the needs and objectives of the smaller subcontractors are frequently overlooked, and that despite exploiting partnering agreements with employers, main contractors still treat their relationships with their smaller subcontractors in an opportunistic and somewhat cavalier fashion.

Nonetheless commentators such as Wood and McDermott (2001) argue that there is significant evidence (Lewicki and Bunker 1996) that, provided the parties are able to establish even a fragile level of trust, then collaborative relationships are viable. Establishment of at least a basic level of trust is vital, and is extremely difficult to achieve. Ekström and Björnsson (2002) explore the use of source credibility theory (Hovland et al. 1953) in building trust in an e-tendering relationship, and Das and Teng (1998) explore the issue of managing relational risk in co-operative alliances. There may be significant benefit in developing these and other social science based approaches to the procurement problem, particularly where a collaborative relationship is sought.

There can, however, be little doubt that, to be successful, partnering is dependent upon assembling a project team which already shares (or may be persuaded to share) a common set of ethical and cultural beliefs and values, and the impact of cultural differences in the project team, both at the organisational level and at the personal level, is a further factor which has been considered by a number of researchers. Ngowi (1996) reviews some experience of innovation in construction procurement within international multicultural teams in ten major projects in Botswana, half of which were carried out by local teams, and the remainder by multinational multicultural teams comprising combinations of British, Chinese, German, Kuwaiti, Belgian and local Tswana organisations. He found evidence that where multicultural teams were involved innovation tended to be less prevalent, whilst miscommunication and disputes tended to increase within a multicultural framework. He concluded that teams sharing the same culture tended to be characterised by lower degrees of uncertainty and anxiety than those that did not. Plainly care needs to be taken to ensure that particular attention is paid in multicultural teams to ensuring not only good communication and a clear understanding of shared language, but also that each team member has a good understanding, appreciation and acceptance of the cultural values and beliefs of the others involved.

Jeffries et al. (2001) take the partnering issue one stage further with a detailed case study examining how a strategic alliance model derived from the oil industry was used to structure an alliance between a number of specialist companies in order that an economically marginal project could be successfully completed.

Supply chain management

The area of supply chain management (Saunders 1997) within the construction project team has also received some attention, with several researchers asserting that construction can learn key lessons from manufacturing and production management

(Koskela 1992; Love and Gunasekaran 1996; Kornelius and Wamelink 1998; Love et al. 1998a; Howell 1999; Green et al. 2002). Lee et al. (2000) describe the development of a supply chain management tool called the Generic Design and Construction Process Protocol (GDCPP) which attempts to provide:

> 'a common set of definitions, documentation and procedures that provides the basis to allow a wide range of organisations involved in a construction project to work together seamlessly.' (Kagioglou et al. 1998)

Root et al. (2003) explore the application of supply chain management techniques to the construction design process. Following Schneider (1993) they contend that organisational boundaries pose a significant barrier to collaboration, concluding that there is a clear need for integration within the design process, and that formal supply chain management techniques can be used to overcome the problems.

Finnemore et al. (2000) argue for improvements in project delivery driven by improvements in management of the process of construction, and report upon the early stages of the development of a tool called the Standardised Process Improvement for Construction Enterprises (SPICE). SPICE follows the philosophies of Deming (1986) and Juran (1988), and is predicated upon putting into place continuous process improvement based upon a series of small evolutionary steps.

The role of the client in the procurement process

Traditional methods of contracting generally involve employers or their agents designing or at least specifying in detail the work required prior to competitive tenders being invited from some group of construction contractors. The employer subsequently chooses which of the tenders submitted appears to represent the best value for money, and enters into some kind of contract for the construction work. The contractual relationship that results from this traditional process is therefore essentially that of supplier and customer, where the employer decides, in theory at least, in detail what he/she wants and the contractor simply constructs the work as designed. To be successful, the method depends upon employers being able to specify their requirements in sufficient detail for the contractor to accurately price the work. Whilst this becomes progressively more difficult the more the design responsibility is passed to the contractor, it is still possible for the process to deliver acceptable results, but choice of the most acceptable tender becomes more problematic as the number of variables involved increases. Clear goal definition has long been known to be a major determinant in management success (see for example the work of McGregor (1960) and Hersey and Blanchard (1982)) and failure by employers to accurately represent their requirements to the tendering contractors usually leads to confrontation and dispute.

Avoiding confrontation and dispute leads directly to the idea of alternative methods of procurement based upon achieving congruence between the objectives of the principal stakeholders, shared risk and return, and the joint resolution of problems. Working in this way, the relationship between employer and contractor becomes less 'supplier/customer' and more like a partnership, and therefore places much more emphasis on the whole project organisation working together in pursuit of common goals.

The way in which the client interacts with the remainder of the construction team has long been seen to be a key issue in project success (NEDO 1988; Potter 1994; Green and Lenard 1999; Barrett and Stanley 2000). Whilst it is of course possible that regular construction clients with a deep understanding of the way the construction industry works can operate a supplier/customer relationship very effectively, indeed some, including some large property developers, appear to prefer to do so, it was pointed out many years ago (see for example Ferry 1978) that 'the uninformed client has an unrealistic idea of what he is letting himself in for'. Most clients interact with the construction team through the appointment of a client representative, sometimes termed a 'project sponsor', 'client's project manager' or 'project director', and it appears that the project success may be significantly influenced by the level of confidence that this client representative inspires in the client team (Walker 1995).

The interrelationship between the various components of the overall procurement approach: strategic procurement management

Strategic procurement management (defined by Cox (1995) as 'the development of an external sourcing and supply strategy which links the total business plan of an organisation (public or private) so as to maintain a sustainable position for that organisation in the total value chain') has attracted comparatively little attention.

The significance of each of the various different elements embodied within the overall strategy (for example contractor and consultant selection processes, contractual framework and tender evaluation models) in enhancing project success has, however, been the subject of some research and debate. Kumaraswamy (1994) points out the need to develop synergy through informed choice of the components of the system, and careful alignment of the system with the objectives of the stakeholders, but the evidence as to the particular contribution made to overall project success by individual components seems to be far from clear cut. Walker (1995) for example concluded that contract type did not significantly alter speed of construction, whereas Rwelmalila and Hall (1994) found that lack of adequate management of the 'human aspects' contributed significantly to poorer performance.

Legal and contractual issues, including public sector procurement legislation, and their effect on project performance

Legal and contractual issues in procurement

Common law contractual practice has, in general, provided an adequate framework for 'traditional' client-designed construction projects, but recent years have seen the growth of design and build methods and, more recently, the encouragement of innovation through the process of inviting contractors to submit 'alternative' tenders of various kinds. Craig (1999) presents a detailed critique of the legal theory and practice surrounding the process of tendering for construction work, and discusses the concept of a collateral contract arising during the tendering process, which may constrain both the employer's freedom in deciding which tender he or she should accept, and the contractor's freedom to withdraw his or her tender prior to formal acceptance. Craig further points out that, in the light of established legal precedents, it is increasingly important for those engaged in construction procurement, particularly where innovation is sought, to design tender processes that encourage innovation, but at the same time to place sufficient limits on contractors' freedom to innovate in order to retain proper control of the tender process.

Early researchers (e.g. Emmerson 1962; NEDO 1983) found that the construction contracts in use at the time of their research provided little or no incentive for contractors to look for opportunities to add value to the client. This is not altogether surprising since the construction industry has historically had a reputation for being confrontational and 'claims conscious', and this attitude was largely fuelled by the predominant supplier/customer contractual relationship. There is some evidence that this situation was not altogether unhelpful to contractors, with one 1995 survey apparently indicating that, of the top 50 UK contractors 25% earned as much as 10–15% of their turnover from contractual claims (Cox and Thompson 1998).

In terms of contractual relationships, Cox and Thompson (1998) showed that little had changed in the previous 35 years. Their study of 163 construction clients revealed that, notwithstanding a high level of customer dissatisfaction with the standard forms of contract in common use, amongst predominantly building-biased respondents the JCT range (comprising the JCT 80 Standard Form of Building Contract, CD 81 Standard Form of Building Contract with Contractor's Design, IFC 84 Intermediate Form of Building Contract and the MC 87 Standard Form of Management Contract) jointly accounted for 76% of the sample. In civil engineering, however, whilst some 55% of respondents used the traditional ICE 6th Edition and the Agreement for Minor Building Works, only 9% said that they would use only the ICE forms. Their work led them to believe that '. . . a small minority of clients have developed rigorous techniques for establishing the most appropriate contracting strategy for their needs', whilst the remainder continued to base their choice on tradition and what they had always done before.

The theory of relational contracting was first formulated by MacNiel more than 30 years ago in a seminal essay published in the *Southern California Law Review* (MacNiel 1974), in which he draws attention to the differences in the kind and scope of possible contractual relationships. MacNiel makes a firm distinction between 'transactional' contracts, such as buying a newspaper, which are essentially short term and where the contract is affected by no past events and concerned with no future ones, and upon which the classical law of contract has been developed, and 'relational' contracts where the parties anticipate a lengthy relationship and where the contract attempts, in one way or another, to predict future events. He goes on to suggest a framework for extra-contractual alternative dispute resolution.

Campbell and Harris (1993) went further, maintaining that the foundation of classical contracting is based on the principle of 'presensiation', defined as an assumption that one can predict the future with some degree of accuracy, and that contracts can then be drawn up that adequately apportion the consequent risks and assign strict liabilities for breach. This, they maintain, is a somewhat unrealistic view of the real world as far as complex long-term relational contracts are concerned. Their research led them to believe that, where such contracts contained strict liabilities in respect of foreseen future breach, the contracting parties, in the private sector at least, frequently compensated for the contractual inadequacy by ignoring them in favour of '. . . a repertoire of extra-legal strategies when liability arises'. They go on to assert that:

> 'Efficient long term contractual behaviour must be understood as consciously co-operative. We see a long term contract as an analogy to a partnership. The parties are not aiming at utility maximisation directly through performance of specified obligations; rather they are aiming at utility maximisation indirectly through long term co-operative behaviour manifested in trust and not in reliance on obligations specified in advance.'

The growth of partnering has given rise to considerable debate about the legal significance of the partnering relationship and the undertakings given by the parties in respect of how they intend to conduct their business. Uher (1999) provides a useful summary of the issues and a detailed review of the major points of contention. Much revolves around the notion of good faith in commercial contracts. In the UK, and in many other common law jurisdictions, the courts have consistently avoided finding any such duty, preferring to stay as far as possible with the common law doctrine of *caveat emptor*. In the USA, however, and in European law, the legal establishment has been much more amenable to the notion of fair trading including an implied promise of good faith in commercial transactions. This issue plainly lies at the heart of any partnering relationship which claims to rely on trust, the absence of hidden agendas and full disclosure of all relevant facts as major planks in its construction. Australian legal commentators have advised against attempting to include 'good faith' clauses in contract on the grounds that, under the doctrine of *contra preferentum*, a court or an arbitrator would be likely to construe them against the party responsible for their insertion (Davenport 1993).

Public sector procurement law

As far as public sector procurement legislation is concerned, Dorée (1997), in a review of the procurement practices adopted by a number of Dutch municipalities, is extremely critical of the European procurement directives, questioning their adequacy 'especially about the emphasis on public tendering combined with price competition'. He concludes that, in the Netherlands at least:

> 'the directives ignore the problematical nature of the transactions in the fields of construction and building, and disregard the value of selective tendering as an effective tool in project control.'

International procurement comparisons and the impact of cultural differences

A number of researchers have presented analyses of construction procurement practice in various parts of the world, and the internationalisation of the construction industry is frequently a cause for concern and debate. In developing countries for example there is always the danger that projects constructed using overseas aid and resources will, whilst proving to be a successful source of income for international consultants and contractors, nonetheless fail to deliver facilities that provide maximum benefit for the host nation. The question of the impact of culture and organisation on project performance has been considered in depth by several authors including Winch et al. (1997), Winch (1999), Seymour and Fellows (1999) and Rahman et al. (2002).

Bin Abdul Rashid and Morledge (1998) and Bin Abdul Rashid (1999) consider the role of construction in economic growth within the framework of a rapidly developing country, Malaysia. They argue that the effective and productive procurement of construction work is an essential ingredient in maximising the contribution made by the construction industry to overall economic development in attaining the required levels of economic growth, but that this process is typically constrained by a lack of the necessary skills and resources. Bin Abdul Rashid (1999) presents a series of measures that could be taken domestically to address the problems, but there is little doubt that whilst self development is a vital part of the strategy, few countries could attempt to solve these issues by 'home growing' alone. Considerable care therefore needs to be taken to ensure that where external skills and resources are employed, their use is carefully controlled and managed by the host nation in order to ensure that the economic and social benefits are maximised. Technology transfer is plainly an important issue here, and poses serious challenges to the world construction industry as a whole.

An alternative view, again with a Malaysian focus, is put by Abdul-Rashid Abdul Aziz (1998). Here the issue of localisation is considered from the point of view of the company through a study of 26 Japanese contractors operating in Malaysia. Building upon the work of management theorists such as Lawrence and Lorsch (1967), Fayerweather (1969) and Doz and Prahalad (1984), the author points out that while localisation can help by opening up opportunities within national markets whilst generating a minimum of local hostility, the downside for companies is that excessive localisation may lead to the company losing the benefits of integrating its activities in the international marketplace. He concludes that the mingling of cross-cultural concepts in this way needs to be carried out with extreme care if the company involved is to successfully bid for work in local markets whilst at the same time securing its future in the international marketplace.

Ngowi (1996) discusses the changes emanating from increased internationalisation of construction procurement in Botswana. He points out that the local culture in procurement is by nature collaborative due to the fact that 'construction activities were carried out by members of the community who owned the work'. This naturally collaborative state has, he reports, been adversely affected by increasing internationalisation caused by increasing sophistication and complexity of projects, the use of international funding and scarcity of skilled local labour.

Dorée (1997) examines a number of factors affecting public sector construction procurement in several Dutch municipalities. Here the predominant method of procurement has historically been based around limited competition, supported by an 'atmosphere of co-operation between municipalities and contractors' which was characterised by 'attitudes . . . (which) . . . were far less antagonistic, opposing, hostile and conflicting than were predicted in the literature'. His conclusion is that this natural co-operation has again been adversely affected by external forces, in this case the European Union procurement legislation discussed earlier.

Kumaraswamy et al. (2002) adopt a somewhat different approach. Working from a base of recent experience in Hong Kong (Grove 1998; Hong Kong Housing Authority 2000; Rowlinson 2001; Tang 2001), possibly the most 'international' construction industry in the world in terms of the breadth of experience and expertise carried by the construction companies and consultancy practices operating there, they link Hong Kong experience with experience gained elsewhere in the world to develop the hypothesis that major cultural change within the world construction industry must be driven from within through systematic changes in construction procurement practices and processes. Whilst they admit that 'attempts to merely "bolt on" collaborative elements such as partnering to existing cultures and contracts have rarely succeeded', they contend that cultural change can be accelerated through factors such as:

- Better briefing (Green and Lenard 1999; Barrett and Stanley 2000).
- Improved selection systems and a more careful matching of client and contractor objectives (Lowndes 1998; Crane 2001; Bayliss 2002).
- The careful use of incentives (Scheznayder and Ohrn 1997; Wright 1998).
- More effective use of information technology.
- A wider acceptance of 'lean construction' principles.

Finally, in the international arena, Lahdenperä (2001) presents the results of a 10 year study (1989–1998) covering 20 050 projects in Finland, each with a value greater than 1 million Finnish Marks (about 170 000 Euros). Overall, in the public sector, the balance is tilted overwhelmingly in favour of a traditional procurement with separation of the design and construction functions, with minority use (14%) of design and build. In the private sector, however, the use of design and build increases to in excess of 23%.

Risk allocation and reward

Risk allocation in construction projects has long been a subject of intense debate, and there is evidence that inappropriate and unrealistic risk shifting is a major cause of disputes (Wall 1994), and that risk allocation determined on sound commercial principles helps to avoid disputes (NPWC 1990).

It is often said that 'risk should be placed with the party best able to control it', but Cheung (1997), following on from work by Abrahamson (1984) asserts that a more detailed taxonomy of risk distribution is more appropriate. The taxonomy suggested by Cheung suggests that risk should be allocated using the following scale of priorities:

- The party best able to control it effectively
- The party best able to absorb it financially
- The party who has most information to control it
- The party who benefits most from controlling it
- The party for whom the risk is inherent in its commercial role

Cheung then uses these principles together with a set of risks developed from the work of Casey (1979) in the construction of a risk allocation model which was subsequently tested with some success in Australia, although the sample size was rather small.

Also in Australia, Kenley et al. (2000) examine rewards and incentives in the context of the *Construct Australia* report on public sector construction (APCC 1997). They provide a useful review of the literature relating to incentives and rewards, and conclude that appropriate incentive and reward systems can help to change a contractor's behaviour both during the bidding process and during project execution.

Bid evaluation techniques

It has been argued (Trickey 1982; Smith 1986) that seeking the lowest tender price on the basis of common tendering information leads to maximum value for money for the client through free and fair competition. This may well be true in an environment of perfect economic competition, but in the real world, where price is the sole criterion, the bidding problem for the contractor becomes simply to submit the lowest price. Contractors have historically derived some comfort from the fact that the forms of contract employed would provide them with the opportunity to increase their return through claims made as the result of client variations and other changes. Contractors' actions in deciding what figure to bid have therefore been largely governed by their attitude to risk and their expectations of the way the client and his or her representatives will behave during the construction process. There is also evidence that, perhaps not surprisingly

- Lowest initial bid does not necessarily equate with lowest out-turn cost (Pearson 1985; Dawood 1994; Pasquire and Collins 1997).
- Contracts let by open competition are often less successful than contracts awarded by other means (Holt et al. 1995a, 1995b).
- Lowest price exposes the client to an increased risk of post-contract claims and cost over-runs (Crowley and Hancher 1995).

In order to counter these problems, a number of alternative bid models based on price alone have been developed, designed to identify and eliminate 'rogue' bids, thus hopefully increasing the client's chances of project success. Lingard et al. (1999) contains a concise review.

There now appears to be a general acceptance, in the UK at least, that bid acceptance based on price alone is unlikely to lead to improved project success, although as recently as 2001 Holt and Proverbs (2001) found that out of a sample of 120 English local authorities comprising equal numbers of counties, district authorities, metropolitan districts and London boroughs:

- 11% believed that lowest price provided best value for money.
- Only 18% believed that best value would be delivered through a combination of price and quality.
- 50% believed that price was the most important aspect of a contractor's tender submission.

Several researchers have attempted to develop quantitative models to aid the process of analysing multicriterion bids, including:

- Multidimensional utility analysis (Diekman 1981)
- Cluster analysis to identify the bid representing overall utility optimisation (Seydel and Olson 1990)
- Models based on fuzzy set theory (Nguyen 1985; Wong and So 1995)
- Hierarchical decision models (Mustapha and Ryan 1990; Russell 1992; Kong and Cheung 1996)
- Expert systems (Russell et al. 1990)
- Discriminant analysis (Tam 1993)
- Performance ranking using a Likert scale (Assaf and Jannadi 1994)
- Composite models involving a combination of processes (Prasertsintanah 1996)

Lingard et al. (1999) contend that contractor selection models ought to be considered in respect of their propensity to reduce the transaction costs involved, and that this can best be achieved by multiparameter models, since they address a broader range of sources of uncertainty and opportunistic behaviour than do the simpler comparative bid evaluation models.

As explained earlier, some clients have attempted to introduce innovation into the procurement process through the introduction of 'alternative tenders', but some commentators believe that such practices may pose significant legal difficulties (Craig 1999).

Bid evaluation is now more commonly carried out using a combination of price and 'qualitative' factors such as the contractor's attitude, proposed method of working, the strength of the contractor's team, etc. This issue becomes even more important where projects are to be conducted under some form of collaborative arrangement, and it has been proposed that in these cases the price and qualitative factors should be supplemented by a series of factors concerned with aligning cultural fit between the employer and the contractor. Examples of such factors might include ethical values and principles, and common and transparent business practices.

Hampson and Kwok (1997) suggest testing the following attributes in order to indicate likely success in collaborative relationships between firms:

- Trust in one's trading partners
- Commitment to a win–win solution
- Interdependence and a recognition of the benefits to be gained from shared mutually agreed goals, shared risks and shared rewards
- Recognition that self-interest requires co-operation and that co-operative working integrates self-interest in order to achieve mutual goals
- Recognition that effective communication is essential in co-operative working
- Commitment to joint problem solving

Diekman et al. (2000) propose the use of similar techniques to help potential developers of privatised infrastructure projects assess the relative merits of alternative projects through a series of discrete clusters of information based on:

- Political issues (legislative status, organisational structures, etc.)
- Public/private partnership attributes (the nature of the partnership agreement, ownership, operational responsibility, length of concession period, etc.)
- Project scope
- Environmental factors
- Construction risks
- Operational risks
- Financing issues (debt/equity, project subsidies, etc.)
- Economic viability
- Developer's financial involvement (including development rights, performance incentives, etc.)

A detailed set of scoring documents are also presented.

Environmental and sustainability issues and the procurement of construction work

Environmental sustainability has long been a topic of intense discussion in the construction industry. Langston (1997) provides a useful review of contemporary issues, and there appears to be general agreement that increased sustainability and mitigation of the environmental impact of major projects form two of the construction and development industries' more important key performance indicators. Note, however, that some (see for example Gilham 1998) believe that, in the wider international arena, for many construction businesses the adoption of strategies for construction sustainability may be too far from their current set of cultural values and business reality to be a practical proposition.

A wide range of tools have been developed to aid in the environmental assessment of both construction materials (see for example the Green Materials Index (Shiers and Keeping 1994), the Building Materials Ecological Sustainability Index (Lawson and Partridge 1995) and the Environmental Preference Method (Anik et al. 1996)) and buildings (see for example Environmental Value Engineering (Roudebush 1992) and the Building Research Establishment Environmental Assessment Method

(BREEM) (Baldwin and Yates (1996)). Furthermore, it has been proposed (Langston 1996) that cost–benefit analysis can also be adapted for use in this way. A combined framework incorporating both environmental assessment and environmental sustainability is presented in the *Green Building Challenge* (Larrson and Cole 1998).

A number of commentators and researchers have suggested that environmental issues should form part of the project procurement process (see for example Baldry (1997) and Walker and Lloyd-Walker (1998)).

Graham and Walker (2000) review a range of environmental performance assessment tools in order to explore the extent to which they were able to provide a robust process to enable assessments of overall environmental sustainability to be made. On the basis of their analysis they propose a 'triple bottom line' model for project assessment based upon financial viability, environmental performance and social sustainability, which they then suggest should be incorporated into project procurement strategies.

Conclusion

So what general lessons can we learn from the above, admittedly brief and to some extent selective, review of research into the process of construction procurement? It is evident that there is now considerable expertise around the world in many aspects of the discipline, and there are many instances of good practice. On the other hand, there are areas where there are considerable differences of opinion, and where considerably more work is required in order to establish a consensus view.

It is also heartening to note a significant shift in construction management research as a whole away from the simple evaluation of practical experience and towards a process of integrating the observation of practical experience with more mature and generally accepted existing theoretical constructs. It is only through the development of a body of robust research data linked to established theory that we can develop a truly scientific body of reliable construction procurement knowledge.

It is true that construction researchers have historically attempted to link their work to elements of general management theory, and to some extent to generally accepted theories of organisational behaviour, but what has become increasingly evident in recent years is a more generally accepted understanding that the construction procurement process is largely, perhaps mostly, about people and the way they interact with each other and their environment. This has led to a considerable increase in interest in the application of theoretical constructs taken from the 'softer' disciplines such as communication, social and behavioural science, cognitive and industrial psychology and the influence of culture, ethics and personal beliefs. There are, we believe, strong reasons for believing that it is from these areas that the major improvements to existing construction procurement practice will be derived in the future.

This chapter has attempted to introduce some of the major issues in construction procurement research, many of which will be explored in greater detail in subsequent chapters.

References

Abdul-Rashid Abdul Aziz (1998) Adapting procurement practices to suit host country as a probable localisation tactic. *Journal of Construction Procurement*, **4** (1), 45–58.

Abrahamson M.W. (1984) Risk management. *International Construction Law Review*, **1** (Part 3), 587–598.

Akintoye S.A. (1994) Design and build procurement method in the UK construction industry. In: *Proceedings of CIB W92 Procurement Systems Symposium, East Meets West*, Hong Kong (ed. S. Rowlinson), pp. 1–10.

Al-Tabtabi H.M. (2002) Construction procurement selection strategy using analytical hierarchy process. *Journal of Construction Procurement*, **8** (2), 117–132.

Ambrose M.D. and Tucker S.N. (2000) Procurement system evaluation for the construction industry. *Journal of Construction Procurement*, **6** (2), 121–134.

Anik D., Boonstra C. and Mak J. (1996) *Handbook of Sustainable Building – An Environmental Preference Method for Selection of Materials for Use in Construction and Refurbishment.* James and James (Science Publishers).

APCC (1997) *Construct Australia – Building a Better Construction Industry in Australia.* Australian Procurement and Construction Council.

Assaf S. and Jannadi M.O. (1994) A multi-criterion decision making model for contractor prequalification selection. *Building Research and Information*, **22**, 332–335.

Baden Hellard R. (1995) *Project Partnering Principle and Practice.* Thomas Telford.

Baldry D. (1997) The role of project management in the environmental impact of the building process. In: *Proceedings of the 2nd International Conference on Buildings and the Environment.* CIB Task Force Group 8, Paris.

Baldwin R. and Yates A. (1996) An environmental assessment method for buildings. In: *Proceedings of CIB Working Commission Meetings*, RMIT Melbourne.

Barrett P. and Stanley C. (2000) *Better Construction Briefing.* Blackwell Science.

Bayliss R.F. (2002) Partnering on MTR Corporation Ltd's Tseung Kwan O Extension. *Transactions Hong Kong Institute of Engineers*, **9** (1), 1–6.

Bennett J. (1991) *International Construction Management General Theory and Practice.* Butterworth Heinemann.

Bennett J. and Grice A. (1990) Procurement systems for building. In: *Quantity Surveying Techniques: New Directions* (ed. P.S. Brandon). BSP Professional Books.

Bennett J. and Jayes S. (1995) *Trusting the Team: The Best Practice Guide to Partnering in Construction.* Centre for Strategic Studies in Construction, University of Reading.

Bennett J. and Jayes S. (1998) *The Seven Pillars of Partnering.* Thomas Telford.

Bin Abdul Rashid K. (1999) Strategies to remove or alleviate constraints affecting the processes of construction procurement in Malaysia. *Journal of Construction Procurement*, **5** (1), 27–39.

Bin Abdul Rashid K. and Morledge R. (1998) Constraints in resources and functions within the process of construction procurement in Malaysia. *Journal of Construction Procurement*, **4** (1), 27–44.

Bowen P.A. and Edwards P.J. (1996) Interpersonal communication in cost planning during the building design phase. *Construction Management and Economics*, **14**, 395–404.

Bowen P.A., Hindle R.D. and Pearl R.G. (1997) The effectiveness of building procurement systems in the attainment of client objectives. In: *Proceedings of CIB W92 Procurement*

Systems Symposium, Procurement – The Key to Innovation (eds C.H. Davidson and T.A. Meguid), Montreal, pp. 39–49. IF Research Corporation.

Brandon P.S. (1990) Expert systems – after the hype. In: *Proceedings of CIB 90 Conference, Building Economics and Construction Management*, Sydney, vol. 2, pp. 314–344.

Brandon P.S., Basden A., Hamilton I.W. and Stockley J.E. (1988) *Application of Expert Systems to Quantity Surveying*. Royal Institution of Chartered Surveyors.

Building EDC (1985) *Thinking About Building*. Economic Development Committee for the Building Industry.

Campbell D. and Harris D. (1993) Flexibility in long term contractual relationships: the role of co-operation. *Journal of the Law Society*, **20** (2), 166–191.

Casey J.J. (1979) Identification and nature of risks in construction projects: a contractor's perspective. In: *Proceedings of Conference on Construction Risks and Liability Sharing*, vol. 1. American Society of Civil Engineers.

Chan A.P.C., Tam C.M., Lam K.C. and So A.T.P. (1994) A multi attribute approach for procurement selection – an Australian model. In: *Proceedings of the 10th Annual Conference of the Association of Researchers in Construction Management*, Loughborough University of Technology, pp. 621–630.

Cherns A.R. and Bryant D.T. (1984) Studying the client's role in construction management. *Construction Management and Economics*, **2**, 177–184.

Cheung S.O. (1997) Risk allocation: an essential tool for construction project management. *Journal of Construction Procurement*, **3** (1), 16–27.

CIB (1997) *Partnering in the Team*. Thomas Telford.

Cook W.D., Johnston D.A. and Mosche A. (1993) Selecting a new technology strategy using multilevel decision trees. *Journal of Multi-Criteria Decision Analysis*, **2**, 129–144.

Cox A. (1995) Strategic procurement in the public and private sectors: the relative benefits of competitive and collaborative approaches. In: *Strategic Procurement Management in the 1990s* (eds R. Lamming and A. Cox). Earlsgate Press.

Cox A. and Thompson I. (1997) 'Fit for purpose' contractual relations: determining a theoretical framework for construction projects. *European Journal of Purchasing and Supply Management*, **3** (3), 127–135.

Cox A. and Thompson I. (1998) Has contracting lost its customer focus? *Journal of Construction Procurement*, **4** (1), 5–15.

Craig R. (1999) How innovative is the common law of tendering? *Journal of Construction Procurement*, **5** (1), 15–26.

Crane A. (2001) Re-thinking construction – its implementation in the UK construction industry. In: *Proceedings of the International Conference on Construction*, Hong Kong, pp. 53–60.

Crowley L.G. and Hancher D.E. (1995) Risk assessment of competitive procurement. *Journal of Construction Engineering and Management*, **121**, 230–237.

Das T.K. and Teng B.S. (1998) Between trust and control: developing confidence in partner co-operation in alliances. *The Academy of Management Review*, **23** (3), 491–512.

Davenport P. (1993) Good faith clauses are not good. *Australian Construction Law Newsletter*, **33**, 21–23.

Dawood N.N. (1994) Developing an integrated bidding management expert system for the precast concrete industry. *Building Research and Information*, **22**, 95–102.

Dell'Isola M.D., Licameli J.P. and Arnold C. (1998) How to form a decision matrix for selecting a project delivery system. *Design-Build Strategies*, **4**, 2.

Deming W.E. (1986) *Out of the Crisis*. MIT Centre for Advanced Engineering Study.

Diekman J.E. (1981) Cost plus contractor selection; a case study. *Journal of the Technical Council, American Society of Civil Engineers*, **107**, 13–25.

Diekman J.E., Ashley D., Bauman R., Carroll J. and Findlayson F. (2000) Viability of privatised transportation projects: an evaluation tool for design/build teams. *Journal of Construction Procurement*, **6** (1), 33–43.

Dorée A.G. (1997) Construction procurement by Dutch municipalities. *Journal of Construction Procurement*, **3** (3), 78–88.

Doz Y. and Prahalad C.K. (1984) Patterns of strategic control within multinational corporations. *Journal of International Business Studies*, **15**, 55–72.

Egan Sir J. (1998) *Rethinking Construction.* UK Government Department of the Environment, Transport and the Regions.

Ekström M.A. and Björnsson H.C. (2002) Trust building tools supporting electronic bidding in construction. *Journal of Construction Procurement*, **8** (2), 133–147.

Emmerson H. (1962) *Survey of the Problems Before the Construction Industry.* HMSO.

Fayerweather J. (1969) *International Business Management: A Conceptual Framework.* McGraw-Hill.

Ferry D. (1978) Client–consultant expectations: fact or fantasy. *The Building Economist*, June, 8–10.

Finnemore M., Sarshar M., Haigh R., Goulding J., Barrett P. and Aouad G. (2000) Can process capability be used to manage the construction supply chain? *Journal of Construction Procurement*, **6** (2), 184–201.

Franks J. (1984) *Building Procurement Systems.* Chartered Institute of Building.

Gilham A. (1998) Strategies for change – understanding sustainable development from a construction industry perspective. In: *Proceedings of CIB 98 Symposium, Construction and the Environment*, Gavle.

Gordon C.M. (1994) Choosing appropriate construction contracting method. *Journal of Construction Engineering and Management*, **120** (1), 196–210.

Graham P.M. and Walker D.H.T. (2000) First steps towards achieving environmental sustainability for developed projects – an holistic life cycle procurement objective. *Journal of Construction Procurement*, **6** (1), 66–83.

Green S.D. (1999) Partnering: the propaganda of corporatism? *Journal of Construction Procurement*, **5** (2), 177–186.

Green S.D. and Lenard D. (1999) Organising the procurement process. In: *Procurement Systems: A Guide to Best Practice in Construction* (eds S. Rowlinson and P. McDermott). E. & F.N. Spon.

Green S.D., Newcombe R., Williams M., Fernie S. and Weller S. (2002) Supply chain management: a contextual analysis of aerospace and construction. In: *Proceedings of CIB W92 Procurement Systems Symposium, Information and Communication in Construction Procurement.*

Grove J.B. III (1998) *Consultant's Report on Review of General Conditions of Contract for Construction Works for the Government of the Hong Kong Special Administrative Region.* http:/www.constructionweblinks.com/resources/industry_reports_newsletters/nov_6_2000/grove_report.htm.

Hampson K. and Kwok T. (1997) Strategic alliances in building construction: a tender evaluation tool for the public sector. *Journal of Construction Procurement*, **3** (1), 28–41.

Hersey P. and Blanchard K. (1982) *Management of Organisational Behaviour: Utilizing Human Resources*, 4th edn. Prentice-Hall International.

Hibberd P.R. and Djebarni R. (1996) Criteria for choice of procurement systems. In: *Proceedings Cobra '96*, University of the West of England, Royal Institution of Chartered Surveyors.

Holt G.D. and Proverbs D.G. (2001) A survey of public sector procurement in England. *Journal of Construction Procurement*, **7** (1), 3–10.

Holt G.D., Olomolaiye P.O. and Harris F.C. (1995a) A review of contractor selection practice in the UK construction industry. *Building and Environment*, **30**, 553–561.

Holt G.D., Olomolaiye P.O. and Harris F.C. (1995b) Application of an alternative contractor selection model. *Building Research and Information*, **23**, 255–264.

Hong Kong Housing Authority (2000) *Quality Housing: Partnering for Change*. Hong Kong Housing Authority (HKHA).

Hovland C.I., Janis I.L. and Kelley H.H. (1953) *Communication and Persuasion*. Yale University Press.

Howell D. (1999) Builders get the manufacturers in. *Professional Engineer*, May, 24–25.

Jeffries M., Gameson R. and Chen S.E. (2001) The justification and implementation of project alliances – reflections from the Wandoo B development. *Journal of Construction Procurement*, **7** (2), 31–41.

Juran J.M. (1988) *Juran on Planning for Quality*. The Free Press.

Kagioglou M., Cooper R., Aouad G., Hinks J., Sexton M. and Sheath D. (1998) *Final Report: Generic Design and Construction Process Protocol*. The University of Salford.

Kashiwagi D.T. and Mayo R.E. (2001) Best value procurement in construction using artificial intelligence. *Journal of Construction Procurement*, **7** (2), 42–59.

Kashiwagi D.T., Halmrast C.T. and Tisthammern T. (1996) 'Intelligent' procurement of construction systems. *Journal of Construction Procurement*, **2** (1), 56–65.

Kenley R., London K. and Watson J. (2000) Strategic procurement in the construction industry: mechanisms for public sector clients to encourage improved performance in Australia. *Journal of Construction Procurement*, **6** (1), 4–19.

Kong W.K. and Cheung S.M. (1996) A multi-attribute tender evaluation model. In: *Proceedings of CIB W89 Symposium, Construction Modernisation and Education*, Beijing. Architecture and Building Press.

Kornelius L. and Wamelink J.W.F. (1998) The virtual corporation: learning from construction. *Supply Chain Management*, **3** (4), 193–202.

Koskela L. (1992) *Application of New Production Philosophy to Construction*, CIFE Tech Rep 72. Centre for Integrated Facility Engineering, Stanford University.

Kumaraswamy M.M. (1994) New paradigms for procurement protocols. In: *Proceedings of CIB W92 Procurement Systems Symposium, East Meets West*, Hong Kong (ed. S. Rowlinson), pp. 143–148, 399–400.

Kumaraswamy M.M. (1998) Industry development through creative project packaging and integrated management. *Journal of Engineering, Construction and Architectural Management*, **5** (3), 228–238.

Kumaraswamy M.M. and Dissanayaka S.M. (1996) Procurement by objectives. *Journal of Construction Procurement*, **2** (2), 38–51.

Kumaraswamy M.M., Rowlinson S. and Phua F. (2002) Accelerating cultural changes through innovative procurement processes: a Hong Kong perspective. *Journal of Construction Procurement*, **8** (1), 3–16.

Lahdenperä P. (2001) An analysis of the statistics on project procurement methods in Finland 1989–1998. *Journal of Construction Procurement*, **7** (1), 27–41.

Langston C. (1996) The application of cost benefit analysis to the evaluation of environmentally sensitive projects. In: *Proceedings of CIB Working Commission Meetings*, RMIT Melbourne.

Langston C. (ed.) (1997) *Sustainable Practices: ESD and the Construction Industry*. Envirobook Publishing.

Larrson N. and Cole R. (1998) GBC 98: Context, history and structure. In: *Proceedings of an International Conference on the Performance Assessment of Buildings, Green Building Challenge '98*, Vancouver.

Larson E. (1995) Project partnering: results of a study of 280 construction projects. *Journal of Management in Engineering*, **11** (2), 30–35.

Latham Sir M. (1994) *Constructing the Team*. The Stationery Office.

Lawrence P. and Lorsch J. (1967) *Organisation and Environment*. Harvard Business School.

Lawson W.R. and Partridge H. (1995) *Building Materials Ecological Sustainability (BES) Index – Assessing the Ecological Sustainability of Building Materials*. Partridge Partners.

Lee A., Kagioglou M., Cooper R. and Aouad G. (2000) Production management: the process protocol approach. *Journal of Construction Procurement*, **6** (2), 164–183.

Lenard D.J. (1999) Future challenges in construction management: creating a symbiotic learning environment. *Journal of Construction Procurement*, **5** (2), 197–210.

Lewicki R.J. and Bunker B.B. (1996) Developing and maintaining trust in work relationships. In: *Trust in Organisations* (eds R.M. Kramer and T.R. Tyler). Sage Publications.

Lingard H., Hughes H. and Chinyio E. (1999) The impact of contractor selection method on transaction costs; a review. *Journal of Construction Procurement*, **4** (2), 89–102.

Liu A.M.M. (1994) From act to outcome – a cognitive model of construction procurement. In: *Proceedings of CIB W92 Procurement Systems Symposium*, Hong Kong, pp. 169–178. Department of Surveying, Hong Kong University.

Love P.E.D. and Gunasekaran A. (1996) Towards concurrency and integration in the construction industry. In: *Proceedings of an International Conference on Concurrent Engineering*, Toronto.

Love P.E.D., Gunasekaran A. and Li H. (1998a) Concurrent engineering: a strategy for procuring construction projects. *International Journal of Construction Management*, **6** (6), 177–185.

Love P.E.D., Skitmore M. and Earl G. (1998b) Selecting a suitable procurement method for a building project. *Construction Economics and Management*, **16**, 221–233.

Lowndes S. (1998) *Fast Track to Change on the Heathrow Express*. IPD.

Lynch R.P. (1989) *The Practical Guide to Joint Ventures and Corporate Alliances*. Wiley and Sons.

MacNeil I.R. (1974) The many futures of contracts. *Southern California Law Review*, **47**, 691–816.

Maloney W.F. (1997) Improvement example from the 1990s: increased use of design-build as a project delivery system. In: *Proceedings of CIB W65 Symposium, Transfer of Construction Management Best Practice Between Different Cultures, Organisation and Management of Construction*, Publ 205, pp. 183–187.

Masterman J.W.E. (1996) *An Introduction to Building Procurement Systems*. E. & F.N. Spon.

McGregor D. (1960) *The Human Side of Enterprise*. McGraw-Hill.

Miller C., Packam G. and Thomas B. (2003) Inter-organisational relationships and their effect upon small construction enterprises in South Wales: co-operation at a cost. *Journal of Construction Procurement*, **9** (2), 17–28.

Mohsini R.A. and Botros A.F. (1990) PASCON an expert system to evaluate alternative project procurement processes. In: *Proceedings of CIB 90 Conference, Building Economics and Construction Management*, Sydney, vol. 2, pp. 525–537.

Mohsini R.A. and Davidson C.H. (1992) Determinants of performance in the traditional building process. *Construction Management and Economics*, **10**, 343–359.

Moore D.R. and Dainty A.R.J. (2000) Work-group communication patterns in design and build project teams: an investigative framework. *Journal of Construction Procurement*, **6** (1), 44–54.

Mustapha M.A. and Ryan T.C. (1990) Decision support for bid evaluation. *International Journal of Project Management*, **8**, 230–235.

Nahapiet J. and Nahapiet H. (1985) *The Management of Construction Projects*. Chartered Institute of Building.

Naoum S.G. and Mustapha F.H. (1995) Relationship between the building team, procurement methods and project performance. *Journal of Construction Procurement*, **1** (1), 50–63.

National Audit Office (1999) *Modernising Procurement*. The Stationery Office.

National Economic Development Office (NEDO) (1983) *Faster Building for Industry*. HMSO.

National Economic Development Office (NEDO) Construction Industry Safety Group (1988) *Faster Building for Commerce*. Report prepared by the NEDO Commercial Building Steering Group.

Newcombe R. (1992) Procurement paths – a power paradigm. In: *Proceedings of CIB W92 Procurement Systems Symposium*, Hong Kong.

Newcombe R. (1999) Procurement as a learning process. *Journal of Construction Procurement*, **5** (2), 211–220.

Newcombe R. (2000) An investigation into simulating the procurement process in the United Kingdom construction industry. *Journal of Construction Procurement*, **6** (2), 104–120.

Ngowi A.B. (1996) Impact of culture on construction procurement. *Journal of Construction Procurement*, **3** (1), 3–15.

Nguyen V.U. (1985) Tender evaluation by fuzzy sets. *Journal of Construction Engineering and Management*, **111**, 231–243.

NPWC (1990) No-dispute strategies for improvement in the Australian building and construction industry. In: *Proceedings of National Public Works Conference*, Australia.

Ogunlana S.O. (1999) Procurement lessons from Solomon's temple project. *Journal of Construction Procurement*, **5** (2), 187–195.

Pasquire C. and Collins S. (1997) The effect of competitive tendering on value in construction. *RICS Research Papers*, **2** (2).

Pearson G. (1985) Tender assessment. *Chartered Quantity Surveyor*, **8**, 194–195.

Potter M. (1994) Procurement of construction work: the client's role. In: *Proceedings of Centre for Construction Law and Management 7th Annual Conference on Risk Management and Procurement in Construction*, London.

Prasertsintanah P. (1996) An evaluation of a premium scheme in tendering as an incentive strategy. In: *Proceedings of CIB W89 Conference, Construction Modernisation and Education*, Beijing. China Architecture and Building Press.

Rahman M., Palaneeswaran E. and Kumaraswami M.M. (2002) Transformed culture and enhanced procurement: through relational contracting and enlightened selection. In: *Proceedings of CIB Symposium, Procurement Systems and Technology Transfer* (eds. T.M. Lewis), Trinidad and Tobago, pp. 383–401.

RICS (1996) *The Procurement Guide*. Royal Institution of Chartered Surveyors.

Root D.S., Thorpe A., Thompson D.S., Austin S.A. and Hammond J.W. (2003) Design chains: introducing supply chain management into the construction design process. *Journal of Construction Procurement*, **9** (2), 29–39.

Roudebush W.H. (1992) *Environmental value engineering (EVE): a system for analysing the environmental impact of built environment alternatives.* PhD dissertation, Graduate School of the University of Florida.

Rowlinson S. (2001) Matrix organisation structure, culture and commitment – a Hong Kong public sector case study of change. *Construction Management and Economics*, **9** (7), 669–673.

Russell J.S. (1992) Decision models for analysis and evaluation of construction contractors. *Construction Management and Economics*, March, 117–135.

Russell J.S., Skibniewski M.J. and Cozier D.R. (1990) QUALIFIER-2; knowledge based system for contractor prequalification. *Journal of Construction Engineering and Management*, **2**, 17–25.

Rwelmalila P.D. and Hall K.A. (1994) An inadequate traditional procurement system? Where do we go from here?. In: *Proceedings of CIB W92 Procurement Systems Symposium, East Meets West*, Hong Kong (ed. S. Rowlinson), pp. 107–114.

Saaty T. (1980) *The Analytical Hierarchy Process.* McGraw-Hill.

Saunders M. (1997) *Strategic Purchasing and Supply Chain Management.* Pitman Publishing.

Scheznayder C. and Ohrn L.G. (1997) Highways specifications – quality vs pay. *Journal of Construction Engineering and Management*, **123** (4), 437–443.

Schneider M.E. (1993) Mastering the interfaces – construction contracts drafting for dispute avoidance. *International Construction Law Review*, **10**, 403–424.

Seydel J. and Olson D.L. (1990) Bids considering multiple criteria. *Journal of Construction Engineering and Management*, **116**, 609–623.

Seymour D.E. and Fellows R.F. (1999) Towards a culture of quality in the UK construction industry. In: *Proceedings of Conference on Profitable Partnering in Construction Procurement* (ed. S.O. Ogunlana). E. & F.N. Spon.

Sharif A. and Morledge R. (1996) Procurement strategies and national organisations: the dependency linkage. In: *Proceedings of CIB W92 Symposium, 'North Meets South' Procurement Systems*, Durban (ed. R.G. Taylor), pp. 566–577.

Shiers D. and Keeping M. (1994) The Green Materials Index – development of an environmental auditing system for building materials. In: *Proceedings of the 1st International Conference of CIB TG16, Sustainable Construction*, Tampa.

Shoesmith D. and Langford D. (1991) Research into specialist building services procurement and its impact on project performance. In: *Refereed Proceedings of the 7th Annual Conference of the Association of Researchers in Construction Management*, pp. 79–90.

Skibnewski M.J. and Chao L. (1992) Evaluation of advanced construction technology with AHP method. *Journal of Construction Engineering and Management*, **118** (3), 577–593.

Skitmore R.M. and Marsden D.E. (1988) Which procurement system? Towards a universal procurement selection technique. *Construction Management and Economics*, **6**, 71–89.

Smith A.J. and Wilkins B. (1996) Team relationships and related critical success factors in the successful procurement of health care facilities. *Journal of Construction Procurement*, **2** (1), 30–40.

Smith R.C. (1986) *Estimating and Tendering for Building Work.* Longman.

Smyth H.J. and Thompson N.J. (1999) Partnering and the conditions of trust. In: *Proceedings of CIB W55 and W65 Joint Triennial Symposium, Customer Satisfaction: A Focus for Research and Practice* (eds P. Bowen and R. Hindle), Cape Town.

Stephenson R.J. (1996) *Project Partnering for the Design and Construction Industry.* John Wiley & Sons.

Tam C.M. (1993) *Discriminant analysis model for predicting contractor performance.* Unpublished PhD thesis, Loughborough University.

Tang H. (2001) *Construct for Excellence.* Report of the Construction Industry Review Committee, Hong Kong Government.

Tavistock Institute (1965) *Interdependence and Uncertainty.* Tavistock Publications.

Trickey G.G. (1982) Does the client get what he wants? *Quantity Surveyor,* **38**, 86–87.

Uher T.E. (1999) Partnering performance in Australia. *Journal of Construction Procurement,* **5** (2), 163–176.

Walker A. (1989) *Project Construction Management.* BSP.

Walker D.H.T. (1995) The influence of client and project team relationships upon construction time performance. *Journal of Construction Procurement,* **1** (1), 4–20.

Walker D.H.T. (1997) Construction time performance and traditional versus non-traditional procurement methods. *Journal of Construction Procurement,* **3** (1), 42–55.

Walker D.H.T. and Lloyd-Walker A. (1998) Organisational learning as a vehicle for improved building procurement. In: *Procurement Systems: A Guide to Best Practice* (eds S. Rowlinson and P. McDermott). E. & F.N. Spon.

Wall C.J. (1994) Dispute prevention and resolution for design and build contracts in Hong Kong. In: *Proceedings of CIB W92 Procurement Systems Symposium, East Meets West,* Hong Kong (ed. S. Rowlinson).

Walter M. (1998) The essential accessory. *Construction Manager* (CIOB), **4** (1), 16–17.

Weston D.C. and Gibson G.E. (1993) Partnering – project performance in the US Army Corps of Engineers. *Journal of Management in Engineering,* **9** (4), 410–425.

Wilkins B. and Smith A.J. (1995) Organisational change, technological innovation and the emerging role of the owner's project manager in the delivery of health care facilities. In: *Proceedings of the 1st International Conference on Construction Project Management* (ed. K.T. Yeo), Nanyang Technological University, Singapore, pp. 149–156.

Winch G. (1999) Innovativeness in British and French construction: the evidence from Transmanche-link. *Construction Management and Economics,* **18**, 807–817.

Winch G., Millar C. and Clifton N. (1997) Culture and organisation: the case of the Transmanche-link. *British Journal of Management,* **8**, 237–249.

Wong K.C. and So A.T.P. (1995) A fuzzy set expert system for contract decision making. *Construction Management and Economics,* **13**, 95–103.

Wood G. and McDermott P. (2001) Building on trust: a co-operative approach to construction procurement. *Journal of Construction Procurement,* **7** (2), 4–14.

Wright G. (1998) Tapping the team synergy. *Building Design and Construction,* **39** (8), 52–56.

Zavadskas E. and Kallauskas A. (1996) Determination of an efficient contractor by using the new method of multicriteria assessment. *The Organisation and Management of Construction: Shaping Theory and Practice,* **2**, 95–104.

UNIVERSITY OF CENTRAL
LANCASHIRE LIBRARY

3 Principles of strategic procurement

Introduction

The aim of this chapter is to introduce the basic concepts of strategic procurement, and to review the need for a strategic approach to procurement issues across the whole construction industry supply chain. The concept is developed to provide the basis of a cohesive framework for construction procurement, which in turn provides an anchor for more extensive study of the various elements involved.

The continuing search for maximum value for money in construction work has, in recent years, increasingly focused attention upon the procurement process. The impetus for change, in the UK at least, has been driven by a number of factors, not least of which has been a growing dissatisfaction among construction clients with the quality and predictability of construction work, coupled with the development of an increasingly knowledgeable and empowered body of influential corporate and public sector clients. Similar pressures are now becoming evident elsewhere in the world, particularly in South Africa, Australia and in the Far East including Hong Kong, China, Malaysia and Singapore.

The need for a more strategic and cohesive approach to construction procurement has long been proposed by construction commentators and academics as a significant prerequisite for improved project success, and the publication of *The Procurement Guide* by the Royal Institution of Chartered Surveyors (RICS) in 1996 (RICS 1996) provided much valuable advice to practitioners. The issue was given significant emphasis in Egan's *Rethinking Construction* report (Egan 1998), and has since been adopted as a major plank in the UK Government's procurement policy through guidance issued initially by the Treasury and latterly by the Office of Government Commerce.

Considerable effort has also been devoted to examining the ways in which both the procurement process and project supply chains are managed in other industries such as car and aeroplane manufacture, shipbuilding and oil exploration, with a view to transferring the technology to construction. A number of relevant case studies are included in Cox and Townsend (1998). Such efforts, however, have met with limited initial success, and quickly led to the view that whilst these techniques may have considerable potential there is a need for them to be significantly modified and adapted in order to meet the special needs of the construction industry.

In addition to these largely external influences, there has , in very recent times, been a significant shift in the way both commercial companies and public sector clients view their interaction with the construction and development industries. An increasing understanding by clients of their own generic supply chain relationships and the factors that drive them has increasingly led enlightened clients to recognise that the improved supply chain management techniques that they have developed for the direct benefit of their own core business can and should be adapted and applied to the construction supply chain. Consequently we are beginning to see experienced and enlightened clients taking a much closer interest in the way their construction supply chains, once the exclusive purview of the construction contractors, are being organised and managed.

Procurement: a review of theory and practice

It seems to be generally agreed that the historical development of construction procurement systems across the world has been a somewhat ad-hoc process. The gradual evolution in the UK of the 'traditional' design-led approach occurred largely as a natural development of the rise of the architectural profession throughout the seventeenth and eighteenth centuries. Initially, although predominantly design led, work was commonly procured as separate trade-based packages often carried out by individual tradesmen, in much the same form as the modern-day French *lots séparés* approach. However, this approach led to many disputes, and the traditional system as we know it today was finally brought to fruition with the development of the practice of 'contracting in gross' by the Government Office of Works in the course of procuring a substantial programme of military construction required as a result of the Napoleonic Wars around the turn of the nineteenth century. 'Contracting in gross', coupled with the nineteenth century building boom, caused primarily by the development of the railways, the rapid expansion of the civil engineering profession and subsequent suburban expansion, led to the development of the modern general contractor.

Whilst the traditional approach held sway for more than 150 years, by the early 1960s problems arising from the separation of the design and construction functions, often evidenced by the difficulty of apportioning responsibility for defects in completed buildings, led to the development of what have been called 'integrated procurement systems' (Masterman 2002) generally based upon a design and build approach. The design and build family has been relatively slow to take off, but Masterman (2002) estimates that by the turn of the twenty-first century design and build based approaches were used in probably 50% of UK construction projects.

The use of management oriented procurement approaches, where the primary emphasis is placed on the management of the process, dates initially from the 1920s, but the approach was progressively developed in the USA in the late 1960s, and more rapidly during the 1970s, as a direct result of government pressure to improve predictability and reduce construction delays through better process management. This development was fuelled as much as anything by contemporary developments in major projects in other industries such as military shipbuilding, oil exploration and delivery, and the American space programme. The approach achieved prominence in the UK in 1969 with the construction of the Horizon project for John Player Ltd in Nottingham, and has continued to be used, mainly by experienced clients, with varying degrees of success ever since, but its use in the UK has never achieved truly popular support.

The development of the use of private sector funding to secure public sector infrastructure has a long history, and has had significant impact upon construction procurement. The use of privately financed solutions is considered in detail elsewhere, but the development in the UK of the Private Finance Initiative (PFI) and other forms of public private partnership (PPP) has had significant effects upon the way the supplier/customer relationship has been structured, and upon the way risk distribution and value for money have been perceived by employers and contractors alike.

Finally, in the closing decade of the twentieth century, we began to see a resurgence of interest in examining the structure of the relationship between the construction client and the contracting team. This process, driven largely by the desire of some clients to take a more active role in the ownership of their projects during development and construction, and particularly in the UK through a central government drive to improve the value derived from the expenditure of public money, led to the development of a class of collaborative procurement approaches.

The development of partnering, strategic alliances and other collaborative methods has led to a substantial redefinition of the procurement process as a whole, and has concentrated interest upon study of a large body of management theory and practice dealing with team motivation, behaviour, performance measurement, incentivisation and similar issues within the broader constructs of construction procurement and project management.

The cumulative impact of these developments has caused some to look more critically at the management of the whole construction supply chain.

In light of the above, it is clear that the development of construction procurement approaches has historically been driven primarily by perceived deficiencies in previously popular approaches, a process described by Cox and Townsend (1998) as 'barefoot empiricism'. Despite the attempts of Masterman (2002) and others to construct some kind of theoretical framework within which to position the different approaches, and by researchers such as Skitmore and Marsden (1988), Bennett and Grice (1990), Liu (1994) and many others to link various parts of the process to an accepted theoretical base, and to formalise the process of selection of the most appropriate procurement route, Cox and Townsend (1998) still felt able to write, at the end of the twentieth century, that:

> 'Essentially there has been no theoretical framework on which to derive either an ideal or an optimum approach to procurement, only a reactive evolution of modus operandi.'

A strategic approach to procurement

From the above discussion it is clear that, to be effective, a strategic and inclusive approach to construction procurement must provide a cohesive framework within which all of the various project objectives required by the client may be addressed in the most effective way. This then implies that the client's needs and wants must be sufficiently well articulated, analysed and communicated for an appropriate procurement process to be developed. It must also be clearly recognised that a strategic approach to project procurement must be dynamic rather than static, in that 'formula-based' standardised approaches are unlikely to be satisfactory as projects become more complex, and as client needs change over time. It is therefore unlikely that any kind of 'one size fits all' approach will be entirely satisfactory in many, perhaps the majority of, cases.

Every new project is to some extent unique, and is a product of both the client's needs and desires and the external influences operating on the client and the project at that particular point in time. It therefore follows that the generic procurement processes must evolve and develop in order to meet particular sets of circumstances at specific points in time. In light of this it can be convincingly argued that there is no such thing as universal 'best practice' in construction procurement in the sense of simply choosing one overall procurement methodology (e.g. traditional, design and build, management contracting, etc.) from a limited and predetermined menu of all-inclusive options. This then leads one to conclude that what is required, particularly for projects that are themselves complex, or those that are to be procured in a complex environment, is a bespoke approach designed specifically to address each particular set of circumstances. If this argument is accepted, then instead of viewing procurement approaches simply as a predetermined and limited menu of ready-made solutions, we must instead consider the overall procurement strategy to be the sum of a set of processes, some of which will be 'off the shelf' and some bespoke, all of which are interrelated within an overall strategic framework designed to provide the best possible chance that the employer's objectives will be at least satisfactorily achieved and if possible exceeded.

This then leads us to view the individual techniques that might be used to address specific aspects of the procurement problem simply as elements of a comprehensive 'tool kit'. The development of an acceptable overall procurement methodology then comprises selection of the right combination of tools and techniques from the practitioner's toolbox to meet specific client and project needs.

Components of the procurement process

The basic components of any procurement process will include the following:

- A functional needs analysis.
- Selection and development of an overall procurement philosophy, which will normally be based upon either a design led, construction led or management led approach.
- Analysis of the most suitable form of relationship between the demand and supply sides of the procurement equation (i.e. a supplier/customer relationship or some form of collaborative relationship).
- A detailed design of the specific procurement approach to be used (i.e. selection and development of the appropriate tools).

- Formalisation of the contractual relationships in the most appropriate way.
- Selection of the most appropriate supply chain (i.e. contractors and consultants).
- Implementation.

The functional needs analysis

For construction projects the starting point for the functional needs analysis will usually be the client's brief (see Chapter 6). This issue is explored in some detail elsewhere, but experience shows that historically both clients and their professional advisors and designers have, in too many cases, exhibited poor briefing skills and that for complex projects, especially where clients are large and multiheaded, considerable skill, time and effort may be required to derive an acceptable brief.

Deciding precisely what the project objectives should be and the characteristics to be expected from the completed scheme is frequently difficult, and whilst the overriding objectives may be derived from the business case, compromises frequently need to be made in order to fit the objectives with the available resources. Assessing and weighting the relative 'worth' of competing objectives within the overall scheme therefore frequently requires some serious consideration.

Modern construction clients in both the public and private sectors are also beginning to think more strategically about their interaction with and expectations from the construction industry, particularly in terms of the value added by construction services to their core business. Standard project briefing procedures will therefore frequently need to be supplemented by an externally facilitated project strategy workshop incorporating input from all of the major client stakeholders. For a private sector project such a workshop would typically include end users, budget holders and those responsible for maintenance of the completed facility. For public sector projects this list would typically be supplemented by representatives from audit and legal departments in order to ensure that the project procurement mechanisms chosen are acceptably robust in terms of public accountability.

Development of an overall procurement philosophy

The relative importance of project success

The functional needs analysis should provide a set of prioritised objectives which the project is required to fulfil, complete with sets of key performance indicators and benchmarks defining how project success is to be measured. The needs analysis should also show how important the project is to employers within their overall business plan and development strategy, and will thus help to focus attention on the amount of effort that should be put into the project procurement process.

There are many ways in which this relationship between client needs and procurement strategy can be modelled.

Figure 3.1 shows a very simplistic approach, modelling the relationship in terms of the likely impact on the employer if the contractual relationship between employer and contracting partner were to fail. The vertical axis indicates the effect in terms of impact upon non-cost related factors such as reputation, whereas the horizontal axis denotes impact in terms of cost.

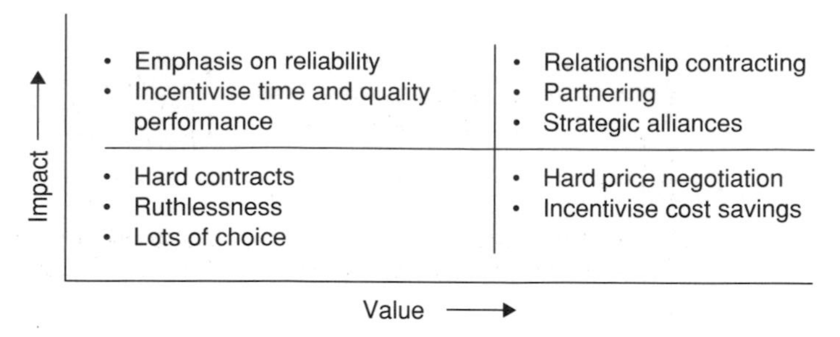

Figure 3.1 Procurement strategy as a function of the impact of failure.

Projects lying in the bottom left-hand quarter of the figure have minimum impact in both financial and non-financial terms, and can therefore be dealt with effectively using conventional procurement approaches involving little management effort. Projects lying in the top right-hand quarter on the other hand are those with maximum impact and are therefore those to which maximum effort should be applied. Projects lying in the top left and bottom right blocks have limited impact, and should be dealt with through the use of a procurement strategy incorporating the appropriate incentives.

Figure 3.2 develops the concept of collaborative methodologies as they might be affected by increasing project complexity and an increasing impact on the employer's core business. Plainly as projects and/or commercial relationships become more complex and more fundamental to the employer's core business, so the nature of the relationship needs to change, and increasing effort needs to be devoted to ensuring that they work properly and support the employer's key business aims and project objectives. As project significance increases, and as the possible consequences of failure become more significant, we might therefore expect to see a progression

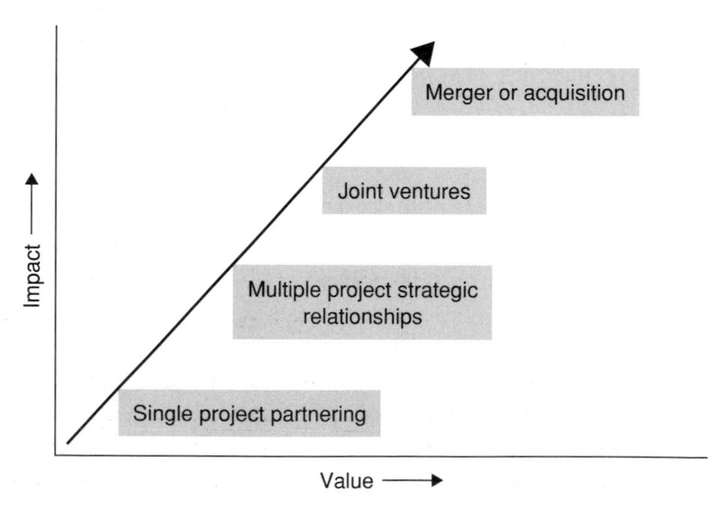

Figure 3.2 Collaborative trading relationships related to project complexity/impact on the employer's core business.

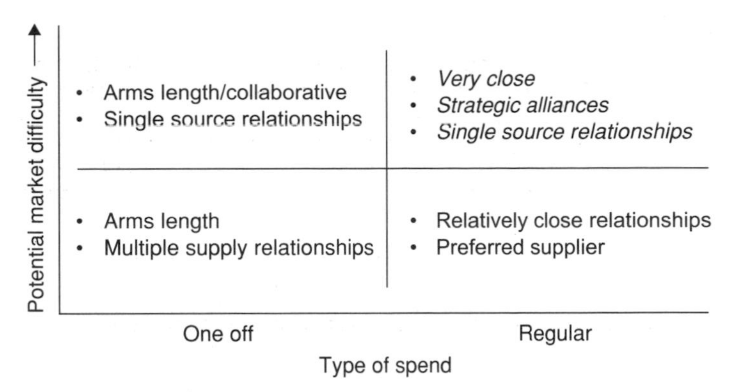

Figure 3.3 Procurement strategy as a function of market difficulty and type of spend. (Adapted from Cox A. and Townsend M. (1998) *Strategic Procurement in Construction*. With permission of Thomas Telford Publishing.)

from the simpler single project partnering relationships, through more complex multiproject strategic alliances, to true joint ventures where both parties have a financial stake. The limit of the model, where projects and/or relationships become absolutely fundamental to the employer's core business, might ultimately involve bringing the operation in house often through either a merger or acquisition route.

Figure 3.3 recognises that other external influences may also have a bearing on the procurement decision, and thus models procurement strategy as a function of potential market difficulty and type of spend.

Here the contention is that any trading relationships incorporating multiple projects ought to be based on collaborative strategies, whilst those involving only a single project might be better based upon more traditional 'arms length' arrangements. Others may, however argue that complex issues involving high degrees of market difficulty should also be based upon collaborative arrangements.

As with Figure 3.1, the most important relationships, and thus the ones to which the most effort should be devoted, are those in the top right-hand quarter of the diagram. Similarly those in the bottom left-hand quarter are the least important, and therefore justify rather less management effort.

Components of a collaborative relationship

The development of successful collaborative relationships will depend upon alignment of the objectives of the various organisations involved in the project, and this will in turn be dependent upon successful agreement of both the commercial and the cultural elements of the arrangement. The cultural and commercial issues upon which attention should be focused are illustrated in Figures 3.4 and 3.5.

It terms of cultural issues (Figure 3.4), the central issues to be addressed are those of trust and teamworking, and these areas are conventionally addressed in a 'workshop' environment, generally using an external facilitator. Many different types of management development techniques have been developed to deal with these issues, ranging from outdoor-based activities where people learn to work together to solve

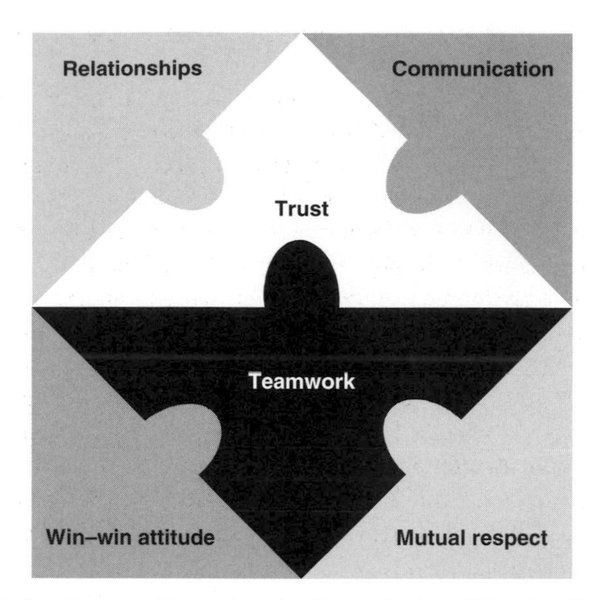

Figure 3.4 Cultural issues. (Reproduced with permission of Knowles Management.)

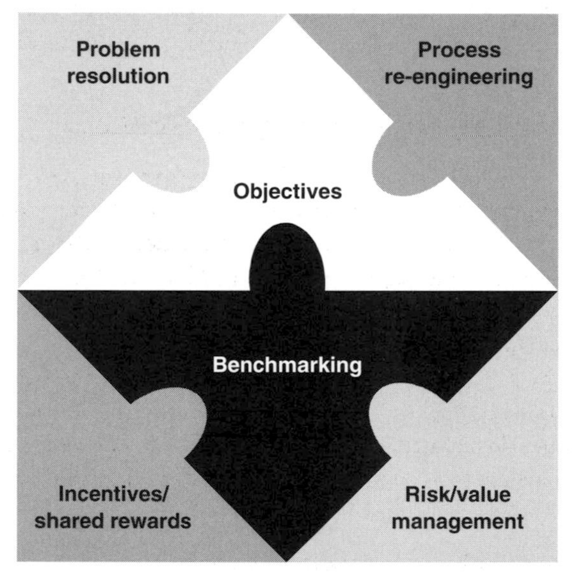

Figure 3.5 Commercial issues. (Reproduced with permission of Knowles Management.)

problems in an environment perceived by the participants to present some physical hazard or challenge, to purely academic desk-based problem solving tasks often incorporating some element of role play.

Development of the other cultural issues illustrated in Figure 3.4 (relationships, communications systems, a 'win–win' attitude and mutual respect) will all follow from the development of a strong team culture. A more detailed examination of these issues is given in Chapter 12, in the section headed 'The management of relationships'.

Equally important is the development of the commercial aspects shown in Figure 3.5. Central to this is the development of a joint set of mutually agreed project objectives, which each party to the agreement is prepared to support and work towards. The objectives must be measurable, and thus need to be supported by a robust set of key performance indicators and benchmarks designed to indicate the extent to which the objectives are being achieved as the project proceeds.

The development and/or selection of robust procedures for both risk management and value management are crucial to project success, and must support the achievement of the mutual objectives. Also important is development of a fair and timely dispute resolution methodology which will allow potential problems to be resolved before they escalate into full-blown disputes. A detailed discussion of dispute resolution methodologies is beyond the scope of this book, but selected aspects of dispute resolution practice are considered in outline in later chapters.

Achievement of the mutual objectives will be further aided by commitment of all of the parties to a critical re-examination of the processes conventionally used for the management of projects, with a view to seeking improvement through innovation and process re-engineering. This will in turn be aided by the development and introduction of suitable incentive mechanisms designed to enable those bringing forward innovative ideas to share in the potential rewards.

Detailed design of the procurement approach to be used

Once the form of the overall procurement methodology has been derived, then development of the detailed design will generally comprise selection and/or development of those tools best suited to achievement of the project objectives. Note that not all projects will require use of the whole tool kit, and that there are significant advantages to keeping the detailed design as simple as possible, consistent with ensuring adequate project performance.

The use of collaborative procurement methods may not be appropriate for all projects, and the use of sophisticated risk and value analysis tools may not be justified in all cases. The important issue is to ensure an appropriate selection of tools to meet the needs of the particular client at that specific point in time.

Formalisation of contractual relationships

An important part of the procurement process is to ensure that the chosen procurement route, and the relationship between the parties, is formalised in an appropriate way. The majority of construction projects use one or other of the generally available standard form contracts, of which there is now a bewildering array available. What is important is that the chosen contractual arrangement should supplement and support the chosen procurement route.

For some clients the form of contract is prescribed, as was until recently the case for central government, where for traditionally procured projects the GC/Works series has long been regarded as virtually the only available option. It remains to be seen whether the current Office of Government Commerce (OGC) advice that '. . . the form of contract used has to be selected according to the objectives of the project to meet the needs of *Achieving Excellence in Construction* principles' (NEC 2005) leads to more variety in the choice of public sector contracts, or whether the OGC endorsement of NEC3 will simply lead to this form replacing GC/Works as the default choice. For the private sector, however, the range is much wider.

Whilst it would appear that the majority of contracts let in the UK outside central government still use one or other of the JCT Standard forms, these forms have in the past been regarded as somewhat unsuitable for collaborative arrangements. Despite the fact that the Joint Contracts Tribunal issued a Practice Note explaining how these forms may be used in a partnering context, they have still been seen by some to be rather confrontational in nature and as encouraging a supplier/customer style of relationship that is at odds with partnering and other similar techniques. It remains to be seen whether or not the 2005 revision of the JCT suite will modify this view.

The partnered style of relationship is generally considered to be better supported by one of the more modern alternatives, of which there are several available. The Engineering and Construction Contract (also known as the NEC Form), published by Thomas Telford on behalf of the Institution of Civil Engineers and now in its third edition, is probably the most popular, but for those wishing to bring the partnering process firmly within the contractual ambit the PPC 2000 Standard Form of Contract for Project Partnering published by the Association of Consultant Architects is becoming a popular choice.

It is important to note here that this is not to say that a partnered relationship or some other collaborative form is generically better or worse than the more traditional supplier/customer approach, merely that once the most appropriate approach is selected for a particular project then it should be supported by the most appropriate form of contract. In this context, and as an example, reference might be made to the JCT Major Project Construction Contract, which is specifically designed for employers who wish to have minimal involvement in their projects. The Guidance Notes to the document state that the form is designed such that '. . . having defined its Requirements, the Employer should permit the Contractor to undertake the project without the Contractor being reliant upon the Employer for anything more than access to the site, the review of Design Documents and payment.' Its use in a partnered context might therefore be questionable.

Selection of the most appropriate partners (i.e. contractors and consultants)

Selection of the most appropriate trading partner is of paramount importance in achieving successful project delivery. Contractor and consultant selection procedures are dealt with elsewhere, but it should be noted that, in the case of more complex projects, particularly perhaps those requiring a collaborative approach, quite complex and detailed selection processes are likely to be required to ensure the necessary 'cultural fit' between the trading partners, and this issue is also explored in more detail in Chapter 12 under Establishing the 'cultural fit'.

Contractor selection processes usually involve bringing the potential partners together in an interactive way. At the lower end of the scale, the interaction has historically consisted simply of an interview lasting perhaps an hour or an hour and a half, during which potential trading partners (contractors or consultants) are given the opportunity to explain their approach to the project, and in return the employer has the opportunity to ask questions. Whilst the concept of interviews of this kind has some merit, experience shows that many contractors have approached the interview primarily as a marketing opportunity. They have therefore put forward their marketing team rather than the team actually responsible for carrying out the work, and this has led in some cases to considerable client disappointment when the project team subsequently fails to fulfil the promises made.

More recent developments in the case of larger projects, and particularly where multiproject strategic alliances are involved, have seen the use of assessment centre techniques largely borrowed from human resources management. Here the contractor's proposed project team is required to work with the employer for an extended period of time (perhaps a whole day) in a structured and highly interactive way. Typical assessment centre activities might include a combination of business games and simulations, small group discussion sessions and mini-workshops, some of which might be designed and run by the contractor, in addition to the more formal contractor's presentation.

Implementation

The successful design and implementation of a workable procurement strategy is plainly a highly skilled process, and the increasing attention being paid to this has led to the development of specialist procurement advisors. Whilst initially these people tended to come from within the construction disciplines, increasingly this area is becoming the focus of attention by procurement specialists from other industries.

In summary, the following might represent the skills and knowledge required of a construction procurement advisor:

- A clear understanding of the generic management issues posed by the organisation and operation of complex supply chains.
- A knowledge of the factors governing the management of relationships.
- Clear understanding of the nature of organisational structures, their boundaries and the factors affecting communication between them.
- Comprehensive knowledge of the nature and function of the various standard contract forms commonly used in the construction supply chain.
- Clear insight into the way the construction industry is structured and the industry-specific factors affecting the working of the construction supply chain.

Conclusion

It is clear that the present advances being made by commercial companies in understanding the ways in which supply chains work will lead to significantly enhanced expectations of those responsible for commissioning, procuring and managing construction work. This will lead directly to the need to develop more sophisticated approaches both to the 'front line' procurement of construction works and services and to the management of construction supply chains by the industry itself. It is also clear that increasingly well informed corporate employers will not shrink from direct involvement in construction supply chain management in the event that the construction industry fails to actively engage with the challenges posed.

There are also serious implications here for those involved in project management. Project success is increasingly being seen to be less to do with detailed management of the technical issues such as time, cost, quality, etc., and more to do with managing the relationships between the organisations involved in the extended supply chain. This extended supply chain is increasingly seen as spanning the complete range of project participants, from end users at one extreme, through client budget holders, main contractors, subcontractors and suppliers to those responsible for facilities management at the other. Indeed some are beginning to consider project success as being directly related to successful relationship management, and the successful project as being the aggregate of a series of successful relationships.

This is not to say that the traditional project management skills will become redundant, rather that employers will, in the future, take such skills for granted and will, in addition, require project managers to be able to understand and implement all of the additional softer management skills inherent in a modern strategic procurement environment.

References

Bennett J. and Grice A. (1990) Procurement systems for building. In: *Quantity Surveying Techniques* (eds. P.S. Brandon). BSP Professional Books.

Cox A. and Townsend M. (1998) *Strategic Procurement in Construction*. Thomas Telford.

Egan Sir J. (1998) *Rethinking Construction*. UK Government Department of the Environment, Transport and the Regions.

Liu A.M.M. (1994) From act to outcome – a cognitive model of construction procurement. In: *Proceedings of CIB W92 International Procurement Symposium, East Meets West*, University of Hong Kong, pp. 71–89.

Masterman J.W.E. (2002) *Introduction to Building Procurement Systems*, 2nd edn. Spon Press.

NEC (2005) *NEC3 Engineering and Construction Contract*. Thomas Telford.

RICS (1996) *The Procurement Guide*. Royal Institution of Chartered Surveyors.

Skitmore R.M. and Marsden D.E. (1988) Which procurement system? Towards a universal procurement selection technique. *Construction Management and Economics*, **6**, 71–86.

4 Public sector projects

Introduction: why should public sector projects be different?

This chapter explores the procurement of construction works and services in the public sector. Note that, in this context, the 'public' sector refers specifically to work carried out wholly or partly with public funds. Public companies, who by and large must only satisfy their own shareholders, therefore constitute part of the 'private' sector. Public/private partnerships (PPP) using some combination of public and private sector funds, such as those being developed under the Government's Private Finance Initiative (PFI), are subject to their own particular body of regulation and are discussed in detail in Chapters 13 and 14.

There are, of course, many common areas between public and private sector projects, and the underlying principles of contract law and of 'best practice' in procurement set out in this book and elsewhere do, of course, apply equally to both. There are, however, further issues to be taken into account when working with the public sector, in that the public sector must be seen to be accountable to the taxpayer in respect of:

(1) *The way in which projects and services are procured*: public sector contracts must be seen to be awarded fairly and without discrimination. The award process must be both transparent and accountable.
(2) *The expenditure of the public funds involved*: public accountability is usually of paramount importance. Taxpayers, in a democracy, have the right to be shown that their money is being spent in accordance with the approved published policies, standing orders, financial regulations, etc., and that adequate safeguards are in place to prevent the misappropriation of funds. The need for good audit control and the provision of a clear audit trail are therefore of paramount importance in the selection of the public sector project procurement strategy.
(3) *Maximising value for money*: value for money has often, in the past, been equated in the layman's eye with lowest price. This is no longer the case in the public sector, and current UK Government advice now supports the principles of good practice elsewhere in advocating seeking the optimum combination of price and quality for each specific project or service. The Government has also imposed a legal duty of 'best value' upon a substantial portion of the public sector, with the aim of securing continuing improvements in the standards

of public services. This issue is discussed in detail later, in the section headed 'Best value'.

These obligations, particularly those in respect of transparency and fair competition, go far beyond those placed upon either private sector companies or individuals contracting between themselves, and a substantial body of legislation and regulation has therefore been developed in an attempt to ensure that these objectives are consistently met. Consultants working with the public sector for the first time are frequently both confused by and contemptuous of the apparent bureaucracy involved. One must, however, understand that the obligations imposed upon the public sector involve employees and consultants in a significantly higher level of responsibility, transparency and accountability than is conventionally the case in the private sector. Failure to comply with the letter of the law may render individuals and/or the client body as a whole liable to actions at civil law, and/or to possible criminal prosecution. In some extreme cases, for example the 'best value' legislation as it applies to local authorities and some other public bodies, failure to meet the requirements could result in administration of the authority's services being taken over by central government.

In addition, much of the legislation, particularly in relation to larger projects, has emanated from the European Union's (EU) drive for a common market in goods and services. Serious breaches of the law may therefore result in the European Commission (EC) taking action against the responsible organisations or governments, and the Commission has the power to impose fines and/or to require organisations to modify their practices to ensure compliance.

A final influence upon the procurement of supplies and services by some sections of central government is the Government Procurement Agreement (GPA) which arises out of Britain's membership of the World Trade Organization (WTO).

The procedures required to ensure compliance with EC Regulations and the GPA are considered in detail in the section on 'Procurement rules: European Community principles and procedures'.

What constitutes the public sector?

For the purposes of EC legislation, contracting authorities are defined in the EC Directives as being all bodies governed by public law. These are then subsequently defined as:

> 'any body established for the specific purpose of meeting needs in the general interest and not having an industrial or commercial character, which has legal personality and is financed for the most part by the State, or is subject to management supervision by the latter.'

The WTO, however, for the purposes of the GPA, defines public authorities as being confined to central government and its agencies.

The public sector in Britain therefore comprises a wide range of organisations including, but not limited to:

- Central government
- Local authorities (including police and fire authorities, etc.)
- The National Health Service
- Public utilities (the majority of which have now been privatised but are still subject to some public sector procurement legislation)
- Other entities, including QUANGOs and higher education institutions

Central government

Central government in Britain has historically been accused of taking a somewhat autocratic attitude towards its role as a client of the construction industry, imposing upon contractors its own unilaterally written standard forms of contract with little attempt to make any rational allocation of risk between the parties to the contract.

Lately, however, following the Latham review of construction industry practice (Latham 1994) and the subsequent requirements of Part 2 of the Housing Grants, Construction and Regeneration Act 1996 (the Construction Act), Government has publicly indicated its intention to become a 'best practice' client. A review of government construction procurement (Cabinet Office Efficiency Unit 1995) recommended that the GC/Works suite of contracts should be revised both to meet the requirements of the Construction Act and to ensure that they reflect a fully integrated approach to roles and responsibilities in accordance with the Latham recommendations. The GC/Works suite at the time of writing (mid 2005) is outlined in Box 4.1.

Some experimentation has been reported (PACE Central Advice Unit 1998) with the use of the Engineering and Construction Contract (NEC ECC), and the latest version, Edition 3 published in the summer of 2005 (NEC 2005), has been endorsed by the Office of Government Commerce (OGC). It therefore seems likely that the use of the GC/Works suite as the exclusive Government contract may be limited, and that Government will in due course adopt the NEC ECC as its preferred standard form.

Government has also been an enthusiastic supporter of, and participant in, bodies such as the Construction Industry Board (CIB) and the Construction Clients' Forum (CCF), both of whom advocate a less adversarial and more collaborative approach to construction procurement. They have also actively encouraged and supported the use of collaborative approaches such as partnering, in the form of

Box 4.1 The GC/Works suite of contracts 1998.

GC/Works/1: Intended for high value major building and civil engineering work

GC/Works/2: Intended for moderately costly building and civil engineering where GC/Works/1 is considered too complex

GC/Works/3: Used for mechanical and electrical installations

GC/Works/4: Intended for small works including building, civil engineering, mechanical and electrical work

GC/Works/5: Used for the appointment of consultants

both single project agreements and long-term strategic alliances with a small number of 'prime' contractors.

Responsible agencies: the Office of Government Commerce (OGC)

Virtually every branch of central government is involved in some way with placing contracts for construction projects and services, and, in general, property management services are provided through executive agencies. The Defence Estates Agency (DEA) provides services to the military estate, whilst the Office of Government Commerce (OGC) provides advice and guidance in respect of the civil (i.e. non-military) estate.

The OGC is an independent Office of the Treasury established as part of the Government's efficiency and modernisation agendas. The establishment of the agency builds largely upon a continuous series of government-sponsored pieces of work ranging over a period of some five years including:

- Sir Michael Latham's 1994 report *Constructing the Team.*
- Sir Peter Levene's 1995 report *Construction Procurement by Government: An Efficiency Unit Scrutiny.*
- A Bath University study commissioned jointly by the Government Construction Client's Panel and the Treasury Procurement Group into Government's performance as a client.
- The Defence Estates Agency *Building Down Barriers* project.
- Treasury benchmarking studies of performance on a sample of government construction contracts carried out during 1998/99, which showed that more than half of all of the projects studied exceeded their pre-tender budgets, and two thirds were completed late.

The OGC's mission is therefore set as 'to work with the public sector as a catalyst to achieve efficiency, value for money in commercial activities and improved success in the delivery of programmes and projects'. This mission statement is supported by the following general objectives:

- To be regarded as a trusted advisor to public sector organisations in supporting the successful and efficient delivery of their objectives.
- To improve public services by working with departments to help them meet their efficiency targets – amounting to £21.5 billion a year by 2007/2008.
- To deliver a further £3 billion saving by 2007/2008 in central government civil procurement (building on the savings achieved since 2000) through improvements in the success rate of programmes and projects and through other commercial initiatives.
- To improve the success rate of mission critical programmes and projects.

The OGC is therefore intended to provide 'best practice' advice to enable and encourage other government departments to meet their requirements in the most cost-effective way. Services are provided in four distinct areas as follows:

(1) *Area 1 – better projects.*
This service area includes the Gateway process (described below), embedding centres of excellence, and mission critical reporting, intervention and support. OGC can provide review teams of various types to assist government departments with the project review process, depending upon the perceived degree of project risk.

(2) *Area 2 – access to skills and know how.*
This service area helps government departments to improve delivery by focusing on programme, project and commercial activities. It also develops best practice and a related skills framework, working with departments and central bodies to strengthen departments' capabilities and capacities to deliver, amongst other things, property and construction projects.

(3) *Area 3 – smarter procurement.*
This service area helps public sector organisations in general to get better value from their procurement activities by:
- improving understanding of EU and UK procurement policies;
- collecting information about demands on key markets;
- developing supplier relationships with key suppliers;
- improving the perception of the public sector as a client by reducing bureaucracy and barriers to competition;
- promoting modern procurement methods such as eProcurement and eAuctions.

(4) *Area 4 – the Efficiency Programme.*
The Efficiency Programme aims to drive forward and coordinate implementation of the Government's Efficiency Review by enabling departments and the wider public sector to make fundamental step changes in the way they carry out their business. The Efficiency Programme aims to create a culture of continuous improvement and a reduction in bureaucracy, thus releasing more resources for 'front line' activities.

In addition to the above mentioned service areas, the OGC Corporate Services Directorate (CSD) provides a range of services to both internal and external customers. The external services include the promotion and delivery of a programme of events and conferences, the OGC Service Desk and the website at www.ogc.gov.uk.

OGC also operates an agency, OGCbuying.solutions, the aim of which is to assist OGC to deliver value-for-money gains for both UK central government and the wider public sector. The agency helps customers achieve major savings through the provision of a comprehensive range of products and services that have been through a rigorous competitive tendering process and quality evaluation. It aims to make the procurement process as efficient and effective as possible by offering value for money, fitness for purpose and compliance. It delivers services in the fastest and most appropriate way possible by making optimum use of technology. It provides a platform for suppliers, which allows customers to order products and services online through several catalogues.

The OGC is also responsible for the Achieving Excellence in Construction initiative. This initiative, which is at the heart of Government's drive to become an 'intelligent client', was launched in March 1999 to improve the performance of central government departments, executive agencies and non-departmental public bodies as clients of the construction industry. It put in place a strategy for sustained improvement both in construction procurement performance and in value for money achieved on construction projects, including those involving maintenance and refurbishment. Key aspects include:

- The use of partnering and development of long-term relationships
- The reduction of financial and decision-making approval chains
- Improved skills development and empowerment
- The adoption of performance measurement indicators
- The use of tools such as value and risk management and whole-life costing

Key findings from the first three years of the Achieving Excellence in Construction initiative and the future strategy are set out in OGC (2003).

Procurement guidance

The drive for Government to become a 'best practice' client has given rise to the issue of a considerable volume of procurement guidance and administrative advice, initially from the Treasury and agencies such as the Procurement Advisors to the Civil Estate (PACE), and latterly from the OGC, aimed at providing a modern framework to guide those involved in commissioning construction work and placing contracts. The documents comprising the present set of *Achieving Excellence Procurement Guidance* are shown in Box 4.2. These documents consolidate and build upon other previous literature such as the *PACE Guide to the Appointment of Consultants and Contractors (GACC)* and a series of documents produced by the Construction Industry Board including *Selecting Consultants for the Team*, *Code of Practice for the Selection of Main Contractors* and *Briefing the Team*, reflecting developments in construction procurement over recent years and building upon OGC experience of implementing the Achieving Excellence initiative. The considerable library of

Box 4.2 OGC Procurement Guidance.

PG01	Initiative into Action
PG02	Project Organisation
PG03	Project Procurement Lifecycle
PG04	Risk and Value Management
PG05	The Integrated Project Team
PG06	Procurement and Contract Strategies
PG07	Whole-Life Costing
PG08	Improving Performance
PG09	Design Quality
PG10	Health and Safety
PG11	Sustainability

working documentation and procedures now available in the public sector, largely accessible through the OGC website, coupled with a major drive to educate and train civil servants to effectively perform the client's role has enabled the partnership of Government client, consultants and contractors to be much more productive than may have been the case in the past.

The OGC Gateway process

The OGC Gateway process, initially developed from the PRINCE2 project management system, is a structured project management methodology which examines a programme or project at critical stages in its life cycle in order to provide assurance that it can progress successfully to the next stage. The process is based on well-proven techniques, and if correctly followed should lead to more effective delivery of benefits together with more predictable costs and outcomes. It is designed to be applied to a wide range of project types, and the intention is that all major central government schemes are now subject to Gateway review.

Although the Gateway process was initially developed for central government schemes, the OGC is increasingly encouraging its use by other public sector organisations such as local authorities.

In simple terms, the Gateway process consists of a series of project reviews carried out at key decision points in the project life cycle by an experienced team of specialists independent of the project team. It therefore provides a useful and unbiased reality check on project progress and likely project performance. A 'traffic light' system is used at the end of each Gateway review process. A green light allows the project to go ahead to the next stage, an amber light indicates that although the project may proceed to the next stage there are serious concerns which the project team must address as a matter of urgency, and a red light means that the project will not be allowed to proceed until the review panel's concerns have been addressed and the project reassessed.

There are six Gateways at which reviews may typically be carried out:

(1) *Gateway 0 – strategic assessment.*
Establishes the business need for the project, and considers issues such as:
- Does the project comply with sustainability criteria?
- Is the designer clearly defined?
- Is the allocated timescale adequate to meet the need?

(2) *Gateway 1 – business justification.*
Examines the business case for the project, and considers issues such as:
- Is the brief sufficiently well developed?
- Are stakeholders' requirements likely to be met?
- Is there an outline risk allocation and management strategy in place?
- Is the project team clear and agreed on what the project is to deliver?

(3) *Gateway 2 – procurement strategy.*
Examines whether an appropriate procurement route has been chosen to procure the integrated project team, and whether the proposed commercial arrangements will deliver value for money. Also considers questions such as:

- Is sufficient attention being paid to health and safety?
- Are sustainability requirements adequately addressed?
- Is appropriate weight being given to design quality?
- Does the procurement route comply with government guidance, i.e. PFI, prime contracting or design and build?

(4) *Gateway 3 – investment decision.*
Examines whether the project makes sense as an investment. At this stage the scope of the project is finalised, and a maximum budget based on a whole-life cost basis is fixed. Key questions will include:
- Does the project team have appropriate health and safety provisions in place, together with a clear commitment to zero accidents?
- Are on-site processes sustainable?
- Does the design optimise whole-life cost and time as far as possible?
- Is the project buildable?
- Is the project team, including the contractor, integrated and committed to continuous improvement?

(5) *Gateway 4 – readiness for service.*
Carried out after construction, when the facility has been commissioned and is ready for use. Key questions include:
- Were health and safety risks properly managed?
- Does the completed project comply with health and safety requirements?
- Is a health and safety file available for use by the facility manager?
- Does the completed project meet or exceed sustainability targets?
- Has continuous improvement been achieved?
- Is all necessary operating documentation available?
- Have key lessons been disseminated to inform subsequent projects?

(6) *Gateway 5 – benefits evaluation.*
Checks whether or not the expected business benefits have been/are being achieved. There may be several Gateway 5 reviews during the life of the project, and sensible use of this provision should be used to inform facilities management (FM) decisions and future investment decisions during the operating life of the facility. Key issues will be concerned with the following:
- Is the facility safe and compliant with health and safety legislation?
- Are sustainability targets being achieved?
- Does the completed facility meet stakeholder expectations?
- Has the project team completed a post project review against the agreed project KPIs?
- Has a post implementation review been carried out and the results appropriately acted upon?

Local government

The ability of local authorities to enter into contracts is governed generally by the provisions of their standing orders and financial regulations. These regulations, which may vary significantly in detail from one authority to another, although generally not in substantive content, are drawn up to guide both officers and elected members in the performance of their duties, and to ensure that the authority meets the requirements of all applicable legislation and of the district auditor.

In many cases in the past standing orders and financial regulations have been extremely prescriptive, leaving little scope for officers or elected members to use their professional judgement. However, the new obligations imposed by Government under the best value regime (see below) and other associated legislation now require authorities to adopt much more flexible ways of delivering their services and of dealing with their contractual partners. Accordingly many local authorities have embarked upon major revision and updating of both their standing orders and financial regulations. In most cases a primary objective of the review has been to move away from a prescriptive set of regulations which must be slavishly followed, towards a minimum set of regulations designed to uphold the basic tenets of transparency, public accountability and value for money, whilst at the same time allowing the authority's officers and members to use their professional skill in developing innovative ways of improving performance.

Local authorities have historically used the Joint Contracts Tribunal (JCT) suite of contracts, generally unamended, and many local authority property departments have historically prided themselves on their even-handedness in the way their contracts have been administered. However, the endorsement by the OGC Edition 3 of the New Engineering Contract Engineering and Construction Contract (NEC 2005) is now persuading many authorities and other public sector bodies to move away from their historical JCT allegiance.

'Best value'

In 1997 the British Government replaced the previous Compulsory Competitive Tendering (CCT) regime by a strategic 'best value' initiative designed to encourage public sector organisations to maximise value for money in their organisations. In short, the best value regime imposes upon local authorities and many other public bodies a duty to demonstrate that they are achieving 'best value' for the taxpayer, in accordance with published and auditable criteria.

In the case of property and construction, government advice states that:

- Procurement should support the principles established by the 1994 Latham review (Latham 1994), and should aim to achieve the targets set out in the subsequent Egan review (Egan 1998).
- Procurement on the basis of price as the only choice criterion is unlikely to secure best value.
- Overall value for money should be assessed on the basis of whole-life cost not initial capital cost.
- Project delivery mechanisms should encourage innovation by all concerned in the search for more flexible and cost effective project delivery mechanisms.
- Delivery mechanisms should incorporate partnering principles supported by good risk and value management.

The imposition of the best value regime has given rise to considerable changes in the way construction work is procured, and has led directly to the development of projects based upon partnering, target cost contracts, the use of initiative and pain share strategies, and the use of benchmarking and key performance indicators to promote continuous improvement.

The National Health Service (NHS)

The National Health Service has long maintained detailed procedures for the procurement of construction work. Following the various health service reviews that have taken place in recent years, these procedures have been radically updated and modernised to reflect the increased use of privately financed procurement options. Standard procedures have now been developed for PFI schemes and for the construction of publicly funded hospitals.

Of particular interest is the ProCure21 approach developed for the provision of major hospital projects exceeding £1 million in value.

ProCure21

NHS ProCure21 has been developed to deliver publicly funded schemes with a works cost of above £1 million across the NHS. The programme, launched nationally in April 2000, is estimated to have a value of £1.2–1.4 billion per year (at 2004 prices).

The main objectives of the programme are to promote better capital procurement in the NHS through the use of modern techniques based upon the ideas put forward in the Latham and Egan reviews. The approach is therefore based upon a series of long-term strategic partnering arrangements with carefully chosen partners, termed 'principal supply chain partners' (PSCPs), established under framework agreements. Projects are delivered using a design-and-build approach based upon functional output specifications, and PSCPs are much more than simply construction contractors. Each PSCP consists of a complete supply chain including designers, constructors and facilities managers, and at the time of writing a total of 12 PSCPs have been chosen through an extremely rigorous selection process.

At the same time work has been carried out to ensure that the NHS achieves 'best client' status, through the adoption of the Achieving Excellence in Construction and Sustainability agendas.

The key benefits of the ProCure21 initiative are stated to be the provision of improved health care facilities for all stakeholders through:

- Improved scheme management as the result of better understanding between PSCPs and the NHS
- Improved cost and time certainty
- Consistent high quality in design and construction including the fostering of 'zero defects', leading to improved stakeholder satisfaction

- Improved financial security
- Joint skills development
- Improved value for money over the life of the programme through reduced transaction costs, embodied learning, better risk and value management and the optimisation of whole-life costs
- Faster start on site due to pre-accredited supply chains, avoiding the need for separate EU tender notices for each project and thereby saving between 6 and 12 months on the pre-construction programme

Procurement rules: European Community principles and procedures

Non-discrimination and transparency: the role of the EU

A key objective of the original European Economic Community, enshrined in the Treaty of Rome and confirmed in the Maastricht Treaty, was the expansion of trade both within the community and by the community members with the rest of the world. Achievement of this objective required the creation of an internal market which presented a 'level playing field' for all community members, and accordingly the European single market was finally established in 1993. The primary aim was to create 'an area without frontiers in which the free movement of goods, persons, services, and capital is ensured'. This in turn brought forward the need to abolish restrictive trading practices, to encourage the free movement of capital and labour within Europe, and to establish closer ties between the Union members. In economic terms it was envisaged that harmonisation of the market in this way would bring about benefits to the public sector amounting to around 0.5% of GDP, or 20 billion ECU (Commission of the European Communities 1988).

Public sector construction output has historically constituted a significant component of the total value of public sector procurement, and the economic significance of the construction industry has therefore ensured that it has received considerable attention. Harmonisation of the procurement of public sector construction works and services has thus long been seen as a major plank in the successful achievement of a common market.

The principal objective of the European Commission in attempting to regulate the procurement of public services and works has consistently been to eliminate unfair competition within the European Union through the establishment of a legal framework which abolishes: (1) discrimination on the grounds of nationality and (2) preferential treatment through the favouring of 'national champions' or the use of narrowly defined national standards.

Regulation has therefore concentrated purely upon the demand side of the market, by attempting to force all purchasing authorities into a consistent procurement framework whilst completely ignoring the supply side of the market place. This strategy is not without its critics (see for example Bovis 1998).

The Directives

Problems of disparity between the public law systems of the members of the EU meant that simply introducing primary legislation to regulate procurement issues would have been highly unlikely to achieve the required result, and the Commission therefore introduced the measures through the use of Directives. EU Directives are unique legal instruments which carry binding force within each member state to which they are addressed, and which must, under the EU 'rules for membership', be incorporated into member states' legal systems. They are, however, highly flexible in terms of their implementation. In essence, although the Directives specify the objectives to be achieved and the time frame within which they must be brought into force, individual states are allowed considerable freedom to determine the precise form in which the required legislation is incorporated into their own individual legal processes.

The EU has therefore published a number of Directives to regulate public sector procurement, the primary objective of which is to ensure that the European market for public sector works and services operates in an acceptably open and transparent way without favouritism or discrimination on the grounds of location, race or nationality. In addition to the Procurement Directives, other Directives that have a major impact upon the construction industry are those concerned with: (1) the free movement of labour and mutual recognition of qualifications; and (2) the harmonisation of construction standards and specifications.

It must be clearly understood that the EU's purpose in introducing the legislation is solely concerned with the need to uphold the functioning of the single European market. The legislation is not intended to ensure that customers necessarily get maximum value for money for any specific project. Indeed questions have been raised as to whether compliance with the rules aids or obstructs public sector clients in this respect, either with individual projects or across their construction programmes as a whole.

EU Directives relating to the procurement of public sector works and services

The Directives concerning public procurement have as their prime objective the co-ordination of national procedures for the award of public contracts. However, the intention is not to prescribe every detail of a rigid procurement process to which everyone should conform, but rather to put in place sufficient safeguards to ensure that the common market principles of fair trade, transparency and non-discrimination are enforced.

The procurement Directives collectively and individually have at their core three fundamental principal requirements:

(1) EU-wide advertisement of all public contracts above a predetermined threshold value
(2) Harmonisation of technical specifications such that discrimination on the grounds of national standards alone is abolished

(3) Consistent application of transparent and objective criteria in tendering, assessment and award procedures

The principal Directives in force at the time of writing (September 2005) governing the procurement of public sector construction works and services are each considered below.

The Supplies Directive and subsequent amendments

Public supplies contracts are defined as:

> 'contracts for pecuniary consideration concluded in writing between a supplier (natural or legal persons) and public authorities, which include the state, local and regional authorities and bodies governed by public law, having as their objective the delivery of goods.'

The Works Directive and amendments

Public works contracts are defined as:

> 'contracts for pecuniary consideration concluded in writing between a contractor (natural or legal persons) and public authorities, which include the state, local and regional authorities and bodies governed by public law, having as their objective the completion of works/construction projects.'

Note that the Works Directive also includes management contracts for the design and execution, or the execution only, of works projects. The relevant requirements were enacted in Britain by the Public Works Contract regulations 1991. For construction projects, 'contractors' comprise any legal or natural person engaged in construction activities, whilst 'contracting authorities' include all bodies governed by public law, and the UK Government has decided that this should include publicly funded bodies responsible for social housing such as housing associations and arms length management organisations (ALMOs). Note, however, that the WTO provisions define 'public authorities' simply as central government and its agencies.

Note also that the project does not have to be completely funded by the public sector. For certain civil engineering work, hospitals, sports, recreation or leisure projects, school or university buildings and administrative buildings there is an obligation to comply with the Directives in respect of any contracts that are directly subsidised by more than 50%. However, the rules only appear to apply to direct financial subsidies. Projects that benefit from subsidies in kind, for example the free provision of land, are excluded.

There are a number of other exclusions including:

- Works contracts in the defence sector that are declared to be secret or where the state's basic security interests may be compromised.
- Contracts awarded under certain international agreements or in accordance with the established procedures of an international organisation such as NATO.
- Concession contracts, which are subject to a special procedure and only need to comply with the regulations for advertising.

Note also that this Directive does not attempt to create a common public procurement procedure within the EU; instead it establishes a common set of rules to govern the advertisement and notification of qualifying projects, tendering procedures, selection of contractors and the award of contracts.

The Utilities Directive and the Utilities Remedies Directive

In construction terms these Directives are concerned with public sector supplies and works contracts in the water, energy, transportation and telecommunications sectors, and with ensuring that affected parties can seek redress in cases where the Utilities Directive has been breached.

The Utilities Directives are in many ways the most complex. In theory, construction works and services in these sectors could have been covered by the Works and Services Directives, but the large number of private sector operators in this area would have meant that the definition of 'contracting authorities' would have become unacceptably broad. On the other hand the public service nature of this sector of the market meant that to leave it unregulated was not acceptable either. Hence a separate directive was developed simply to control the utilities. Some major problems with this are as follows:

- The legal status of many public utility undertakings is often unclear in that they may act both as statutory undertakings and as private companies. The nature of their contractual obligations may therefore vary depending upon whether or not they are acting in the capacity of statutory undertaker.
- There are many exemptions from the Directives including, for example, radio and television broadcast systems, bus, rail and similar transport operating companies, research and development organisations, companies engaged in prospecting for gas, oil, coal, etc., and utility services companies supplying less than 30% of their output to the public.

The Services Directive

Public services contracts are defined as:

> 'contracts for pecuniary consideration concluded in writing between a service provider (natural or legal persons) and public authorities, which include the state, local and regional authorities and bodies governed by public law, having as their objective the provision of services, as defined in the United Nations Product and Service Classification nomenclature.'

This classification incorporates virtually everything, including 'design contests' such as architectural competitions with or without prizes.

Services are classified into priority and non-priority services. Priority services are subject to the full rigour of the public procurement regime, whereas non-priority services are subject only to the basic rules of non-discrimination and publicity of the results of the award. Maintenance and repair services, architectural, engineering, urban planning, landscape design and associated scientific and technical consultancy services are classified as priority services. Note that management contracts for the design and execution of projects are excluded and fall under the scope of the Works Directive.

Specific exemptions relating to the construction and property industries include:

- Contracts for the acquisition of land, existing buildings or other non-moveable assets or rights therein. Note, however, that any associated financial services contracts are subject to the regulations.
- Contracts for arbitration or conciliation services.

The Public Sector Directive

The Directives as described above have been heavily criticised for their complexity, and the fact that certain provisions, for example framework agreements, were included in some Directives and not in others. Consultation on provisions to reform the Directives has therefore been ongoing since 1996, and new draft Directives were published in the *Official Journal of the European Union* in April 2004, with the provision that the new Directives must be transposed into national legislation by 31 January 2006.

The major revision included in the new provisions is the consolidation of the three previous Directives for public works, supplies and services into a single new text termed the Public Sector Directive.

Many of the basic provisions in the existing Directives, including overall methodology, timescales and thresholds, remain in place, although the actual threshold values and timescales have been altered by the new provisions. These issues are considered later in this chapter. The new Directive does, however, introduce major new provisions in several specific areas relevant to construction including:

- Framework agreements
- Central purchasing bodies
- Electronic auctions
- Sustainability and environmental issues

In addition, a new tendering procedure, termed competitive dialogue, has been introduced to supplement the existing open, restricted and negotiated procedures. Each of these issues is considered in detail later in this chapter.

Tendering

All qualifying public sector construction projects, and the procurement of services to the public sector, must be advertised (free of charge) in the *Official Journal of the European Union* (*OJEU*). Individual contract notices are required for each project as it arises, and timescales are specified for various phases of the tendering process. The timescales are considered in detail later. Contracting authorities may, however, publish a single prior indicative notice (PIN) giving an annual forecast by category of service of all of their qualifying requirements for the following 12 months, and where this is done the timescales for individual schemes may then be reduced.

Advertisements must include basic details of the project, the form of contract to be used and the identity of the employer, and also information such as key dates, an estimate of the likely cost, details of the tender award criteria and the documentary evidence that will be required in order to make the necessary financial and technical appraisals. Once a tender has been accepted a tender award notice must be published within a specified time.

There are standard model forms of advertisement and tender award notification which can be downloaded from the EU website.

Note that the regulations do not prohibit authorities from advertising projects whose value is below the threshold in the *OJEU* if they wish to do so in order to improve the competition by seeking tenders from European sources. Provided the project value is estimated to be at least £66 000, authorities may advertise the project through the issue of a voluntary notice.

In addition to the printed *OJEU*, the EU offers the tenders electronic daily (TED) service. This subscription-based service provides an on-line database of current contract notices updated daily.

Value thresholds

Only projects in excess of certain specified values are subject to the EU requirements. EU threshold values are expressed in Euros and are reviewed every two years. The Treasury calculates annual conversion to sterling values, which are available on the OGC website from 31 January each year.

Note that, in general, the rules apply to single contracts where the value exceeds the threshold. It is, however, illegal to subdivide a large project into a number of small contracts simply to avoid the regulations. Where a contract is let in this way, the value of all of the separate contracts must be added together in order to ensure that the rules are not circumvented, and the general principle is that each of the separate contracts must be let in accordance with the rules.

A typical example would be the use of a construction management procurement route where the project as a whole is subdivided into a series of work packages, each of which is let as a separate contract. Similar arrangements occur in a number of European countries, the most common being variations of the French *lots séparés* approach. In this case the criterion for assessing whether or not the project falls within the regulations is the value of the whole project not the value of each of the individual packages. It is, however, recognised that the use of this type of approach may give rise to some small contracts which it would plainly not be sensible to advertise in a pan-European arena, and the regulations make special provision that such work packages may be exempted provided that the total of all such contracts in any specific project does not exceed 20% of the estimated value of the project as a whole.

Tendering procedures

The present Directives permit open, restricted and negotiated tendering arrangements, and from January 2006 the new Public Sector Directive will permit the use of the competitive dialogue procedure. However, the Directives limit the circumstances in which the restricted and negotiated procedures may be used for the procurement of works. The Directives do though allow prequalification of contractors, leading to competitive tendering on the basis of a selected list.

Open tendering

The Directives stipulate that open tendering should, wherever possible, be the norm since it maximises competition. Contracting authorities are, however, free to opt for restricted procedures if they wish, provided that the decision can be justified on the grounds of the nature of the work and the administrative cost involved in assessing tenders. Construction work is therefore most often let under the restricted arrangements.

Restricted tendering

Restricted tendering permits contracting authorities to invite only specified contractors to submit final tenders. The process takes place in two rounds. In the first round all interested contractors may express an interest, and from this group the contracting authority may then select a smaller number of contractors to submit detailed tenders. In principle the minimum number of bids invited is five, but there may be occasions when a smaller number can be justified.

Negotiated tendering

Negotiated tendering, where a contracting authority negotiates directly with one or more bidders, is permitted by the Directives, but is subject to a more rigorous set of procedures and conditions. Negotiated procedures may take place with or without prior notification.

Negotiation with prior notification must be justified on the grounds of irregular or unacceptable tenders received in response to an earlier call for bids. In this case the selection process takes place over two rounds. In the first round any interested contractors may submit an expression of interest, and from this pool the contracting authority chooses one or more companies with whom to negotiate. In principle the minimum number of firms passing through to the second round is three, provided that there are sufficient acceptable firms arising from the first round.

The regulations governing a contracting authority's ability to negotiate directly with a single firm with no prior notification are very restrictive. The cases where single tendering is allowed are limited to:

(1) Where no suitable supplier was found in a previous open or restricted tender because no, or only irregular, bids were received, or because the bids submitted were unacceptable under national provisions that are consistent with the Community rules on public sector procurement. There is also a proviso that the original terms of the contract (e.g. financing, delivery dates and in particular the technical specifications of the products) are not substantially altered. If these conditions are not met then the whole procedure must be recommenced with a re-advertisement of the tender.

(2) Where, for technical or artistic reasons, or because of the existence of exclusive rights, there is only one supplier in the Community able to supply the product.

(3) Where the product is manufactured purely for the purposes of research, experiment, study or development.
(4) In cases of extreme urgency resulting from unforeseen circumstances not attributable to the action of the purchaser, where the time-limits laid down in tendering procedures cannot be observed.
(5) For additional deliveries by the original supplier required either as part replacement of regular supplies or equipment, or to extend existing supplies or equipment, where a change of supplier would compel the contracting authority to purchase equipment having different technical characteristics which would result in incompatibility or disproportionate technical difficulties of operation or maintenance.
(6) For goods quoted and purchased on a commodity market in the EU.
(7) Where supplies are classified as secret or where their delivery must be accompanied by special security measures under the law of the member state of the purchaser, or where the protection of the basic interests of that state's security so requires it.

Competitive dialogue

Competitive dialogue is a new procedure intended for use when contracting authorities believe that use of the open or restricted procedure will not allow them to award a satisfactory contract. Examples of such a case may include particularly complex contracts, where authorities may know what their needs are, but may be unable to decide the most appropriate technical, legal or financial solutions. Contracting authorities may also wish to encourage innovative solutions, or may be unable to objectively assess what the market has on offer.

In cases such as these a competitive dialogue between authorities and providers may be useful to identify the most appropriate solution. Since such a dialogue would be prohibited under the old Directives, the new competitive dialogue procedure has been introduced. The Commission also envisages the competitive dialogue procedure being used for many public private partnerships where contracts are complex and the legal and financial structure cannot be determined without dialogue with suppliers.

The procedure works by allowing contracting authorities to call for proposals on the basis of a functional specification of their needs. Submitted proposals are then assessed, following which authorities may choose to enter into a dialogue with a number of selected providers. The purpose of the dialogue is to enable authorities and bidders to achieve agreement over precisely what is to be provided, and once this position is reached potential providers are invited to submit tenders. Tenders are then assessed in the normal way using the assessment criteria specified in the original contract notice.

Note that the competitive dialogue process may be carried out in successive stages in order to progressively reduce the number of potential solutions by applying the particular award criteria, and clarification may be sought by either party either pre or post final assessment, provided that this does not distort competition and is not discriminatory.

The procedure may only be used with 'most economically advantageous' award criteria, and a statement of the award criteria to be used, together with the statement of the contracting authority's needs, must be included either in the contract notice or in the accompanying documentation. Where a prior indicative notice has been issued, then use of the competitive dialogue procedure should be considered before authorities consider use of the negotiated procedure.

Framework agreements

Framework agreements in construction are typically used in the case of continuing programmes of work where the contracting authority wishes to enter into an arrangement with one or more contractors to set up a kind of standing offer under which packages of work may be procured at some time in the future but where the scope, value or timing of the work packages cannot presently be defined with any degree of accuracy. The primary advantage of framework agreements is that the contracting authority can greatly reduce the transaction costs that would otherwise be incurred in a number of different tendering exercises.

Under the existing Directives only the Utilities Directive mentions framework agreements, but in recent years their use, based upon the Utilities Directive model, has become well established. This has now been recognised by the Commission, in the new Public Sector Directive.

Framework agreements in the context of the Directive are agreements to agree, and they may be used in conjunction with any procedure. They are not therefore contracts in themselves, but, primarily in construction contracts, set out the terms of price and quality under which future contracts, which will be subject to EU rules, will be formed at the call-off stage. Note that this does not mean actual prices should be fixed, but merely that the framework should be capable of establishing a *pricing mechanism*. It should also be possible to establish the scope and types of goods and/or services that will need to be called off.

The UK has, in the past, taken the view that the only sensible approach to such agreements is to treat them as if they are contracts in their own right and to apply the EU rules accordingly. The usual practice has therefore been to advertise the framework itself, in the OJEU, and to follow the EU rules for selection and award. The framework therefore establishes the fundamental terms on which subsequent contracts will be awarded with an acceptable degree of transparency, and removes the need to advertise and apply the award procedures to each call-off under the agreement. This practice is now essentially confirmed in the new Directive, and normal timescales and threshold values will apply, but there are some important points to note, as set out below:

(1) *Framework duration.* The duration of a framework agreement is limited to 4 years unless there are exceptional circumstances that would allow a longer term to be justified. In most cases this will depend upon the subject matter of the framework agreement. It has been proposed by some that framework agreements for construction projects that are in themselves of comparatively

long duration might meet this criterion, but the issue of what constitutes 'exceptional circumstances' has not, at the time of writing, been definitively defined. The contract notice must state the planned duration of the framework.

(2) *Number of framework suppliers.* Where there is only one framework supplier, work may be awarded directly provided the terms of the framework agreement are applicable. Where a single appointment has not been made, then the minimum acceptable number of framework providers is three, or the number passing the selection criteria if less.

(3) *Call-offs.* Where only a single framework supplier is appointed, or where the framework is sufficiently precise to allow one particular supplier to be identified to fulfil a particular need, then call off the work from that supplier. If, however, there are a number of framework suppliers all of whom could meet the need then a mini-competition is required to identify the most appropriate supplier.

(4) *Framework conditions.* In general the original framework terms and conditions must be used to award work. If, however, in a particular case, there is the need to refine or supplement the original conditions, a further mini-competition will be necessary involving all those companies in the framework who are capable of providing the goods or services. Existing framework suppliers must be consulted in writing, and bid periods must be reasonable, taking account of the particular circumstances. The award criteria for the mini-competition must be based on the criteria used to set up the original framework agreement.

Framework agreement call-offs can extend beyond the life of the framework agreement itself provided that the period is reasonable and is consistent with the trend established throughout the life of the framework agreement itself.

However, there is no obligation on contracting authorities to place any work with framework suppliers. They may use the framework to procure specific parcels of work if they wish, but they are at liberty to go elsewhere, provided that if they do they then comply with the normal EU rules to place the contract for the work.

Note also that the new Directive explicitly recognises that contracting authorities may purchase through central buying organisations. Contracting authorities that act as central buying organisations (for example consortia of housing associations or local authorities) may therefore set up framework agreements on behalf of other contracting authorities. The contracting authorities entitled to call-off under the terms of the framework agreement must, however, be identified in the documents used to set up the agreement. Authorities permitted to use the agreement can be named or a generic description may be used, for example government departments, local authorities, utilities or all-UK contracting authorities.

If the framework establishes contractual rights and obligations, for example by guaranteeing a specific volume of work to a supplier, then it is a framework *contract* not a framework *agreement* as defined by the Directive. Framework contracts are permissible and must be treated as any other contract for the purposes of the EU rules.

Electronic auctions

Electronic auctions are on-line auctions where selected bidders submit offers electronically against the purchaser's specification. Such systems have been used in industries such as oil exploration and development, but their use in mainstream construction has, to date, been limited. They do, however, offer substantial promise for certain types of construction work provided that the various elements of the auction can be reduced to numerical values. Electronic auctions may also be used on the reopening of competition within a framework (see 'Framework agreements' above).

Electronic auctions are supported by the Public Sector Directive for goods, services and works, and any of the basic tendering mechanisms may be used. It is, however, a requirement of the Directive that all communication following and including the invitation to prequalified bidders to submit new prices and/or values must be electronic and instantaneous. The Public Sector Directive discourages the conduct of electronic auctions for intellectual services.

Where contracting authorities elect to use the most economically advantageous selection criteria, the invitation to participate must be accompanied by the outcome of a full initial evaluation of the first expressions of interest and the mathematical formula to be used in the auction to determine automatic rerankings. The formula used must be based on the declared weightings which, if they were initially expressed as a range, must be reduced to a single value. Separate formulae must be provided for variants where these are permitted.

The auction process is basically carried out in two stages:

- *Stage 1*: a contract notice is placed, inviting expressions of interest in accordance with the selected tendering procedure, setting out the criteria against which they will be assessed. The contract notice must state that an electronic auction is to be held. All expressions of interest are then evaluated, and admissible bidders then proceed to stage 2.
- *Stage 2*: electronic invitations to participate in the auction are sent to the bidders including the criteria to be subject to auction. It is self-evident that, for electronic auctions to work effectively, only criteria that can be expressed as numerical values suitable for incorporation within a formula can be used. The invitation should also set out:
  - the assessment formula;
  - details of the process and the conditions of bidding, including arrangements for bringing the auction to an end. The most common arrangement for closing the auction is to set a fixed closing date and time, but other arrangements, for example closing the auction when minimum price difference criteria between consecutive bids are reached, are also permissible;
  - connection details and date and time of the auction. Note that the auction cannot start sooner than two working days after transmission of the invitation.

Bidders then bid electronically against each other on all of the bid criteria. During each phase of the auction, sufficient information to enable bidders to ascertain their relative ranking in relation to the predefined selection values must be communicated instantaneously. Other bidders' prices and values may also be communicated provided this is stated in the specification, and provided that they remain anonymous. The number of participants bidding may, however, be given. Bidding continues until the conditions set for closing the auction are reached.

Sustainability and environmental issues

The Public Sector Directive addresses a number of issues relating to sustainability and the environment. Those of particular interest in construction are discussed below.

Specifying sustainability requirements and the selection of tenderers

Relevant environmental and social requirements may be specified in the tender documents and used in the assessment of tenderers provided that:

- They are sufficiently well defined to allow both bidders to understand the requirement and the factors to be appropriately assessed during contract award.
- They are compatible with Community law.

Typical environmental issues may include things like the need for contractors to be ISO 14000 accredited, specification of environmentally sustainable materials or the incorporation of appropriate principles within design-and-build schemes. Typical examples of social issues within construction contracts which have been used in the past include requirements to promote the use of local labour, to provide local training opportunities or to provide community facilities, but contracting authorities need to be careful that such provisions do not fall foul of EU antidiscrimination legislation.

Environmental and social issues may also be relevant to the tenderer's track record, grave misconduct findings and technical capacity and ability.

Sheltered workshops

Provided that appropriate provision is made in the contract documentation, relevant contracts or the supply of materials may be limited to sheltered workshops, employing mostly disabled people unable to gain normal employment. At the time of writing, work is under way to define the term 'sheltered workshops' more precisely in order to avoid the possibility of discrimination.

Use of variant bids

Contracting authorities are permitted to encourage tenderers to submit a range of offers through use of the provision for variant bids, which may include comparison of different levels of environmental (and other) performance related to the subject matter of the contract. Minimum requirements including environmental and other criteria must be specified in the tender documentation by the contracting authority, and higher standards of performance are then specified in the form of variants against which a range of offers can be generated. Most economically advantageous award criteria must be used.

Use of contract award criteria

'Most economically advantageous' contract award criteria may include environmental and other characteristics provided that these are linked to the subject matter of the contract and are economically advantageous from the point of view of the contracting authority.

Timescales

The regulations stipulate mandatory minimum time periods for various key activities in the tendering process. Accelerated procedures may be used in certain circumstances where the time limits laid down in tendering procedures cannot be observed. Primarily these will be cases of extreme urgency resulting from unforeseen circumstances not attributable to the action of the purchaser.

The minimum time periods that apply under the Public Sector Directive are shown in Table 4.1, which is adapted from OGC (2004).

Table 4.1 Public Sector Directive timescales.

Procedure	Text	Days
Open	Minimum time for receipt of tenders from date contract notice sent	52
	Reduced when a PIN is published (subject to restrictions) to, generally . . .	36
	and no less than . . .	22
	Electronic transmission reduces all of the above by 7 days so that 52 becomes . . .	45
	and 36/22 becomes . . .	29/22
	Full electronic access to contract documents reduces 52 by 5 days so it becomes . . .	47
	This can also be added to the reduction for 52 days to 45 for electronic transmission so it can become . . .	40
Restricted	Minimum time for receipt of requests to participate from the date the contract notice is sent	37
	Electronic transmission reduces the 37 days by 7 days so it becomes . . .	30
	Minimum time for receipt of tenders from the date the invitation is sent	40
	Reduced when a PIN is published (subject to restrictions) to, generally . . .	36
	and no less than . . .	22
	Full electronic access to contract documents reduces 40 days by 5 days so it becomes . . .	35

Award procedures

All tenderers who meet the specified prequalification criteria must be considered to be suitable for tender award unless there are grounds for exclusion. Acceptable grounds for exclusion are few, and the principal ones are:

- Any supplier who:
  (a) is bankrupt or being wound up, has ceased or suspended trading, or is operating under court protection pending a settlement with creditors, or is in an analogous situation arising from national proceedings of a similar nature;
  (b) is the subject of proceedings for bankruptcy, winding-up or court protection pending a settlement with creditors, or national proceedings of a similar nature;
  (c) has been convicted of an offence concerning his or her professional conduct;
  (d) can be shown by the contracting authority to have been guilty of grave professional misconduct;
  (e) has not fulfilled obligations relating to payment of social security contributions under the statutory provisions of his or her country of residence or of the country of the contracting authority;
  (f) has not fulfilled obligations relating to payment of taxes under the statutory provisions of his or her country of residence or the country of the contracting authority;
  (g) has been guilty of serious misrepresentation in supplying information about his or her current standing or past record or his/her financial or technical capacity.
- Where the burden of proof rests with the tenderer, i.e. all of the above except (d) and (g), the procurement authority is bound to accept as satisfactory evidence:
  — for (a), (b) and (c), the judicial record on the supplier or an equivalent document issued by a judicial or administrative authority in the supplier's country of origin or residence showing that none of these cases applies;
  — for (e) and (f), a certificate issued by the competent authority in the member state concerned.

If such documents are not issued by the country in question, or if they do not cover all the cases referred to, then the supplier may instead produce an affidavit sworn before a judicial or administrative authority, notary or any other competent authority in the member state concerned.

The regulations permit tenders to be awarded on the basis of either lowest cost or most economically advantageous tender (MEAT). Tenders awarded in this way will take into account a wide range of factors, but they must all be stated either in the advertisement or in the tender documents. The EU and European Court of Justice (ECJ) have both consistently supported the view that the range of criteria permitted to be used will be drawn very widely.

Unexpectedly low tenders cannot be rejected without proper enquiries being made to establish the validity of the price quoted. The regulations include a non-exhaustive list of example criteria that might be considered, including price, time and technical quality. Life-cycle cost may also be used as a criterion in the case of projects let on the basis of most advantageous solution.

Once a contract has been awarded for either works or services, a written tender report must be prepared showing the contract details, names of the tendering contractors, reasons for selection and rejection, name of the successful contractor and the reasons for the tender award. This report must be submitted to the EU if required. Unsuccessful tenderers are entitled to be informed of the reasons for rejection of their tender within 15 days of receipt from them of a request for the information. If any bidder feels that he/she has been treated unfairly he/she can sue for damages in the European Court, and if it is found that unfair or discriminatory procedures have been used then the tender award may be set aside by the Court.

Enforcement of EU Directives

Effective enforcement of the EU Directives is plainly a fundamental requirement if the common market is to work as planned. Accordingly the EU framework includes the Compliance Directives which state that member states must ensure that any infringement of public procurement provisions must be effectively and expeditiously reviewed. Anyone who risks being harmed by the alleged infringement and who has an interest in obtaining a public supply or works contract is entitled to seek a review before national courts. The EU has also enforced compliance with the procurement Directives by withdrawing Community funds from defaulting member states, an action that is supported by a decision of the ECJ.

It has also been held, by the ECJ, that failure of individual states to implement Directives may give rise to actions for compensation by individuals. *Francovitch and Bonifaci* v *Italian Republic* (cases 6/90 and 9/90) provided that, in order for such a claim to succeed:

(1) The result of the Directive must be to grant rights to an individual.
(2) The specific rights must be identifiable in the Directive.
(3) There must be a clear link between the failure of the state to implement the Directive and the damage caused.

The EU has taken out proceedings in the ECJ against a number of member states accused of non-compliance with the procurement directives. Of particular interest is the Dundalk Pipeline case (*Commission* v *Ireland*, case 45/87) which concerned the incorporation in a contract of particular Irish specifications. The Court found that this was in contravention of the Treaty of Rome in that compliance with the contractual provisions would effectively exclude non-Irish contractors. It was considered that the inclusion of the words 'or equivalent' in the specification would have been sufficient to comply with the regulations.

References

Bovis C. (1998) *EC Public Procurement Law*, Chapter 4. Longman.

Cabinet Office Efficiency Unit (1995) *Construction Procurement by Government*. HMSO.

Commission of the European Communities (1988) *The Cost of Non-Europe, Basic Findings*. Official Publications of the European Communities.

Egan Sir J. (1998) *Rethinking Construction*. UK Government Department of the Environment, Transport and the Regions.

Latham Sir M. (1994) *Constructing the Team*. The Stationery Office.

NEC (2005) *NEC3 Engineering and Construction Contract*. Thomas Telford.

OGC (2003) *Building on Success: The Future Strategy for Achieving Excellence in Construction*. Office of Government Commerce.

OGC (2004) *The New Public Procurement Directives: Guide to the Changes Introduced by the New Directives*. Office of Government Commerce.

PACE Central Advice Unit (1998) *Information Note 16/98: Use Within Government of the Engineering and Construction Contract (ECC)*. PACE Central Advice Unit.

5 Project initiation

Introduction

Construction projects are responses to 'business needs' identified by clients. They tend to have 'one-off' characteristics with a clear purpose, constraints within which they have to be executed and clear beginnings and ends in the eyes of the client.

The decision to construct a new facility is usually taken as part of a strategy that has considered alternatives such as renting, leasing, extending or adapting existing property.

Male (2002) referred to the corporate value of a project aligned with business mission and the business level value of a project aligned with project needs as delivering primary drivers. The development of a business case is an essential activity for project success. It provides justification for the project in terms of its alignment with the objectives of the organisation; it also evaluates viability and provides the basis for establishing the time, cost and functional parameters for the project.

The business case is an evolving document throughout the procurement processes involved with realising the project outcome, starting as a high-level summary of possible options to meet the business need, then developing into a business case with indicative costs and maturing as project details are developed. The following questions should be addressed as part of business case development:

- Strategic fit – how does the proposed project meet the organisation's objectives, anticipated markets and current priorities?
- Payback – how will the project pay back the client's capital investment?
- Options – has a wide range of options been explored?
- Achievability – can this project be achieved with the organisation's current capability and capacity?
- Affordability – is the budget available sufficient to enable the delivery of a suitable project?

These five key aspects of the business case are considered below.

Strategic fit

This aspect of the business case addresses how the proposed project fits within the client's existing business strategies. It will identify the business need and the completed project's contribution to the organisation's objectives; the appropriateness of its timing; the key benefits to be realised; the anticipated key risks and critical success factors. Whether the project is to contribute to functional efficiency, to provide a solid investment over time and/or to contribute to the fixed asset base of the client company are key aspects of the business case.

Project initiation stems from a belief that construction is the solution to a problem or is in itself a profitable venture, having considered the available options. Construction projects are initiated within a local or national economic environment and there is often a causal link between the construction activity and the economy. For example, where a buoyant economy stimulates property transactions construction can be a way of seeking to benefit through development. Construction itself can also stimulate economic growth, particularly where the construction of infrastructure such as roads, schools and hospitals is required to enable that growth to occur.

Construction of a new facility is not a simple solution. It involves a range of different organisations, from designers to specialist contractors, and is a demanding process. Other solutions should be considered as part of the development of the business case for the specific project, including renting, leasing, buying an existing facility or extending or altering existing premises. Each of these will affect the client's corporate plans in different ways and may not offer the solution achievable by new construction, but they should be considered at an early stage in project development.

Most clients tend to be corporate organisations, large or small. Numerically small-sized clients dominate, but in terms of expenditure on construction the minority of larger clients are responsible for 80% of total expenditure.

Payback

This aspect considers the proposed expenditure on the project against the value of the completed project to the client. In some cases the value contributed to a client will include efficiency and asset value far greater than initial cost and these aspects can be considered against potential risk. The client organisation will normally be able to calculate the nett benefit to the business and the budget available from the value added; the latter being expressed in terms of written down value.

It will address all aspects of the project's direct and indirect costs including to what extent borrowing is necessary and the cost and impact of that borrowing upon current credit rating and cash flow. It will also address the extent of available funding from current reserves.

Corporate worth is a concept that accepts that 'value' can be seen to exist in more than one way. If a construction project is part of a business expansion programme or one targeted at improving business efficiency then its true worth will be to do with greater or better productivity. If the purpose is to build and sell on for a profit then its worth may relate to its market value.

Sometimes the construction of a new facility will also enhance the written down fixed assets of a company and strengthen its balance sheet. Cheaper and more easily obtained corporate borrowing will then be consequently available at times of cash-flow need. As an investment, commercial or residential property has also shown itself to be a largely reliable source of continued income. Thus corporate worth can be related to market value but should also be closely linked to corporate strategy and business need.

No project can logically proceed unless a specific business case has been developed justifying the project's viability and identifying the risks associated with its initiation. Typically, where a project is related to market value, that value should be sufficiently greater than the cost of the project in order to return a profit that reflects the inherent risk of the development process. Where greater or more efficient productivity is the aim then quantification and evaluation of the business gain will be required to justify the commitment being made. As part of this process the period over which the project will pay back its cost is an issue. This will typically be 6–20 years dependent upon the level of risk and stability of the particular market.

Following the verification of business needs and the assessment of options a more detailed business case for the project should be developed to evaluate payback. This part of the process will require return on capital investment to be considered against anticipated benefits and the effects that project realisation will have upon the client organisation. This will require evaluation of the likely benefits to operational efficiency or associated balance sheet benefits of the facility being considered. These considerations will help both to verify the need for the project and set down initial assessments of physical and financial impacts of the project on the client's current business. Independent advice from a construction professional from this early stage should be sought in relation to the costs of the options being considered, the likely programme and any real estate values where these are part of an investment appraisal.

Consider, for example, a client who owns a factory producing 20 000 units per annum at a unit cost of £1000. In addition overhead costs amount to £2 million annually. Resultant profits are £4 million annually. He has assessed improved markets for his product up to 30 000 units per annum which will mean additional overheads of £800 000 per annum, but has estimated that the increased production associated with a more efficient factory layout will result in a decrease in unit costs to £800 per unit.

Current costs: 20 000 @ £1000	20 000 000
Overheads	2 000 000
	22 000 000
Profit	4 000 000
Turnover	26 000 000
Projected costs for more units 30 000 @ 800	24 000 000
Overheads	2 800 000
Profit at current level	4 000 000
Additional profit: 30 000 × £200	6 000 000
New turnover	36 000 000

The additional profit anticipated on the greater output will generate £6 million annually, which can be adjusted by present value calculations to enable strategy decisions to be made. Nonetheless on the basis of, say, a 6 year payback the client may wish to expend some £20–30 million on the development of the new facility depending on his taxation position, capital allowances and any tendency for caution. After the payback period he will then realise the additional profits as well as having any additional fixed asset value on the balance sheet.

In terms of commercial or other property developments the payback may result through an annual rent payable for use of the building. In this case the capital value will be related to the level of the rental recoverable capitalised on the basis of the market for the property in that location.

Assessment of options

Construction of a new facility is not a simple solution to the needs of the client. It involves a range of different organisations, from designers to specialist contractors, and is demanding upon all of them. Other solutions should usually be considered by the client as part of the development of the business case for the specific project, including renting, leasing, buying an existing facility or extending or altering existing premises. Each of these will affect the client's corporate plans in different ways and may not offer the solution achievable by new construction, but they should be considered at an early stage in project development.

Building new will facilitate a bespoke solution developed for the specific need in terms of function and design. It will also enable a company's balance sheet to be strengthened in terms of the proportion of fixed assets against current assets, enhancing company worth even where a significant proportion of loan capital has been used to obtain the new real estate. Ownership (and in particular the cost and extent of borrowing) will constrain cash flow, but it will also enable the extent of available loans to be expanded on the basis of the greater fixed assets. Capability to borrow when needed is increased where balance sheets are stronger, provided borrowing is contained.

Ownership will also enable the benefits of investment where there is growth in the property market and consequential increases in real estate value. It will also enable an income flow to be realised if that property, or a part of that property, is itself leased or rented out.

The downside of ownership is that overall borrowing capacity is reduced where loans are obtained for construction, and also that the costs of ownership, including insurance, maintenance and repair, can be extensive. Additionally flexibility is limited and relocation in the short term is likely to be expensive and disruptive.

Buying existing business premises gives control over a major financial investment without the difficulties associated with the construction process. For owner-occupiers, there is the choice to stay in the premises for ever or, if need be, to raise money through selling, taking out an additional mortgage or letting part of the premises. Putting down a large deposit may seem expensive, compared with the low start-up costs involved in renting, but if tenure is likely to be of a long-term nature, there may be good reasons to buy.

Some clients will not need to have the level of control over real estate that is available through ownership. They may only need to purchase the use of the property, rather than its freehold. The majority of economic activity in the UK is in the service sector where ownership is unnecessary in most cases. This sector of the economy is likely to be attracted to rental or leased property which provides an income flow to investor owners and landlords.

Whilst renting or leasing property is less burdensome on capital or borrowing capacity it usually means that there are fairly strict constraints upon the ability to alter or adapt the property to specific needs. Rental property does not occur as a fixed asset on a company balance sheet and consequently corporate worth and borrowing capacity may be seriously affected.

Following the verification of business needs and the assessment of options a more detailed business case for the project should normally be developed. This part of the process will require return on capital investment to be considered against anticipated benefits and the impacts that project realisation will have upon the client organisation. This will require evaluation of the likely benefits to operational efficiency or associated balance sheet benefits of the facility being considered.

Achievability

The feasibility of the project being completed within the desired time will need to be addressed. Some types of project, if delivered late, can become expensive burdens, particularly if they are constructed to meet particular market conditions.

If possible, evidence of similar projects that have been successful should be obtained, to support the estimates of time for completion. If no similar projects are available for comparison, these estimates will need to be updated from time to time as the design is developed. The Building Cost Information Service (BCIS) provides a useful software package which gives helpful guidance on construction completion times, but this is only a relatively small part of the overall project programme.

Similarly initial estimates of projected total costs will be needed at this early stage to assess whether the project is likely to be achievable within the budget, although initial estimates at this stage are of low reliability. Nonetheless, if a decision to progress is to be made, initial estimates of the time and cost involved are required and their provisional nature accounted for by a risk analysis and some emphasis on contingency allowances.

Affordability

Estimates of the costs associated with the project will indicate whether the initial budget is sufficient and to what extent borrowing is required. Usually clients will need to contribute 10–30% of total borrowing from reserves or available funding.

Fundability anticipates that project expenditure occurs at different times, often not on a regular pattern, and an initial cash flow is helpful to enable the funding pattern to be planned.

Project initiation

The need for a construction project is rooted in the business case for the project which will be developed by the client. Usually the client will see the new project as a solution to the challenge set by the business case and will need to decide quickly whether the project is viable in terms of its long-term 'worth' and achievable in terms of its initial cost and planned duration.

This concept of long-term 'worth' is wider than that of capital value and is unique to each client, whether individual or corporate. It adopts the concepts of functionality, real estate value and return on investment; often all three. For example, if a project is needed to improve the efficiency of a client's production processes then its worth may be initially rooted in its function. Its real estate value will also add to the company's fixed assets on its balance sheet. If the project has investment potential then the client may also be able to realise investment growth at some point in the future.

Consequently a successful construction project outcome can provide increased efficiency, greater borrowing capability and payback, i.e. long-term worth. This is not, however, always the purpose of the project. Sometimes it is there to serve a short-term need or to be sold on as a completed development. In this case short-term value rather than long-term worth may be the primary criteria identified in the business case.

Since the business case underpinning a project will vary from client to client, and each client's experience and willingness to take risk will be different, it is logical that the key parameters and objective should be identified in relation to each project.

This is often referred to as 'defining the project', and can be a most difficult process, particularly for the inexperienced client, since information on all aspects of function, cost and time has to be collated and considered.

Prioritisation of objectives

The definition of the project should clearly address the client's overall business objectives, and indicate the relative priority of each objective. There is a tendency to want a high quality project delivered fast at a cheap price. This is rarely, if ever, achievable and many construction projects suffer from poor definition due to inadequate time and thought being given at an early stage. This is often because, as the Construction Industry Board (1997) has suggested, 'there is a sense of urgency fuelled by the desire for an immediate solution.'

The relative importance of time, cost and functionality in relation to any project will affect the choice of the most suitable procurement strategy for the project.

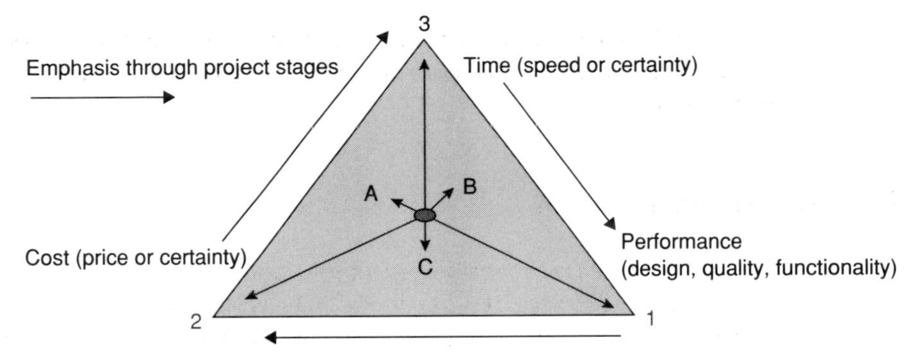

Figure 5.1 The interrelationship of client needs.

Usually, when the business case is developed the most important criterion is identified as that which, if not realised, is the greatest risk to corporate strategy. For example, where building use is most important, functionality in design will be vital; where the building is being constructed to enable a particular market to be served, time may be critical; and in cases where the client has a precise or limited budget, cost control will be most important. Usually, however, more than one criterion is likely to result from the business case. Owner-occupier clients may emphasise function and price certainty, developers cost and speed and investor clients design and speed.

Figure 5.1 illustrates the classic interaction of the three criteria. Where one is emphasised the others will be affected. In the case of the owner-occupier the emphasis on function and price will mean that the process will be more protracted in terms of time; developers may find that design is compromised and the investor client may have less control over the cost.

As the project progresses the emphasis may be affected inappropriately. This is illustrated in Figure 5.1, where position 1 shows the emphasis at an early stage for an owner-occupier. This emphasis may change to position 2 as the design progresses and cost becomes a primary issue and then to position 3 as design is completed and the construction stage commences. Upon completion the client will again concentrate upon the functionality of the design realised in the completed project, which will have to be used and maintained to support the original business case. It is too easy to lose sight of the initial objectives as the project itself becomes a distraction.

Decisions taken at this early stage will drive all future decisions and strategy. It is a key but complex stage carried out at a time when both client and consultants wish to progress quickly but when the pace of co-ordination and synergy of the issues and ideas is dictated by the ability to predict, foresee and design. Often the client will be best advised to appoint a member of his or her own organisation as 'sponsor' to co-ordinate the client function from this stage, or to appoint a project manager to overview the whole process where the project is particularly complex.

Table 5.1 Examples of prioritised criteria by client type.

	Owner-occupier (%)	Developer (%)	Investor (%)
Performance (functionality/quality)	45	20	50
Time (certainty or speed)	25	50	30
Cost (certainty or price)	30	30	20
Total	100	100	100

There is a danger that when the process of defining the project is carried out the primary needs of the business case can, as illustrated above, be given less prominence as short-term objectives relating to cost and time are seen as increasingly important to the project team. The definition of the project should clearly set down the client's overall business objectives, and indicate the relative priority of each objective.

Table 5.1 considers three examples of types of client and makes general assumptions about their likely objectives. These criteria are interdependent and none of the currently adopted procurement strategies will facilitate the delivery of all of them as high priority.

By adopting a technique of initially distributing priorities within parameters (in this case total percentage points) the natural tendency to want performance, time and cost to all be high priority is countered. Consequently prioritisation is necessary. Walker (2002) showed how this can be achieved by further developing this concept (see Figure 5.2).

By discussing with the client organisation the relative importance of the key objectives, the relationship between the initial business case and project priorities can be highlighted. Whilst fast-track solutions deliver speed, other criteria such as cost certainty and functionality may be less achievable, and similarly where cost certainty or performance are considered to be of highest priority the other criteria are affected. Consequently the distribution of prioritisation within parameters is a useful tool. It is a reasonable precept that the greater the concentration on defining the project and linking of the prioritisation of objectives to the business case the greater is the likelihood of the client's objectives being achieved.

Designing and constructing a new building is rarely straightforward. It is subject to a series of risks and uncertainties and involves a number of organisations especially assembled for the project. The way in which the client and the various designers, contractors and suppliers work together as a team is determined by the procurement strategy and the forms of contract entered into between the project participants and the client.

Conclusion

The initiation stage is very important to ultimate success. Time is well spent at this stage to ensure clarity of view by both client and design team.

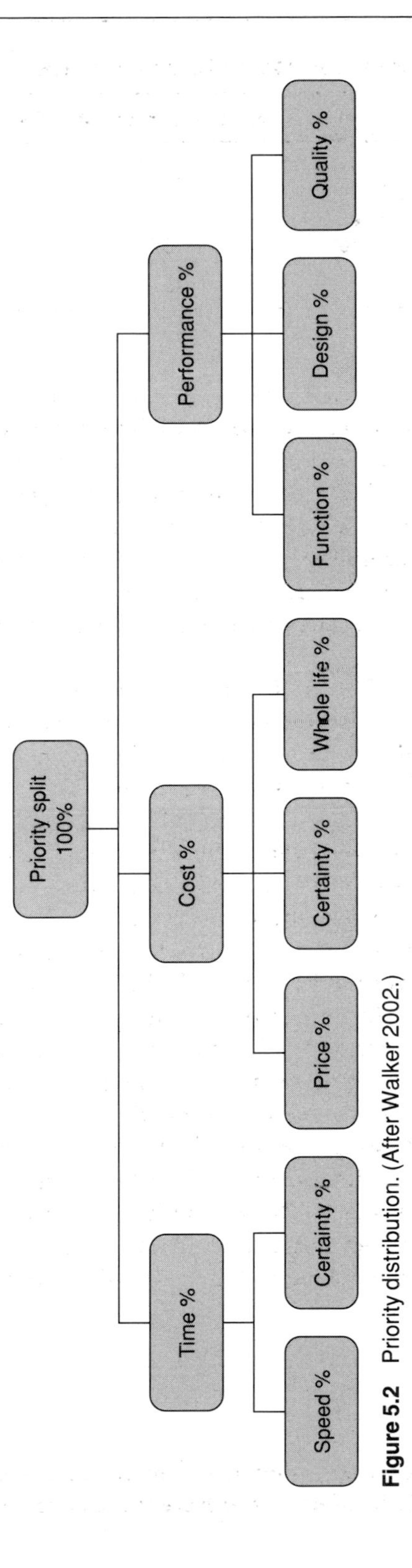

Figure 5.2 Priority distribution. (After Walker 2002.)

References

Construction Industry Board (1997) *Briefing the Team*. Thomas Telford.

Male S. (2002) Building the business value case. In: *Best Value in Construction* (eds J. Kelly, R. Morledge and S. Wilkinson). Blackwell Publishing.

Walker A. (2002) *Project Management in Construction*, 4th edn. Blackwell Publishing.

6 Briefing and the design process

Introduction

Project success or failure may be judged in a large number of different ways. Examples might include:

- Delivery within the agreed time frame and budgets
- By the contribution the development makes to the built environment as a whole
- By the functionality of the building as judged by the end users
- By the extent of critical architectural acclaim, etc.

However, the most important judgement will ultimately be that made by the client in respect of how well the building satisfies the purposes for which it was commissioned – in short, the extent to which the client believes that 'value for money' has been achieved. A fundamental aim of the briefing process must therefore be to identify not only the employer's functional needs and wants, but also the critical success factors that must be achieved in order for the project to be judged a success. This in turn requires a searching and critical evaluation of the employer's value system.

This chapter therefore considers the nature of the briefing process in defining the purpose and scope of the project, and the role played by the client within it.

Project briefing: an overview

Clients of the construction industry worldwide have long believed that the industry is inefficient and untrustworthy, particularly when it comes to the timely delivery of projects within agreed budgets. The construction industry on the other hand has attempted, in recent years, to overcome this perception through a number of innovations designed to improve efficiency, quality, reliability and cost effectiveness through better management and an improved understanding of the client's needs.

Central to this re-engineering process has been a critical re-examination by experienced clients of the way in which buildings have traditionally been procured, and this has led to the development of a variety of alternative procurement philosophies. Among the early developments was a move towards 'one stop' shopping and the use of procurement methodologies in which design and construction are more intimately combined, but more recently perhaps the most significant innovation has been the development of various types of collaborative procurement approach in which the demand and supply elements of the procurement process are more intimately connected than has historically been the case. These factors have directly led to a fundamental reassessment of the traditional roles of the various parties involved, including that of the construction clients themselves.

It has come to be realised that, particularly for large and complex projects, ultimate success or failure is often determined not only by the skill and flair of the design and construction team, but also by the active input of the client body, especially during the initial stages of the procurement process. This philosophy is, however, by no means acceptable to all clients, and we now see a polarisation between those clients who wish to be intimately involved in the genesis of their project and who wish to play a full and active part in the process, and those who wish to have no involvement at all other than to set the initial parameters at the briefing stage. The fact that the JCT has introduced the Major Project Construction Contract, aimed at precisely this group of clients, is testament to the fact that they exist in relatively large numbers. Particular considerations relating to the various types of construction client are considered in more detail below.

Regardless of the philosophies espoused by particular clients, it is surely self-evident that clients are unlikely to judge their projects to be ultimately successful unless the design and construction team and the client share a common view of what the project is expected to achieve, and the initial briefing process is crucial in establishing this understanding. It is therefore clearly essential that clients and their consultants thoroughly understand both the process and their role within it.

History, however, shows all too clearly that in many high-profile cases, often involving complex buildings and complex corporate clients, the necessary 'meeting of minds' has not been successfully achieved, and time after time the post mortem reveals that a major contributor to the failure of the project has been an inadequate or misunderstood briefing process.

Examples are not hard to find. Williams (1989) chronicles the development of Norman Foster's headquarters building for the Hongkong and Shanghai Bank, completed in 1985 in record time but at a cost of some 250% of the original budget, on the basis of a brief that said little more than that the Bank required 'the best bank headquarters in the world' and which was required to provide total flexibility to cope with any future changes. In this case it would appear that the client did actually have a mental picture of what it was it wanted, but was unable to clearly define its needs in a way the architect could understand.

Walker (1994a, 1994b) on the other hand explores the genesis of the Hong Kong University of Science and Technology, another building built in record time but which again considerably exceeded the original budget. In this case the Hong Kong government's original brief clearly stated that it wanted a University 'emphasising science, technology, management and business studies'. Walker contends that 'in academic circles there is no doubt that the perception of a new university concerned with science, technology and management would be in the tradition of Massachusetts Institute of Technology (MIT), Carnegie Mellon and Imperial College, University of London' (Walker 1994b). He goes on to conclude that the implications of such a perception were not fully understood, even by the appointed client body representatives, let alone by anyone else involved in the project's design and implementation, and that therefore the original budgets were grossly inadequate.

The problem is not, however, limited either to Hong Kong or to high-profile projects. In the United States, for example, Stephenson (1996) writes that 'lack of a good statement of the owner's and users' needs is one of the most frequent causes of flawed and troublesome construction efforts', and in the UK Latham (1994) considered the briefing process to be 'crucial to the effective delivery of the project'.

Types of construction client

The construction industry has a very broad spectrum of client types, ranging from highly experienced clients who build regularly and understand the industry very well to very small, inexperienced clients who may only ever build once in their lives.

Plainly, the client's level of experience and understanding will have a serious impact upon the briefing and procurement process. Inexperienced clients need to be led by the hand and given advice and guidance on every aspect of the process, whereas experienced clients may well drive the process themselves on the basis of what they know has worked well in the past. Furthermore, where clients are largely ignorant of the demands of the process then simply understanding the construction terminology might be a serious challenge. Research has shown that the majority of construction clients by number fall into the inexperienced and naive categories.

Looked at from a slightly different perspective, clients may also range from the single owner-occupier with only him- or herself to satisfy (for example, a small-scale manufacturer building an industrial unit for his own use) to complex corporate organisations with a need to satisfy many stakeholders (for example, a hospital trust building a new hospital). In the case of individuals it is easy to see who should be making the decisions, but where complex multiheaded clients are concerned the status of the various stakeholders is often unclear to the outsider, and there is often little attempt made within the organisations to resolve internal conflict and co-ordinate the various inputs to the briefing process. Clearly, a complex organisation will require a very different approach in terms of the development of the brief, and the management of the process as a whole will be much more composite and demanding.

Construction is a complex process carrying more risk for the parties involved than many other activities of similar value, but often rewarding the developer/initiator well for taking the risk. Construction clients tend to have a wide variety of reasons for initiating the process. The rationale will tend to follow the nature of the client and his/her core activity. Each time a construction client identifies the need for space he/she will need also to develop a business case that identifies the nature of the need and the characteristics of the payback underpinning the endeavour.

As indicated in the previous chapter, achieving a successful project outcome depends upon verifying the business needs for the project that is being considered in the context of business strategy. This process will need to involve those stakeholders who are driving that need, those who will take the risks associated with the project and those who will be directly involved in the use of the completed facility.

The decision to build will usually be based upon a business strategy that has reviewed the aims of the organisation, the projected life of the planned building and its corporate 'worth'.

The variety of construction clients is extensive but can be partly categorised in order to assist an understanding of their varied motives or reasons for construction, as well as their likely attitude to project priorities. Typically public sector clients will seek functional projects giving provable value for money at low risk. Some private clients will have the same aims but will not have to justify expenditure in public accountability terms. Alternatively some private clients (usually the more experienced) will be prepared to take some risk to obtain the forecast gain arising from a particular market opportunity or other demand need.

Most clients are inexperienced since they build as a consequence of their primary function, so they build infrequently and depend heavily upon professional advice. Relatively few clients are frequent purchasers of construction and consequently are both more experienced and can use their buying power to demand exactly what they need.

Masterman (1992) categorised clients into public and private sector clients, experienced and inexperienced and into primary and secondary. Primary clients were those who built for a business and secondary clients those who built as the consequence of their business. These categories are probably becoming oversimplistic in the increasingly complex construction market but are still helpful in enabling an initial view to be taken.

Duffy (1992) identified different types of client value propositions:

- *Use-value* building: custom designed for the owner-occupier, maximises the use value for the end user organisation.
- *Exchange-value* building: developed speculatively, and designed to maximise the building exchange value as a commodity to be traded.
- *Image-value* building: designed to maximise the image value of the building (often at the expense of efficiency or other qualities).
- *Business-value* building: where use, exchange and image are synthesised into a building where technology is fully exploited to maximise the range of options for the end user.

For example:

- The major retail chains in the UK have been major spenders on new stores. They can be described as private, secondary, experienced clients, with a business-value proposition.
- Public clients, by the nature of their activity, will be experienced or inexperienced. They will have a use-value approach.
- Property developers are likely to be experienced private primary clients with an exchange-value and/or image-value proposition dependent upon the particular case.

The outcomes that must be achieved in order to satisfy each client will vary with their mission and the primary objectives for the project they have initiated. They can be quite different.

For example, owner-occupiers will want a facility that is functional, aesthetically pleasing and relatively inexpensive to run and maintain, developers on the other hand will aspire to speed and simplicity, whilst investors in property will look for long functional, physical and economic life buildings which retain their marketability. Most clients will also want low initial cost but only if a valuable output is achieved. This variability of criteria will impact upon ultimate client satisfaction and the briefing process is critical in ensuring that individual client needs are satisfied.

Every client presents unique problems, but as a guide the following questions will provide a starting point in determining the project implementation plan. However, it should not be forgotten that securing maximum value for money from a construction project often requires a fundamental re-evaluation of the client's underlying value system. This is frequently a difficult and time-consuming task, which many clients find very challenging, but is nonetheless essential if the project is to be ultimately successful.

Time

What is important about time? Does the client require the project to be completed as quickly as possible? Or does he or she have a date by which the project *must* be completed (for example the start of a school year)? Or does the client simply want certainty of time (that is, the client is not particularly concerned about having the project completed for a specific date, but once he or she has been given a date, it must be kept)?

Cost

Similar questions arise in respect of the project budget. Is the client concerned primarily with initial capital cost (for example, a speculative developer who intends to sell the project on after completion), or is he or she concerned about the costs that will arise throughout the project's life cycle? Does the client have a fixed budget which cannot be exceeded, or does he or she simply want certainty of price once the project budget has been agreed?

Quality

Quality is arguably the most difficult attribute to define and reach a common understanding over. Typical items for discussion might be to do with:

- The client's concern with image and building style (i.e. overall design quality).
- The quality level to be achieved in terms of building finishings and fittings (most often expressed in comparative terms, e.g. 'a five star hotel', 'high quality office block', 'standard factory construction').

- The quality standards required in terms of materials and workmanship.
- The quality assurance procedures required.

Likelihood of post-contract client changes

How firmly is the client able to articulate his or her needs? What is the likelihood that the client's needs will change before the project is completed?

Degree of accountability required

To whom does the client have to justify his or her expenditure? Will the client have to account for every last penny (for example a publicly funded project), or are his/her requirements less stringent?

All clients will be keen to achieve good value for money, but few outside the most experienced group are likely to truly understand what the concept means or how it might be measured. It is common for inexperienced clients to ask for the highest quality at the lowest price to be completed in the shortest time – and the client's project manager may have to devote considerable time and effort to arrive at a workable compromise that meets the client's real objectives.

It is also worth noting here that projects vary widely, both in complexity and in degree of innovation. Innovation is inherently risky, and although little definitive research has yet been reported, experience seems to show that, of the following two extremes, the first might be more difficult to manage, be more likely to run over time and/or budget and generally less likely to be successful than the second:

- *Complex projects* incorporating a high level of innovation, where the client is both inexperienced and complex (for example a large and prestigious company headquarters or a major performance venue).
- *Simple projects* with little or no innovation, constructed by simple and experienced clients (for example the speculative development of industrial units).

The briefing process: an historical perspective

Historically the process of briefing seems largely to have been regarded as almost unworthy of any serious attention. Turner (1945), for example, in the third edition of his *Architectural Practice and Procedure*, ignores the topic almost totally, save to give budding architects the advice that:

> 'One of your most important rules should be to endeavour to give a client what he wants. . . . It is no doubt often the case that a client has but a dim notion as to how his ideas can be carried out, and it is then the architect's privilege to guide him with expert advice.'

Quite how this process is supposed to occur is not explained, and there is no doubt that at the time Turner formulated his advice both buildings and clients were rather less complex, but even so we might speculate that, in many cases, a successful end result came about largely by chance. This attitude towards briefing plainly persisted for some time. O'Reilly (1973) highlighted deficiencies in briefs for housing work, noting not only that the brief commonly lacked essential information, but also that both the client and the architect often failed to appreciate the significance of the omission.

Later work resulted in a number of guides for both clients (Goodacre et al. 1982; Murray et al. 1991) and architects (Newman et al. 1981; Salisbury 1990). Kelly et al. (1992) critically reviewed the briefing process.

The client's role in the briefing process has, however, been seen to be crucial to the success of the project for more than a decade. Salisbury (1990), for example, summed up the issue well:

> '[Briefing] is the most important contribution the client can make to the building project. It is as creative as anything the architect or any other designers subsequently do. This does not mean that it is a short term activity to be got out of the way quickly. It should never be rushed or done in a cursory fashion.'

Both Kometa et al. (1995) and Walker (1996) examined the client's role and responsibilities in communicating project objectives, but Pilling and Lawson (1996) expressed serious doubts about architects' brief-taking skills. Best and Lenard (1997) provide comment upon contemporary developments in Australia.

Recent years have seen the introduction of a number of client guides to getting the best from the modern briefing and procurement process. Examples include *The Procurement Guide* (RICS 1996) and *Briefing the Team* prepared by the Construction Industry Board (1997b). This latter document is described as:

> '. . . written for clients and potential clients of the construction industry. It is designed to help potential clients to work out whether or not they need a construction project, and if construction is the option chosen, to improve the briefing they give so that the project team fully understands their needs and they secure the product and outcome they require.'

The document divides the overall process into five key stages as follows:

(1) Getting started
(2) Defining the project
(3) Assembling the team
(4) Designing and constructing
(5) Completion and evaluation

Advice on each stage is presented in the form of questions and check lists. The document forms part of a co-ordinated portfolio of documents based upon the Board's code of practice for construction clients (Construction Industry Board 1997a).

This approach is valuable in that the process is carefully subdivided into manageable chunks which even inexperienced clients should be able to understand, but research appears to show that understanding the mechanics of the process is only part of the problem. Equally important is understanding how to successfully *manage* the process, and this aspect takes on a particular significance in large and complex projects with complex 'multiheaded' clients.

Developing the project brief

No project is initiated without a clearly identified purpose. This may relate to the need for a more efficient facility in which the client will do his/her business, or the wish to increase the client organisation's fixed asset base, or both. The development of the project brief must therefore begin with a clear statement of the needs to be met. The definitive statement of need arises out of the initial project initiation and option appraisal studies, and represents the agreed 'best option' to meet the client's requirements. It contains, and clearly discriminates between, 'needs' (defined here as fundamental issues which must be satisfied for the project to be successful) and 'wants' (defined here as things that are not essential but which it would be nice to have if possible). The statement of need therefore summarises the client's rationale for the project and contains the essence of the issues that will subsequently become the key indicators of successful project performance.

Once the definitive statement of need is developed, the economic feasibility of the project needs to be checked and further tested through the formulation of a definitive detailed business case to explore the payback from the project in terms of both capital investment and overall value for money. Contemporary concern with sustainability will often mean that the business case will be prepared on a whole-life cost basis, taking into account not only the initial capital expenditure but also the costs and benefits incurred throughout the project life cycle.

An essential part of the business case will comprise an early attempt to identify and evaluate the risks embodied in the project, and a robust business case based upon a clear and unambiguous statement of need will form the basis for the prioritisation of the relative importance of speed, cost and quality within a framework of clearly defined functionality in terms of design and performance. Among the risks to be considered is that arising from innovation. Contemporary thought would appear to advocate that innovation is, of itself, good, but the bespoke nature of construction inevitably means that innovation comes at a price. Anecdotal evidence would appear to show, however, that the potential cost of innovation is rarely appreciated by the client.

Issues arising from the business case may require reconsideration of some elements of the statement of need, and at the end of this process the statement of need and the business case together are combined in a strategic brief, from which the overall procurement strategy, set out in the project implementation plan, will be developed. Experience seems to show that the development of a reliable and comprehensive strategic brief is at the root of achieving a successful outcome, particularly where the project is complex in design or where there are external influences on the design and management process from, for example, end users and funders.

The project execution plan (PEP)

The project execution plan (PEP) sets out how the project which has been defined in the strategic brief will be procured. It is the document that defines the project strategy, organisation, control procedures and responsibilities. It is therefore the highest level project control document and forms a major plank in the project management and control (PMC) system. However, remember that it is a plan – a framework for action. It therefore must, to some extent, evolve as the project proceeds and as decisions are made and confirmed. It should be flexible, not carved in stone.

The function of the PEP is to ensure that the client's objectives as identified in the strategic brief are carried through to reality. It is therefore the primary tool used to ensure that the project is completed to the satisfaction of the users, on time, within budget and to the required quality standards. Remember also that the PEP is intended for the use of all members of the project team – it should therefore be written in an appropriate style and should assume no prior knowledge of the client organisation or administrative procedures.

The issues the PEP needs to address, when and how it should be prepared and who should prepare it need to be considered.

What issues does the PEP need to address?

To some extent the detail of the PEP will be dependent upon the chosen procurement approach, and the format of the plan may well be dictated by the client or the constraints of specific projects. The form and detail of the plan will also change as the project evolves, and the key objective at all times should be to ensure that the plan both documents what has gone before in terms of decisions that have been taken, and looks ahead to set a framework for future action.

The following is a generic list of the topics that any PEP ought to include:

(1) *Overview* – a short summary defining the objectives and the scope of the project in functional terms. It should specify the big problem that the project is intended to address, the project goals and how the project fits into the client's business and his or her future plans.
(2) *Specific objectives* – what are the specific functional objectives to be achieved? These should include time, cost and quality constraints as well as an outline description of the kind of accommodation required and the functional standards to be achieved. The plan should also define any constraints arising from the need for transparency and accountability, and should specify how much flexibility there is in respect of the time and cost constraints.
(3) *Proposed procurement strategy* – most construction projects will present a choice, ranging from a traditional approach, through some variant of the design and build family, to management-based approaches such as construction management or management contracting. Choice of an appropriate procurement route is perhaps the most important decision in the entire process,

and the plan should document not only what procurement strategy has been chosen but also the underlying rationale.

(4) *Project control mechanisms* – this section of the plan provides details of the administrative, contractual and financial mechanisms through which the project will be controlled. In short it defines how the project will be managed. It therefore needs to specify things such as:
— reporting structures, together with any key dates for committee meetings, etc.;
— procedures to accommodate ongoing project reviews, changes, etc.;
— any particular client requirements or restrictions (for example on the use of subcontractors);
— value, quality and health and safety management procedures (note the requirements of the Construction Act with regard to Health and Safety plans, etc.).

Of particular importance are the procedures for project completion and the handing over of the completed facility to the client. The PEP needs to specify things such as who will be responsible for taking possession of the building and its future management, what documentation will be required, what training needs to be provided, etc.

(5) *Project time schedule* – defines all aspects of the project timescale including critical dates and milestone events. The project schedule will plainly evolve as the project proceeds, and will eventually comprise both outline long-term and detailed short-term programmes. The proposed timescales must, of course, meet the client's delivery requirements, but they must also be realistic. Imposing unrealistic time constraints simply increases the risk to the client of failure to meet key targets.

(6) *Project budget* – project budgets will normally include not only capital construction costs but also ongoing and recurrent expenditure during the project life cycle. The budget will normally become progressively more detailed as the project proceeds. The PEP should include details of funding sources, and document cost monitoring and review procedures.

(7) *Personnel and lines of responsibility* – who is responsible to whom and for what? The PEP should define the limits of authority and responsibility for all of the key players.

(8) *Evaluation methods* – the PEP should document the criteria against which the completed project will be measured, and how the measurement will be carried out. It is important that these issues are considered at an early stage in order that the project team know where the goal posts are. Project evaluation is impossible unless the evaluation criteria are fixed, but it is a fact that clients' objectives frequently change between initial project initiation and project handover. If major changes are likely to occur, for example as the result of rapidly changing technology, then the PEP should make provision for them.

(9) *Potential problems* – potential areas of risk which might compromise successful completion of the project must be identified, together with appropriate risk management strategies.

When should it be prepared?

The preliminary feasibility studies which, together with the preparation of the brief, form part of the initial project inception stage must obviously include:

- Preliminary consideration of how the proposed project might be executed, including an outline budget and programme.
- Consideration of the potential risks and uncertainties which might prevent successful completion, together with strategies that could be used to manage them.

All of these issues need to be collected together in a preliminary project execution plan, which should form part of the feasibility report. Many issues at this stage may remain to be resolved, and where this is the case the plan should document potential alternatives. The plan should subsequently be revised as decisions are progressively firmed up.

Once the client is satisfied that the project is feasible in outline, a decision must be made regarding the procurement route. Much of the definitive PEP will be dictated by the procurement route decision. Two questions are important: (1) which procurement route and (2) is a specialist project manager required?

In making these decisions much will depend upon the technical capabilities and expertise of the senior client representative and his/her team. This person must decide whether he/she is competent to adequately advise the client, or whether he/she needs to seek independent professional construction advice (for example from a project manager, quantity surveyor, architect or engineer).

Choice of an appropriate methodology will depend upon factors such as:

- Client constraints in terms of time, cost and accountability.
- The type of project (e.g. the level of technology required, the likely extent of innovation and the client's objectives in terms of project aesthetics and design impact).
- The extent to which the client is able to define what is required and the likelihood of design changes during the construction stage.

Who prepares the PEP?

Again much will depend upon the details of the project, the capabilities of the personnel involved and the chosen procurement route. If an external independent project manager is appointed then preparation of the plan will normally form part of his or her duties. In this case input from the client representative may vary from virtually nothing to co-authorship, depending upon the circumstances of the particular project. If no independent project manager is appointed then the initial plan may well have to be prepared by the client representative him/herself. It should, however, be stressed that the preparation of a PEP will require considerable technical expertise. A workable PEP is of vital importance, and it is therefore imperative that the client should be very aware of the limits of his or her technical skill, and that outside assistance may be required.

How is the PEP prepared?

Preparation of the definitive PEP consists of the following:

- Summarising in clear and unambiguous terms the decisions that have already been reached. This may not be easy in the case of complex multiheaded clients, and in this case one of the senior client representative's most important tasks will be to resolve internal disputes within the client body in order to present a single point interface to the external project team.
- Collecting together and analysing the cumulative knowledge and expertise of the project team, and finally synthesising a feasible PEP.

In order to be considered valid the completed definitive PEP must be explicitly approved by the client and accepted as realistic by the external project team.

Project briefing: the case of hospitals

Hospital projects are generally complex and expensive, and in the public sector modern clients, many of whom are now hospital trusts, tend to be both corporate and often inexperienced. These projects therefore tend to illustrate, and often to magnify, problems with both the briefing process and its management. These projects have, in the past, often tended to enjoy very extended gestation periods, giving apparently ample time for the briefing process to take place, and in many cases detailed manuals of procedure are available. Gray (1994) discusses the process of drawing up the 'ideal brief' for hospital projects. Despite these apparent advantages, however, research would appear to show that the project outcome is not always successful.

Wilkins and Smith (1996) discuss the management of the briefing process on the basis of results from a comparative study of the procurement methodologies used for eleven major hospital building projects in Britain, the United States and Hong Kong. The majority of the projects were in the public sector. Five of the projects studied display significant failures to meet time targets (between +16% and +150%) and/or cost targets (between +3% and +14% of the tender figure). In each of these cases a major cause of the failure was identified as client changes initiated as a direct result of inadequate briefing. This was despite pre-contract gestation periods of between 5 and 8 years. These time spans are not unusual and are confirmed by other researchers; Loosemore and Davies (1994) for example report one project in South Wales with a 10-year pre-contract period.

Wilkins and Smith (1996) do, however, also report several projects, generally those procured using non-traditional procurement methodologies, where the gestation period is much shorter and where the project outcome has been much more successful.

A major factor in determining success or failure appears to be that of effective management of the corporate client body during the briefing stage. Loosemore and Davies (1994) examine the interfaces that commonly occur both inside and outside the client body, and conclude that the main problems lie in communication and co-ordination. Wilkins and Smith (1996) similarly identify the 'multiheaded client' as a major problem. They summarise the problem as follows:

> 'End users are represented by medical consultants each of whom typically takes a narrow and specialist view of their particular unit, and administrative department heads each of whom take a broad and increasingly commercial view of the whole enterprise. Our research shows that both groups expect to act as the client for "their" part of the building, and that individuals may have exaggerated and somewhat idiosyncratic views about the importance of "their" part within the project as a whole. Each therefore expects to have an equal voice in the briefing and design process. This group of end users, which may consist of up to 100 individual experts, is uncoordinated and largely uncontrolled by higher levels of authority, and has traditionally expected to be able to instruct the design team during both design and construction. During briefing, the design team, and in the "traditional" procurement model specifically the project architect, is left to interpret and to arbitrate between the often conflicting requirements of individual consultants and department heads.'

The use of non-traditional procurement approaches such as the design and build family sharply focuses the multiheaded client problem. With a traditional procurement approach the management of the multiheaded client problem takes place largely through the project architect in the conventional briefing stage, but this may no longer be so in the case of a non-traditional procurement route. Yet it is in precisely this situation that the production of a clear, comprehensive and co-ordinated brief is essential in ensuring that tenderers know precisely what is required of them (Dix 1995).

The need for client changes during construction can be seen as merely an extension of the briefing process into the construction phase. Changes introduced at this point have long been known to be unduly disruptive and expensive, particularly where non-traditional procurement methods are employed. Inaccuracies in the client's brief and subsequent changes have been identified as a significant cause of dispute in design and build projects (Ndekugri and Turner 1994). The briefing process therefore plainly needs to be managed by someone close to the client's organisation, and needs to consider not only the competing demands of the various parts of the client body, but also the most appropriate choice of procurement methodology given the likely degree of certainty embodied within the final briefing document.

The role of the client's project manager in hospital projects of this type was explored in Wilkins and Smith (1995). A number of projects were examined using various models of project management, including managers drawn from within the client organisation, external consultants and a combination of the two, and using various traditional and non-traditional methods of procurement. It is unlikely that a project manager drawn from within the client organisation will have sufficient knowledge of the construction industry to be successful acting alone. They therefore conclude that those projects that have had the benefit of skilled project management during their early gestation and that use a combination of in-house and external consultant project managers appear to have greatly improved chances of ultimate success. Some commentators (e.g. the Arts Council of Great Britain 1993) discriminate between the client's representative, termed the 'project sponsor', and the external consultant, or 'project manager'.

Critical factors for success in the briefing process

The Construction Industry Board has identified the following as being the essential qualities of a good brief (Construction Industry Board 1997b):

- Embody the mission and convey this to the reader
- State what is expected, by when and from whom
- Define the context
- Set out the perceived problem unambiguously
- Establish requirements on cost, quality and time
- Be clear about the required lifetime of the project
- Tease out the assumptions of the people involved
- Do not leave any 'big' questions unanswered
- Be honest without raising unrealistic expectations
- Be able to respond to change
- Be flexible so that different options can be explored
- Be explanatory, stating the reasons for the requirements
- Include success factors and measurements that can be used to test the result
- Set out potential conflicts so that the project team can respond

In addition the following are seen to be factors that are critical to success:

- Clear and agreed objectives
- Carefully thought out requirements
- Provision of essential information at each stage
- A flexible approach that balances the requirement for quality with the concern to freeze requirements in order to control costs and meet deadlines
- Trusting relationships

Given that effective management of the briefing process appears to be a major determinant of project success, the following issues appear to be crucial if the 'ideal brief' set out above is to be achieved, particularly in the case of complex multiheaded clients:

- A project management system must be designed and implemented. The principal goals of the system should be to:
  - co-ordinate the various parts of the client body, broker negotiations between competing factions and present a unified and single-point interface between the client and the design and construction team;
  - assist the various parts of the client body to identify their real needs, to consider and evaluate options and to transmute initial 'wish lists' into clearly identified and measurable desired project outcomes;

- advise the client on the most appropriate form(s) of procurement, bearing in mind the degree of certainty embodied in the brief at each stage;
- advise the client on the assessment of competing bids.

- The project manager must be a manager, not merely a co-ordinator. He/she must take the responsibility for guiding the client through the briefing process, but must also be given the necessary authority to enable him/her to broker negotiations between competing factions.
- Establishment of the project management system may require input from an external consultant.

References

Arts Council of Great Britain (1993) *Architecture and Executive Agencies*. Architecture Unit of the Arts Council of Great Britain.

Best R. and Lenard D. (1997) The design brief and integrated design and construction: recent trends in building procurement in Australia. In: *Procurement – A Key to Innovation*. CIB publication no. 203. Construction Industry Board.

Construction Industry Board (1997a) *Constructing Success: Code of Practice for Clients of the Construction Industry*. Thomas Telford.

Construction Industry Board (1997b) *Briefing the Team*. Thomas Telford.

Dix A. (1995) Design and build in the dock. *Hospital Development*, June.

Duffy F. (1992) *The Changing Workplace*. Architectural Design and Technology Press.

Goodacre P., Pain J., Noble B.M. and Murray J. (1982) *Client Aid Program*. Occasional Paper no. 5. Department of Construction Management, University of Reading.

Gray C. (1994) Building confidence – part 1: the project sponsor. *Hospital Development*, April, 16–17.

Kelly J., MacPherson S. and Male S. (1992) *The Briefing Process; A Review and Critique*. Paper no. 12. RICS Publications.

Kometa C.T., Olomolaiye P.O. and Harris F.C. (1995) An evaluation of clients' needs and responsibilities in the construction process. *Engineering, Construction and Architectural Management*, **2** (1). RICS Publications.

Latham Sir M. (1994) *Constructing the Team*. HMSO.

Loosemore M. and Davies M.G. (1994) Hospital development briefing. In: *Proceedings of CIB W92 Procurement Systems Symposium, East Meets West* (ed. S. Rowlinson). CIB publication no. 175. University of Hong Kong.

Masterman J.W.E. (1992) *An Introduction to Building Procurement Systems*. E. and F.N. Spon.

Murray J.P., Hudson J. and Gameson R.N. (1991) *A Model for Guiding Clients and the Design Team During Briefing for Construction Projects*. Unpublished SERC Research Project final report. Science and Engineering Research Council.

Ndekugri I. and Turner A. (1994) Building procurement by design and build approach. *Journal of Construction Engineering and Management*, **120** (2), 243–256.

Newman R., Jenks M., Bacon S. and Dawson S. (1981) *Brief Formulation and the Design of Buildings*. Research report. Oxford Polytechnic.

O'Reilly J.J.N. (1973) *A Case Study of Design Commission: Problems Highlighted; Initiatives Proposed*. Current paper CP27/73. UK Building Research Establishment.

Pilling S. and Lawson B. (1996) The cost and value of design. *The Architects' Journal*, 7 March.

RICS (1996) *The Procurement Guide*. Royal Institution of Chartered Surveyors.

Salisbury F. (1990) *Architect's Handbook for Client Briefing*. Butterworth Architecture.

Stephenson R.J. (1996) *Project Partnering for the Design and Construction Industry*. John Wiley & Sons.

Turner H.H. (1945) *Architectural Practice and Procedure: A Manual for Practitioners and Students*. Batsford.

Walker A. (1994a) *Building the Future: The Controversial Construction of the Campus of the Hong Kong University of Science and Technology*. Longman Asia.

Walker A. (1994b) The lessons of HKUST. In: *Proceedings of CIB W92 Procurement Systems Symposium, East Meets West* (ed. S. Rowlinson), pp. 379–387. CIB publication no. 175. University of Hong Kong.

Walker D.H.T. (1996) Characteristics of a good client's representative. In: *Proceedings of CIB W92 Procurement Systems Symposium, North Meets South*, pp. 614–622. Department of Property Development and Construction Economics, University of Natal.

Wilkins B. and Smith A.J. (1995) Organisational change, technological innovation and the emerging role of the client's project manager in the delivery of health care facilities. In: *Proceedings of the 1st International Conference on Construction Project Management*, pp. 149–156, Nanyang Technological University, Singapore.

Wilkins B. and Smith A.J. (1996) The management of project briefing: the case of hospitals. *Australian Institute of Building Papers*, **7**, 8–21.

Williams S. (1989) *Hongkong Bank; The Building of Norman Foster's Masterpiece*. Jonathan Cape.

7 Procurement strategies

Introduction

UK strategies for the procurement of new construction projects have not significantly changed in the last 25 years. It is probable that the post-war dominance of the public sector in terms of capital spend established a 'normal' practice where a design was completed for the planned project, tender documents were prepared and then bids submitted based upon those documents, usually from a selected list of contractors. In most cases the tender documents included a bill of quantities identifying precisely the extent of work to be done for the lump sum bid. Until the 1980s fee scales for consultants were mandatory, but once these were outlawed the selection of consultants was also based upon submitted bids.

This design-bid-build process adopted the nomenclature 'traditional procurement' and was for many years the mainstay of procurement activity, the default approach, but latterly the traditional approach has become less appropriate for some projects with the advancement of fragmentation of specialist trades and the emphasis upon subcontracting.

Resistance to change to reflect modern commercial frameworks appears to have come primarily from the established construction professions. Whilst the professions have co-operated in *discussions* about change, the primary drivers in *achieving* change appear to have arisen as the result of client pressure caused by increasing frustration at the apparent inability of the construction industry as a whole to deliver the projects needed on time, to budget and to acceptable quality standards.

Consequently there is evidence of some change in the procurement of construction projects, but this is slow and seemingly driven by a combination of major private sector clients and central government.

The latest data available to indicate the relative usage of different procurement strategies show that in overall terms there has been a reduction of about 10–15% over time in the use of traditional design-bid-build and a concomitant increase in design and build. There has also been, over the last decade, evidence of a significant reduction in the use of bills of quantities in design-bid-build projects and a reduction in the quality of the bills of quantities that are prepared (Kings 2002). There are, however, signs of a resurgence, with major quantity surveying practices and construction contractors complaining publicly that they are unable to recruit graduate quantity surveyors with adequate measurement skills.

There appears to be little general movement in terms of other strategies, except that it appears likely that government pressure has led to an increase in the number of projects procured through collaborative routes such as partnering, although there appears to be no reliable research data to support this.

The picture is quite different if an analysis is carried out of the adopted strategies of the top 50 UK construction clients (reported in *Building* (Builder Group 2003)), where traditional methods are in the minority, and partnering is the most commonly reported strategy.

There is also increasing evidence of the adoption of projects procured through franchising opportunities to capture income flow, such as PFI schemes, driven no doubt by Government's view that every major project should be tested for suitability as a PFI scheme before any other procurement routes are examined. Both PFI and PPP are considered in detail in Chapters 13 and 14.

In an interview given to the Royal Institution of Chartered Surveyors (RICS) *Business* magazine (February 2003), Digby Jones, Director General of the Confederation of British Industry (CBI), commented that the construction industry:

> 'is not yet exploiting its chance to reinvent itself and become an example to the rest of business . . . [it] talks the talk about change but walking the walk is a lot more difficult. And because it's such a disparate business full of self-employed, temporary staff and conservative people, change is exceedingly hard. It brings insecurity and a feeling of isolation.'

In the same interview the 8% skills shortage in business generally is compared with the 15% skills shortage in construction. Indeed it is reported (Adult Learning Inspectorate 2005) that in 2005 there were 300,000 too few skilled workers to complete projects already planned.

The change in procurement terms mainly affects the way that capital projects are purchased. Traditional methods reflected a time when the public purse dominated construction expenditure and those methods facilitated excellent accountability and relative price certainty. Deterioration of standards seems to have become an acceptable reality, with bills of quantities being prepared and deemed to be 'firm quantities' in contractual terms when in reality they are not (Kings 2002). This may be because of the incomplete state of design when detailed drawings are required but at the same time clients are pressing for earlier and earlier start on site with faster and faster completion. Extensive post-contract changes also appear to have become the norm.

A move to non-traditional methods that are faster but give some level of price and time certainty was perhaps inevitable given the decreasing confidence created by lower standards. This may not, however, be in the long-term interests of clients whose primary criteria or objectives are often related to building performance or functionality.

Egan, in *Rethinking Construction* (1998) and *Accelerating Change* (2002), argued that integrating design and construction would facilitate improved value in projects for clients. Whilst it is probably true that an increase in the use of design and build will enable integrated design and construction, whether this will translate into improved value for money or functional performance is a matter of judgement over time. The shift to design and build also means a shift of the responsibility for the design function away from direct client control to organisations whose core businesses are profit-focused.

Other ways to achieve integration seem to generate less enthusiasm from the client sector or the construction professions, as evidenced by the very low adoption of strategies enabling such integration.

The construction industry is complex and the demand for construction is variable and fickle, with an experienced minority of clients accounting for most of the expenditure and an inexperienced majority accounting for least spend and most projects. The former experienced group are able to lever change using the advantage of their spending power, while the inexperienced latter group are often at the mercy of construction professionals who have historically appeared largely to resist change and to defend traditional practices.

These problems are compounded by the fact that the supply side of the construction industry is fragmented in terms of both design and construction. Advancements in technology have resulted in specialisation structured in such a way that traditional practices do not enable specialists to contribute to the design, costing and programming elements of a project. For example, those involved in constructing the specialist elements of a building, often very significant elements, do not become involved until so late in the process that their ability to contribute is severely curtailed.

This is counter to the way most integrated manufacturing industry processes work, but the construction industry has historically displayed few similar characteristics to manufacturing industry, being mainly project-focused rather than process-focused. There are, however, indications that collaborative procurement practices and improved supply chain management are slowly changing the way in which things are done.

A further problem has historically been that the project team was, in general, temporary and, with a few exceptions, projects have therefore been largely unable to benefit from prototype testing. There are indications that this too is changing. The increasing use of long-term collaborative arrangements and an increase in the use of framework agreements, again largely but not exclusively driven by central Government, is beginning to change the way continuous development is viewed. Nonetheless, these are all important factors that need to be considered when reviewing construction procurement.

Outside Government and the major private sector clients, comparison of the industry key performance indicators (KPIs) with perceived performance over the last 10 years (such as the 1995 *Gallup Survey of Construction Clients* (Gallup 1995)) will produce evidence of little or no change or improvement, even after many years of self-examination and attempts to promote innovative processes.

This may result from the fact that most clients (by number) are inexperienced and have unreasonable expectations. These expectations are often not discouraged by a project team setting targets with little or no involvement of those specialists whose work will have a major effect on costing and programming. Alternatively the perceived poor performance may result from design teams agreeing to unreasonable client targets in ignorance, or in order to encourage the client to progress, or an acceptance of such targets by construction companies unwilling to point out unrealistic outcomes, if that may result in losing the opportunity to obtain the contract.

There is clearly a challenge to seek improvement. In *Rethinking Construction* (Egan 1998) a number of targets were set for such improvement, allied to the adoption of techniques that had achieved improvement in the manufacturing sector. These included supply chain management techniques, including the adoption of strategic alliances and strategic and project partnering with a focus on value improvement.

As observed above, however, the construction industry is still largely project-focused, not process-driven, and whilst desirable, the opportunities to adopt and benefit from these techniques are probably limited to those clients who have a continued relationship with particular constructors, specialists or design teams. The majority of clients, who buy one-off projects occasionally, are unlikely to benefit from the adoption of such techniques. Other strategies will need to be developed to improve perceived construction performance in this demand sector.

The Construction Clients' Forum (CCF) (latterly the Confederation of Construction Clients and now the Construction Clients Group) published, in 1998, a document entitled *Constructing Improvement* which included a suggested 'Pact' with the construction industry, setting down what was felt to be a way forward (Construction Clients' Forum 1998).

This 'Pact' sought improvement from the industry by suggesting that the construction industry:

- Present clients with objective and appropriate advice on the options and choices to meet their needs
- Introduce a 'right first time' culture, with the projects finished on time and to budget
- Eliminate waste, streamline processes and work towards continuous improvement
- Work towards standardisation in components where this provides efficiency gains
- Use a properly trained and certificated workforce and keep skills up to date
- Improve the management of supply chains
- Keep abreast of changing technology by innovation and investment in research and development

An objective view of these demands from customers in other industries would probably result in the view that they are not unreasonable.

Clients, on the other hand, were to accept as part of the 'Pact' that they had responsibility to become 'intelligent clients', and since then the concept of a client's charter has progressed, even if it is only understood and progressed by the experienced clients.

Most of the focus, then, over the last few years has been on strategic issues with the industry and its professionals 'talking the talk'. There is indeed (as Digby Jones suggested) much less evidence of them 'walking the walk' in relation to the adoption of improved strategies, as indicated by the actual strategies adopted and the relatively low use of innovative processes or products.

Progress *can* be achieved and measured through the implementation of known tools and techniques. Where and when measured improvements follow they will be used again. Typical examples may include:

- The adoption of simple techniques to enable project objectives to be prioritised and client needs to be clearly defined
- An increased focus on preplanning to reduce late changes
- The adoption of value management at an early stage
- A balance between capital cost and long-term 'worth' in terms of the business case for the project
- An increase in the early appointment of specialists through collaborative strategies to enable their input into design and costing

Such techniques can very quickly improve predictability and value for money. Their adoption moves aspiration towards implementation.

One key area identified by clients (Construction Clients' Forum 1998) was that they did not like unpleasant surprises, although there was clear evidence from client satisfaction surveys (also by the CCF) in 1999 and 2000 that these were regularly experienced by the majority of clients (Construction Clients' Forum 1999, 2000).

Constructing Excellence (CE), funded by the Department of Trade and Industry, has begun to take up the challenge of seeking methods of achieving continuous improvement by emphasising the need for inexperienced clients to seek independent advice and to ensure that they (clients) appreciate the need to adopt a more strategic approach to their proposed projects. Through its website (www.constructingexcellence.org), CE has emphasised the importance to clients of verifying the need for the project and the development of a project business case. Clients will then be better able to identify and prioritise key project objectives. They will also be able to appreciate the importance of establishing realistic parameters, analysing project risk and developing with their project team an appropriate procurement strategy. The website links to a range of available advice in both business and construction, and examples of successful projects are given.

The simple message being promoted by CE could be effective in reducing unpleasant surprises and improving output satisfaction. It is a message that was identified by Latham in 1994 (Latham 1994) and re-emphasised by Egan (1998) and the Confederation of Construction Clients as of considerable importance.

In summary the industry is unique and complex. It is predominantly project-focused in producing its output but frequently adopts process-manufacturing terminology, often in an inappropriate way. Performance is seen as sub-optimal and whilst this may be misleading, opportunities for improvement in procurement processes do exist by increasing the objective advice available for clients and the extent of adoption of simple techniques already available within the sector.

Other improvements cannot be achieved through the application of techniques and this must be trusted to the representative groups and institutions at the heart of the industry. Continuing to design buildings requiring traditional skills whilst at the same time observing a constant reduction in the availability of skilled tradesmen and training places requires intervention at a national level. Innovative processes and products may be an alternative solution if the skilled workforce continues to shrink, but investment in research and development in the construction industry is about 10% of that of other industries. The innovative products and processes available are frequently resisted by the conservative attitude of many designers and professionals and are unpopular with real-estate funders.

These and more strategic procurement issues may be resolved over time. In the meantime clients will expect some of the improvements that are achievable by the professional application of available techniques in the procurement of their projects.

Procurement strategy

A construction project is one way of delivering a solution to the particular business needs of clients, whether for investment, expansion or improved efficiency. When new build solutions are selected, rather than renting, leasing or purchasing existing real estate, there is usually the need for a bespoke solution which aims to meet particular objectives.

New build projects are complex one-offs with unique designs on unique sites. Successful outcome is achievable only where the complexity of the processes involved are recognised and addressed appropriately. Unlike the processes adopted in manufacturing, construction activities are not on-going and the management of projects requires different skills. The team drawn together for the project will disperse at its completion and are unlikely to form the same team again. If they do, the project will be different.

Design is largely segregated from the project construction process which itself will need to integrate specialists from a wide range of fields of activity.

In such a scenario the role of the client is vital and includes developing a strategic brief for the project based upon a functional needs analysis and rooted in a clearly established business case. Also the client must ensure that necessary decisions are made and resources are available throughout the project. Procuring construction as a business solution can involve high risk to the client organisation. Risk will require managing and will involve client intervention in both strategic and implementation terms with the assistance of a team of construction professionals.

Most clients will want to ensure as far as possible from the outset that they can achieve the solution they require within affordable cost and by an acceptable date in the future. This will be best achieved if the client seeks independent advice on these matters from the outset from an experienced construction professional who will not become part of the project team. Where, in meeting the needs of the business case, there is particular focus upon building function or running costs, or speed to completion or capital cost, an appropriate strategy for procuring the building while reflecting this focus must be adopted.

The establishment of a procurement strategy that identifies and prioritises key project objectives as well as reflects aspects of risk, and establishing how the process will be managed, are key to a successful outcome.

The unique and bespoke nature of most construction projects does increase the inherent risks. These risks include completing a project that does not meet the functional needs of the business, a project that is delivered later than the initial programme or a project that costs more than the client's ability to pay or fund. All of these risks potentially could have a high impact on the client's core business, and a procurement strategy should be developed that balances the risks against those project objectives established at an early stage.

Most clients will want a high-quality functional building, fast, for a low price. This is rarely, if ever, achievable since it is likely that these requirements are often in conflict with each other, and unless there is a priority of criteria, disappointment or worse may follow. The nature of the client's business and the business case underpinning the project will enable consideration to be given to which of the criteria – time certainty, price or function – is of the greatest importance. The identification of which factor will constitute the greatest risk to the business if it fails to be achieved will assist in the development of a weighted list of priorities.

The strategic brief will enable the development of a procurement strategy for the project that establishes how the team will be selected, how the project will be designed, what the parameters are and how project delivery will be implemented. Because the strategy is based upon the unique needs of the client the key parameters will be clearly communicated to the project team. A key role for the client is to ensure that the strategy established at the outset is not lost sight of as the immediate priorities of the design and construction processes are progressed.

Establishment of an appropriate project team to deliver the right project at the right time for the right cost given the adopted strategy is a vital role for the client, who again should take independent advice. In the selection process better out-turns are usually experienced where selection is based upon quality as well as price. Where the running costs of the building or the design itself are of particular complexity or importance, methods of procurement and selection enabling the maximum amount of collaboration within the project team are desirable.

The role of the client in construction is a vital one to success and involves a number of functions, which most clients have to adopt infrequently:

- Developing a vision and confirming its viability
- Funding a single project throughout its progress
- Buying the skills of construction professionals
- Responding to requests for decisions at key stages

Whilst each of these functions are complex in themselves and will need to be carried out in parallel with the usual activities of the client organisation they are achievable if an appropriate strategy is developed at an early stage and adopted throughout the project.

There are procurement strategies that will achieve:

- Certainty of cost and time for a design developed by an architect employed by the client – but this traditional procurement process is sequential and consequently slow.

- Relative speed and cost certainty – but the design will be to a greater or lesser extent the responsibility of a contractor (develop and construct or design and build respectively) and consequently the client will lose some degree of control over the design process.

- Speedy completion for a design developed by an architect employed by the client (either management contracting or construction management) – but in this case cost is uncertain until close to completion.

The establishment of an appropriate procurement strategy is a key decision if project success is to be achieved. Mistakes made at this stage may have unfortunate consequences later and it is worth taking time to ensure that the prioritisation of objectives and attitude to project risks are understood and accepted. As soon as possible after this initial stage the general parameters associated with the project will need defining, such as time constraints, budgets for all associated costs and issues that are critical to the suitability and functionality of design.

For complex buildings, and particularly where complex and multiheaded clients are involved, a successful procurement process may require the appointment of a project sponsor. Project sponsors are typically drawn from within the client organisation, and their role will be to ensure that the client gets the project he or she really needs. The role of the project sponsor is therefore to manage the client, to resolve disputes within the client organisation and to present a single point of contact to the construction team. He/she must therefore be invested with sufficient authority to manage the client function on behalf of the client organisation.

The project sponsor will need to consult with all stakeholders at an early stage to establish the specific functional needs that the project must fulfil (there is an important distinction to be made here between real needs, things the project must do and wants or 'nice to haves'), to develop realistic time and cost budgets against which the client organisation will have to make on-going project decisions, and to plan the resourcing and funding of the project. (see Chapter 6, Briefing and the design process). Often, and particularly with complex projects, this function is delegated in part or in whole to an experienced project manager, who will see the project through all its phases from initiation to commissioning. Though some experienced clients will have such skills within their organisation, for most this will constitute a consultant appointment.

Procurement options

When the project has been defined, all the factors influencing the project have been identified and the project requirements analysed, the final strategy for the project can be developed.

It is likely that there will be more than one strategy that can be adopted to achieve the requirements of the project. It is important to consider carefully each option, as each will address the various influencing factors to a different extent. In developing strategies, a potential danger is that only the most obvious course of action may be considered – this is not necessarily the best in the longer term.

Common strategies differ from each other in relation to:

- The financial risk to which the client is exposed
- The degree of control that the client has over the design and construction processes
- The information required at the time construction work can commence
- The extent of involvement of the contractor in the design stage when the contractor may be able to contribute to the design and planning of the project
- The organisational arrangements which distribute risk, responsibility and accountability
- The sequential character of the process

Selecting a procurement strategy

Clients who can be described as experienced will be able to select a procurement approach that has worked for them before or which they know to be suitable, taking into account their prioritised objectives and their attitude to risk. Inexperienced clients (the majority of clients in the UK fall into this category) will need to seek advice from experienced professionals to help them through the process.

It is most important that the strategy is reviewed at key times in the progress of the project, such as when planning approval is given, before the contract strategy has been finally decided and before the construction contract(s) has (have) been let. Reasons for this include that the development of design is not maintaining the pace anticipated, or that the programme is otherwise affected by unexpected occurrences.

The selection of an appropriate procurement strategy has two components:

(1) Analysis – assessing and setting the priorities of the project objectives and client attitude to risk.
(2) Choice – considering possible options, evaluating them and selecting the most appropriate.

It may be necessary to seek specialist advice from other consultants, for example, in relation to expected costs for the project. Specialist advice should always be sought when developing the strategy for novel or especially difficult projects.

In 1996 the RICS published *The Procurement Guide* (RICS 1996), now updated as part of RICS's *Surveyors' Construction Handbook* (RICS 2000). This is a guide intended to provide both clients and their advisers with a code of procedure to help them in selecting an appropriate procurement strategy for a building project. The guide is intended to be used as a prompt and focus for the issues to be addressed during the development of the procurement strategy, while setting the standards to be achieved by chartered surveyors when advising their clients.

The strategy should be developed from an objective assessment of the client's needs and project characteristics. A best-fit solution is looked for, with an informed client making the decision based on sound judgment, giving due regard to the identified criteria and the acceptable distribution of risk. *The Procurement Guide* included a

	Client's priority: Essential 5, 4, Desirable 3, 2, Do without 1	Traditional				Design & build						Management				Design & manage			
		Sequential		Accelerated		Direct		Competitive		Develop & construct		Management contracting		Construction management		Contractor		Consultant	
		Utility	Score	Utility	Score	Utility	Score	Utility	Score	Utility	Score	Utility	Score	Utility	Score	Utility	Score	Utility	Score
Time Is early completion required?		10		50		100		90		60		100		100		90		80	
Cost Is a firm price needed before any commitment to construction is formed?		90		40		100		100		90		20		10		30		20	
Flexibility Are variations necessary after work has begun on site?		100		90		30		30		40		80		90		60		70	
Complexity Is the building highly specialised, technologically advanced or highly serviced?		40		20		20		10		40		100		100		70		80	
Quality Is high quality important?		100		60		40		40		70		90		100		50		60	
Certainty Is completion on time important?		40		50		100		90		90		40		30		100		80	
Is completion within budget important?		30		30		100		50		70		60		90		100		90	
Division of responsibility Is single-point responsibility wanted?		30		30		100		100		70		30		10		90		90	
Is direct professional responsibility wanted?		100		100		10		10		50		70		100		30		30	
Risk Is transfer of responsibility for the consequence of slippages important?		30		30		80		100		70		30		10		100		80	
Results																			

Figure 7.1 Strengths and weaknesses of procurement systems. (Source: R.M. Skitmore and D.E. Marsden (1998) *Which Procurement System? – Towards a Universal Procurement Selection Technique.*)

check list to be considered in the selection process. Figure 7.1 shows the various strengths and weaknesses of different procurement systems.

Factors to be considered in selecting procurement strategy

At a primary level the client may wish to address whether collaborative strategies can be adopted, including the designers and constructors agreeing to a partnering approach with the client. This is more likely to be adopted where the client has a series of projects through which the performance of the contributors can be plotted through measures such as key performance indicators. This approach is seen as particularly beneficial where mutual objectives can be agreed and a largely open book approach taken to payment, including the disbursement of incentives where appropriate.

There are other strategies enabling collaboration and involvement by constructors and these are referred to in the descriptors of each procurement strategy. The choice of such strategies will depend upon the nature of the business case and the client's prioritised objectives.

Once the primary strategy has been established the factors listed below should be considered when evaluating the most appropriate procurement strategy:

- Factors outside the control of the project team
- Client resources
- Project characteristics
- Ability to make changes
- Risk management
- Cost issues
- Timing
- Quality and performance

Inevitably, some of these requirements will be in conflict and priorities need to be decided. The choice of strategy should ensure that control is maintained over those factors that are of most importance to the client.

Factors outside the control of the project team

Consideration must be given to the potential impact of economic, commercial, technological, social, political and legal factors which influence the client and the project team or are likely to do so during the lifetime of the project. These may include:

- Changes to interest rates where borrowing levels for the project are relatively high.
- Increases in the level of inflation, affecting the products or trades used in the works.
- Changes in the local or national demand for construction, affecting tender price levels.
- Changes to legislation, affecting the design of the works or the methods of achieving that design.

By adopting simple risk analysis, a judgment can be made whether contingency sums should be budgeted in the scheme in order to cope with occurrences of this nature that are likely and where their impact may threaten the project.

Client resources

The client's knowledge, the experience of the client company's organisation and the environment in which it operates are vital in assessing the appropriate procurement strategy. Project objectives are influenced by the nature and culture of the client organisation, external influences and the expectations of individuals affected by the project. The extent to which the client is prepared to take a full and active role is a major consideration.

Project characteristics

The size, complexity and location of the project should be carefully considered and particular attention given to projects with novel elements. For example, if the building is especially large or complex there may be a bigger risk of cost or time overrun. Novel projects present special risks. The novelty potential factor means that estimates of time, cost and performance are all subject to the possibility of greater error with an increased risk of one or more of the project's objectives failing.

Ability to make changes

It is preferable to identify the total needs of the project during the early stages but this is not always possible. Rapidly changing technology often means late changes. Changes in the scope of the project very often result in increased costs, especially if they arise during construction. Changes introduced after the design is well advanced or construction has commenced often have a disproportionate effect on the project, in terms of cost, delay and disruption, compared with the change itself. The design process goes through a progressive series of 'freezes' as it develops, but the client or project team should set a final design freeze date after which no significant changes to requirements or design are allowed.

Some procurement strategies are better than others at handling the introduction of changes later in the project and reducing the possibility of having to pay some form of specific premium.

Cost issues

The following cost issues are important:

- Estimates of the cost of future buildings made at an early stage are notoriously inaccurate but are necessary for decision purposes. This is the inevitable consequence of the unique nature of the project and the lack of design development. Where there is the need for price certainty this influences both project timing and the procurement strategy to be used. Generally, design should be complete if price certainty is required before construction commences. Since design takes time this will cause delay except where design and build strategies are adopted.

- Inflation will inevitably affect total cost where there is a level of variance, and increased price clauses exist which compensate the contractor for such a variance.
- Changes to design should be avoided if cost certainty is to be maintained during the course of construction. Changes, whether initiated by the client or design team, often have cost and time implications on the project well in excess of the change itself. It is therefore important for the client to fix a date after which no significant changes should be introduced, or to select a strategy that enables such changes and accepts the consequences.

Project timing

Most projects are needed within a time frame or by a specific date. Timing will influence whether subsequent activity can occur as planned and in many cases may severely affect those factors identified as critical or high priority in the business case.

Self-evidently, setting unachievable programmes will result in overruns and the UK construction industry has a reputation for delivering projects 'late'. Overoptimism or lack of reliable data on which to rely can be the cause. Realism may frustrate an impatient client but unexpected lateness may have more severe consequences.

The programme of the project is influenced by many of the factors identified in the above section and a particularly large and/or complex project is likely to require more time for design, specification and construction than would a simple small building.

It is vitally important to allow for adequate design time in terms of the total project. This is particularly the case if design is required to be complete before construction commences (where perhaps cost certainty is required). Design can be a complex and lengthy process, often taking as long prior to the commencement of the works as the works do themselves.

In the process of the appointment of the design team, assurances should be obtained about resource levels and the ability to meet key dates or programmes. It is not usual to impose contractual dates upon designers, although their progress is probably the key to the overall completion date.

Decisions to progress with a project may be influenced by the gaining of planning approval, by the successful operation of compulsory purchase order, by land purchase or by some other non-specific but critical factor (such as obtaining funding approval). Depending on whether these factors occur earlier or later, they may be an influence upon the planned or desired time available for design.

Some procurement strategies enable an overlap between the design and construction stages, so construction can start earlier than sequential strategies and offer the potential for earlier completion. It may be necessary to review planned procurement strategy in the light of design progress at the point where restraints to constructions are removed, bearing in mind the stage of design and the consequence in terms of risk.

Time has both a cost and a value. If the worth of a project is identifiable then the cost of relatively late completion and the value of relatively early completion can be sessed and may form an important factor in the decision-making process. This is en referred to as time/cost trade-off and for relatively early completion may encompass early income flow from a commercial market and will enable reduced interest and insurance charges to be realised. Relatively late completion will attract greater interest and insurance charges (amongst others) and potential loss of opportunity. Developing calculations that identify these sums can be a useful management tool.

Construction times

Total construction time is a consequence of design. More complex structures will almost certainly take longer given the same cost or size, and may require more resources. Although it is possible to work on site for extensive hours or to increase resources, it is not always possible to achieve directly resulting productivity. The law of diminishing returns will have an influence because of the limited space and the nature of traditional construction methods (such as concreting and bricklaying).

Performance

The required performance of the project measured in terms of both its response to the needs of the client as expressed in the business case and the quality of individual elements should be clearly identified. If performance is overspecified, a premium will be paid for exceeding actual requirements, thereby affecting the cost objective. Overspecification will also lead to time overruns. Conversely, failure to recognise the true performance objective leads to an unsatisfactory product.

If quality and performance are particularly important the client will probably want to keep direct control over the development of the design. This can be achieved by employing the design team directly.

Procurement strategies: a review

Hughes (2004) has argued that since there are so many variables to each of the commonly adopted procurement strategies, notwithstanding the commonly adopted nomenclature, there is a very wide range of strategies available. This may be so, but there are a range of commonly adopted basic procurement frameworks, each of which is discussed below.

Design-bid-build (also known as 'traditional')

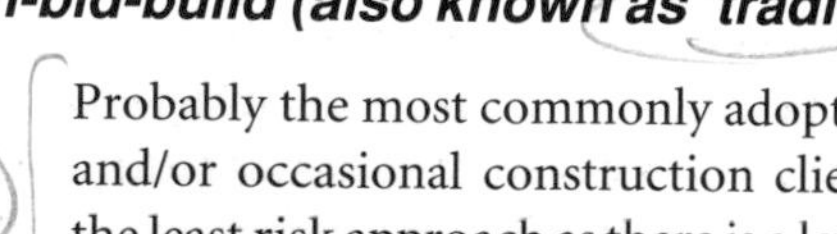

Probably the most commonly adopted UK strategy, particularly for inexperienced and/or occasional construction clients, is that of design-bid-build. It is seen as the least risk approach as there is a level of certainty about design, cost and duration inherent in the strategy if it is properly implemented. The sequential nature of the strategy which is necessary to assure low risk does mean that it can be relatively slow prior to the commencement of construction.

Under a traditional procurement strategy, design should be completed before competitive tenders are invited and before the main construction contract is let. As a result, and assuming no changes are introduced, construction costs can be determined with reasonable certainty before construction starts. This may be particularly attractive to clients with a strictly limited budget or a limit to their borrowing powers.

The tender documents will include drawn designs and a specification of workmanship and materials which the contractor must use in the preparation of his/her price. In many cases the client will arrange to appoint a quantity surveyor to measure the quantity of work to be done in order to satisfy the demands of the design. In such cases each contractor will submit a price upon the same work extent and the client will be responsible for the accuracy of these quantities prepared in the form of a bill of quantities. Whilst such documents reduce the contractors' tendering costs the use of such documents has significantly reduced in the last 20 years, shifting the risk for the extent of work to be done back to the contractor.

Alternatively the tender documents supplied to the contractors bidding for the work may include drawings and a specification, but no quantities are provided. Consequently the costs of the bidding process are increased as all tenderers will have to carry out some form of quantification for which they have to take the risk of error.

The contractor assumes responsibility and financial risk for the construction of the building works to the design produced (usually) by the client's architect, for the contract sum agreed and within the contract period, whilst the client takes the responsibility and risk for the design and the design team performance. Therefore, if the contractor's works are delayed by the failure of the design team to meet their obligations, the contractor may seek recompense from the client for additional costs and/or time to complete the project. In turn, the client could seek to recover these costs from the design team members responsible, if negligence could be proved.

Clients are able to influence the development of the design to meet their requirements because they have direct contractual relationships with the design team. When construction begins, they usually have a single contractual relationship with a main contractor, but are usually only able to influence (not control) the construction process through their architect acting as their agent for this purpose.

The strategy may fail to some degree if any attempt is made to appoint a contractor for the work before the design is complete. Such action will probably result in many post-contract changes which could delay the progress of the works and increase the costs. Where a traditional strategy is chosen because of its particular advantages, the temptation to let the work before the design is complete should be strongly resisted.

It is possible to have an accelerated traditional procurement strategy where some design overlaps construction. This can be achieved by letting a separate, advance works contract, for example, by allowing ground works (site clearance, piling and foundations) to proceed to construction once planning permission has been obtained and whilst the design for the rest of the building is completed, and the above-ground construction tendered separately. This reduces the total time to complete the project at the risk of losing certainty of cost before construction starts. More importantly, a

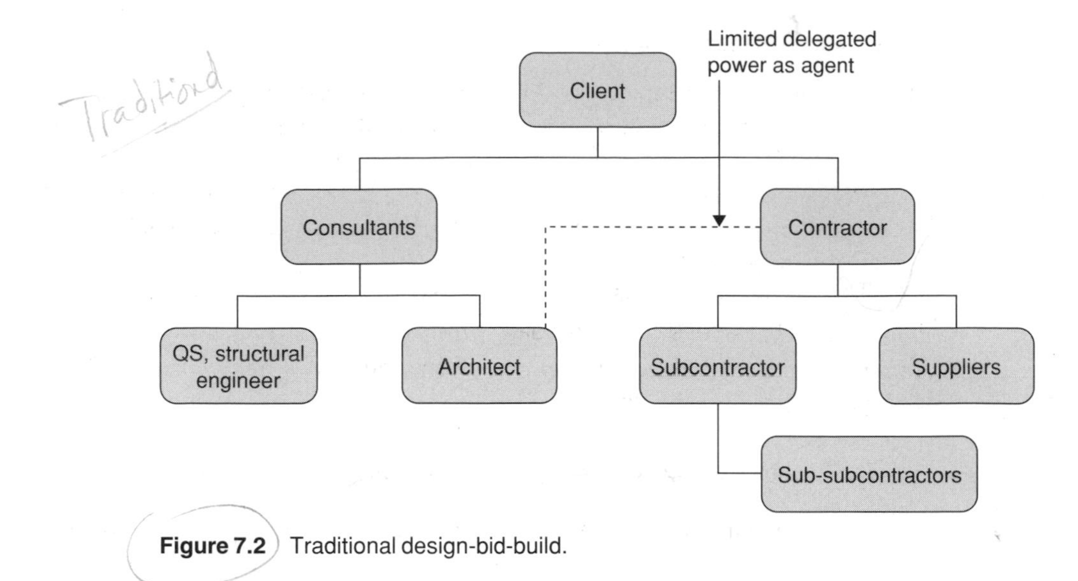

Figure 7.2 Traditional design-bid-build.

substantial risk is created in that the contractor who builds the superstructure has no responsibility for the foundation works carried out by another contractor.

Another alternative is to let the work by a two-stage process or by negotiation, thus reducing the pre-construction time involved. In each of these cases the certainty of end cost is lost in the search for a faster start but the other risks are still contained. Usually price is based upon the predicted cost of known major works elements and the detail negotiated later. This can be done in competition or by negotiation with a chosen contractor. A two-stage process also enables the chosen contractor to be consulted as the design develops.

The organisational structure of a traditional strategy is shown in Figure 7.2. Standard forms of contract are available, enabling a common understanding of rights and responsibilities. These are occasionally updated.

The main advantages of a traditional contract strategy are as follows:

- Competitive fairness, since all tendering contractors are bidding on the same basis.
- Design-led, with the client able to have direct influence, thus facilitating a high level of functionality and bespoke quality in the design.
- Reasonable price certainty at contract award based upon market forces (subject always to design changes or client-led changes which will have cost implications).
- Where public expenditure or audit demands are rigid the strategy is satisfactory in terms of public accountability since it is transparent and based upon competition.
- The procedures are well known, enabling confidence to be assured in those involved throughout the supply chain.
- Changes are reasonably easy to arrange and value where the design needs vary due to changes in client need or technology (though this ease can prove a disadvantage – see below).

The main disadvantages are that:

- It is possible to attempt to speed up the process by producing tender documents from an incomplete design, but this will usually result in less cost and time certainty and can be the cause of expensive disputes.
- The overall project duration may be longer than for other strategies as the strategy is sequential and construction cannot be commenced prior to the completion of design with no parallel working.
- There is no input into the design or planning of the project by the contractor, who will not be appointed at the design stage.
- The strategy is based upon price competition and this can result in adversarial relationships developing.

Measurement (remeasurement or measure and value)

This procurement strategy is only occasionally used, except for civil engineering projects where there is a level of uncertainty in terms of the ground conditions. With a measurement contract the contract sum is only established with certainty on completion of construction, when remeasurement of the quantities of work actually carried out takes place, and is then valued on an agreed basis. Measurement contracts are sometimes referred to as remeasurement or measure and value contracts and are based upon the principle that the work carried out is measured and valued at prices for each type of work tendered by the contractor.

The contract is not a lump sum arrangement in that there is no contract sum. Instead the bill of quantities effectively constitutes a schedule of rates for each unit or item.

The most effective use of a measurement contract is where the work has been substantially designed but final detail has not been completed. Here, as with civil engineering projects, a tender based on drawings and a bill of approximate quantities will be satisfactory.

Measurement contracts allow a client to shorten the overall programme for design, tendering and construction but usually with the result of some lack of early price certainty because the approximate quantities reflect the lack of information on exactly what is to be built at tender stage. The scope of the work, the approximate price and a programme should be clear at the contract stage. Measurement contracts provide more risk than the sort of lump sum contracts achieved through design-bid-build for the client but probably with programme advantages.

The organisational structure of a measurement contract strategy is identical to the design-bid-build (traditional) approach except that it is not a lump sum contract. Standard forms of contract are available for this strategy.

The main advantages of the measurement strategy include the following:

- Pre-construction time-saving potential, as the later aspects of the design are still on-going as the works are progressing on site.
- Competitive prices, as the work is tendered on standard approximate quantities which will be used to value the completed work on site.

- Some public accountability where this is required due to the competitive selection process.
- The procedures are well known, particularly in civil engineering projects.
- The strategy enables changes to be made easily as the later stages of the design are still progressing.
- There is some parallel working possible as the contractor is selected before the design and project planning processes are completed.

The main disadvantages of the strategy are that:

- The strategy offers poor certainty of price since the cost to the client will not be accurately known to the client until the works are complete. This will depend upon the level of certainty in the approximate quantities used in the tendering process.
- There is no contractor involvement in the early planning or design stage when most expensive or time-impacting decisions are made.
- There is a potential for adversarial relationships to develop as with all strategies that are price dependent. This may be more likely in this case unless a willingness to negotiate is evident by both the contractor and design team/client.

Construction management

Under a construction management strategy, the client does not allocate risk and responsibility to a single main contractor. Instead, the client employs the design team and a construction manager is engaged as a fee-earning professional to manage, programme and co-ordinate the design and construction activities and to facilitate collaboration in order to improve the buildability of the design.

Construction work is carried out by trade contractors through direct contracts with the client for distinct trade or work packages. The construction manager supervises the construction process and co-ordinates the design team.

Consequently this is a strategy with little certainty for the client at the outset and one usually adopted where the primary objective for the client is relative speed to completion.

The construction manager, who has no contractual links with the design team or the trade contractors, provides professional construction expertise without assuming financial risk, and is liable only for negligence by failing to perform the role with reasonable skill and care, unless some greater liability is incorporated in the contract. On appointment, the construction manager will take over any preliminary scheduling and costing information already prepared and will draw up a detailed programme of pre-construction activities. Key dates are normally inserted at which client decisions will be required. In adopting the construction management system, the client will be closely involved in each stage of design and construction. The client must have administrative or project management staff with the time and ability to assess the recommendations of the construction manager and take the necessary action. The client needs to maintain a strong presence through a project management team that

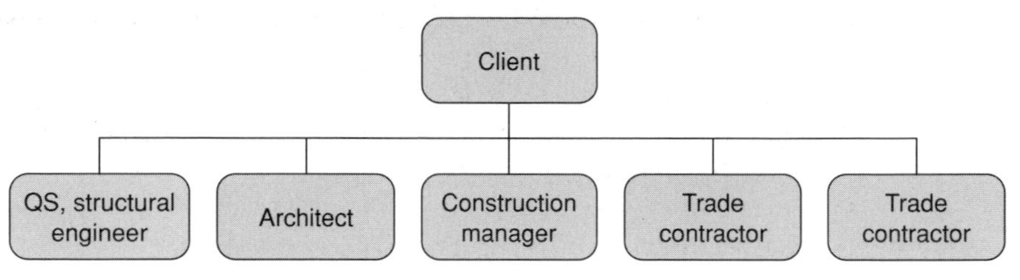

Figure 7.3 Construction management.

is technically and commercially astute. This strategy is not, therefore, suitable for the inexpert or inexperienced client.

With this contract strategy, design and construction can overlap. As this speeds up the project, construction management is known as a 'fast track' strategy. Although the time for completion can be reduced, price certainty is not achieved until the design and construction have advanced to the extent that all the construction work (trade) packages have been let. Also, design development of later packages can affect construction work already completed. The construction manager should, therefore, have a good track record in cost forecasting and cost management. A package is made up of work for which one of the trade contractors is responsible, e.g. foundations, concrete, electrical installation or decorating. These packages are tendered individually, for a lump sum price, usually on the basis of drawings and specification.

Construction management has been used in the UK predominantly for large and/or complex projects, but there is no intrinsic reason for this. Indeed, it is particularly recommended for projects where there is a high degree of design innovation, where the client wants 'hands on' involvement.

The organisational structure of a construction management strategy is shown in Figure 7.3. Standard forms of contract are available for this strategy.

The main advantages of a construction management strategy are as follows:

- The strategy offers relative time saving potential for overall project time due to the overlapping of design and construction procedures.
- The strategy enables constructor contribution to the design and project planning.
- Roles, risks and relationships for all participants are clear.
- Changes in design can be accommodated later than some other strategies, without paying a premium, provided the relevant trade packages have not been let and earlier awarded packages are not too adversely affected.
- The client has direct contracts with trade contractors and pays them directly. (There is some evidence that this results in lower prices because of improved cash flow certainty.)

The disadvantages are that:

- Price certainty is not achieved until the last trade packages have been let. Budgeting depends heavily upon design team estimates.
- An informed, pro-active client is required in order to operate such a strategy.

- The client must provide a good quality brief to the design team as the design will not be complete until the client has committed significant resources to the project.
- The strategy relies upon the client selecting a good quality and committed team.
- Close time and information control is required.

Management contracting

With this strategy, a management contractor is engaged by the client to manage the whole of the building process and is paid a fee for doing so. Unlike construction management, the management contractor has direct contractual links with all the works contractors and is responsible for all the construction works. The management contractor, therefore, bears the responsibility for the construction works without actually carrying out any of that work. The management contractor may provide some of the common services on site, such as office accommodation, tower cranes, hoists and security, which are shared by the works contractors, but in pure management contracting such works are let as a self-contained work package. The client employs the design team and, therefore, bears the risk of the design team delaying construction for reasons other than negligence.

Management contracting is a 'fast track' strategy. All design work will not be complete before the first works contractors start work, although the design necessary for those packages must be complete. Consequently it is a strategy with little certainty for the client at the outset and one usually adopted, as with construction management, where the primary objective for the client is relative speed to completion.

As design is completed, subsequent packages of work are tendered and let. Cost certainty is, thus, not achieved until all works contractors have been appointed. A high level of cost management is, therefore, required. With the agreement of the client, the management contractor selects works contractors by competitive tender to undertake sections of the construction work. The client reimburses the cost of these work packages to the management contractor who, in turn, pays the works contractors. The management contractor co-ordinates the release of information from the design team to the works contractors.

Where the management construction team is not of the highest quality, or where the management fee is inadequate, the management contractor can be less than pro-active and the system can become a reactive 'postbox' approach. It is, therefore, vital to select the management contractor carefully and to ensure that his or her fee is appropriate, bearing in mind market conditions. Similarly, resistance to works contractors' claims can be affected by the same circumstances.

The organisational structure of a management contract is shown in Figure 7.4. A standard form of contract is available for this strategy.

The main advantages of a management contracting strategy are as follows:

- There is a relatively good time-saving potential for the overall project due to the overlapping of the design and construction processes.
- The strategy enables constructor contribution to the design and project planning.

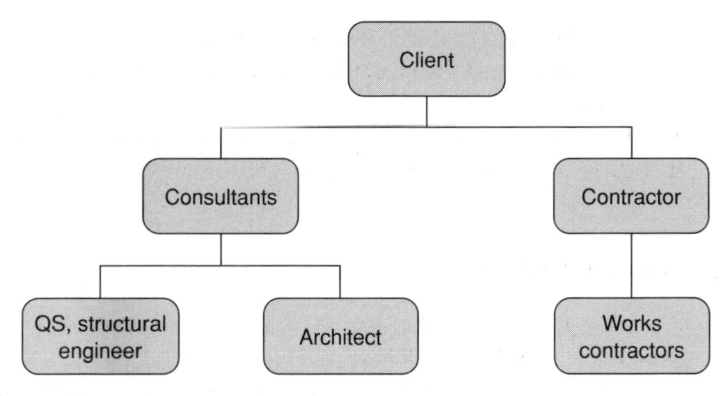

Figure 7.4 Management contracting.

- Changes can be accommodated provided packages affected have not been let and there is little or no impact on those already let.
- Works packages are let competitively at prices that are current at the time the work is let.

The disadvantages are that:

- The client must provide a good quality brief to the design team as the design will not be complete until the client has committed significant resources to the project.
- Poor certainty of price is offered at an early stage and the potential cost commitment depends upon design team estimates.
- The strategy relies on a good quality committed team or it may become no more than a 'post box' system in certain circumstances.
- It reduces resistance to works contractors' claims where such demands are passed on to the client by the management contractor.

Design and manage

A design and manage strategy is similar to management contracting. Under a design and manage contract, the contractor is paid a fee to manage and assume responsibility, not only for the works contractors, but also for the design team.

The main advantages of a design and manage strategy are as follows:

- The client deals with one firm only, enabling improved co-ordination and collaboration between designers and constructors.
- There is a relatively good time-saving potential for the overall project due to the overlapping of the design and construction processes.
- The contractor assumes the risk and responsibility for the integration of the design with construction.

The disadvantages are that:

- Price certainty is not achieved until the last work package has been let.
- The client loses direct control over the design quality which is influenced significantly by the constructors.
- The client has no direct contractual relationship with the works contractors or the design team and it is, therefore, difficult for the client to recover costs if they fail to meet their obligations.

Design and build

Design and build is a fast-track strategy. Construction can start before all the detailed design is completed, but at the contractor's risk. Under a design and build strategy, a single contractor assumes the risk and responsibility for designing and building the project, in return for a fixed-price lump sum.

A variant, known as 'develop and construct', describes the strategy when the client appoints designers to prepare the concept design before the contractor assumes responsibility for completing the detailed design and constructing the works. By transferring risk to the contractor, the client loses some control over the project. Any client requirement that is not directly specified by the client in the statement of requirements provided in the tender documents will constitute a change or variation to the contract. Changes are usually more expensive to introduce after the contract has been let, compared with other types of strategy.

It is very important, therefore, that the brief and performance or quality specifications for important requirements in the project are fully and unambiguously defined before entering into this type of contract. If requirements are not specific, the client should provide contractors with a performance specification at tender stage. The contractor develops the design from the specification, submitting detailed proposals to the client in order to establish that they are in accordance with the requirements of the specification. Clients are, therefore, in a strong position to ensure that their interpretation of the specification takes preference over the contractor's. Specification is a risky area for inexperienced clients: overspecification can cut out useful specialist experience; under-specification can be exploited.

The client will often employ a design team to carry out some preliminary design and prepare the project brief and other tender documents. Sometimes, the successful contractor will assume responsibility for this design team and use them to produce the detailed design. In many cases the contractor agrees in his or her tender to novate (effectively contractually to switch) the contract the client has with his/her designers to the contractor.

If a design and build strategy is identified as a possibility at an early stage, then the basis of the appointment of the design team should reflect the possibility of novation. If it does not, the client may have to pay a termination fee to the design team. The client may wish to retain the independent services of a cost consultant throughout the contract for early cost advice involvement in the bidding process and cost reporting during construction.

Consideration may need to be given to the inclusion of a special clause in design, or design and build type, contracts to ensure that the responsibility for design performance is properly allocated or shared. Such a clause should identify specific obligations that are absolute, that do not require the designer to be an expert in the client's business and, as a consequence, are reasonably insurable.

It has now been clearly established that a design and build contractor has a legal duty to provide the employer with a building that is fit for its purpose. This is a significantly higher duty than that assumed by an architect under a traditional strategy, where the requirement is simply one of due skill and care. Design and build contractors can, however, exclude their fitness for purpose obligation; hence under the JCT Standard Form of Contract with Contractor's Design, the contractor undertakes merely to design to the same standard 'as would an architect if the employer had engaged one direct'. Some clients are now, however, insisting on fitness for purpose in their design and build contracts, and also that where there is a discrepancy between the Statement of Employer's Requirements and the Contractor's Proposals the Statement of Employer's Requirements will prevail. To be effective, however, the Statement of Employer's Requirements must be clear, complete and unambiguous. Few clients appear to be capable of writing such a document unaided.

The imposition on contractors of fitness for purpose in design is a matter of judgement for clients and their professional advisers, even though tenderers in recessionary markets are likely to agree to undertake such risks. The requirement for insurance to cover a higher than normal risk should be weighed against the financial ability of contractors to meet design default claims. It will normally be preferable and represent better value for money to impose a lesser, yet insurable, liability, which will be the subject of an insurance payout in the event of a design fault, rather than a fitness for purpose requirement on a contractor of limited financial assets.

Increasingly, collateral warranties are being used to place additional responsibility on designers or subcontractors. Moreover, the client may take out latent defects insurance to cover post-construction liability, usually covering defects due to faulty design, workmanship and materials. The policy usually is for 10–12 years for a single premium, and is assignable, but this will be at a relatively high cost. This is a matter for specialist advice from insurance experts.

In short, design and build provides a range of options from a package deal or turnkey where the client has little involvement in the design development or building procurement process (effectively, a complete hands off approach), to develop and construct where the client appoints designers to develop his or her brief to a level of sophistication which will leave the design and build contractor to develop detailed or specialist design elements.

The project organisation structure for design and build is shown in Figure 7.5. Contractors may use their own firms' resources for undertaking the design (in-house design and build), or more likely may outsource these to one or more consultancy firms, whilst retaining control. Standard forms of design and build contract are available.

The main advantages of a design and build strategy are as follows:

- The client has only to deal with one firm, significantly reducing the need to commit resources and time to contracting with designers and contractors separately.

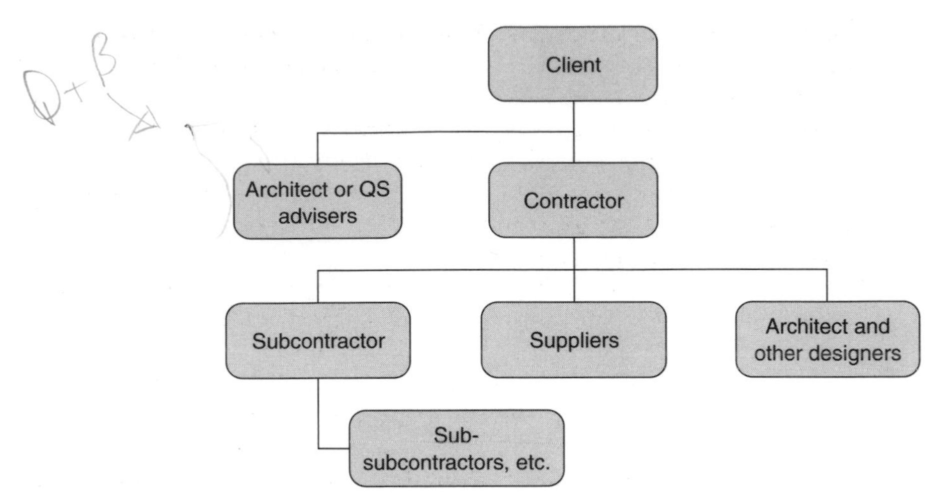

Figure 7.5 Design and build.

- The strategy enables an integrated constructor contribution to the design and project planning.
- Price certainty is obtained before construction starts provided the client's requirements are adequately specified and changes are not introduced.
- Reduced total project time due to early completion is possible because of overlapping activities.

The disadvantages are that:

- Difficulties can be experienced by clients in preparing an adequate and sufficiently comprehensive brief.
- The client is required to commit to a concept design at an early stage and often before the detailed designs are completed.
- Bids are difficult to compare since each design will be different, the project programme will vary between bidders and prices for the project will be different for each different design.
- There is no design overview unless separate consultants are appointed by the client for this purpose.
- Relatively fewer firms offer the design and build service so there is less real competition.
- Client changes to project scope can be expensive.
- Design liability is limited by the standard contracts available.

References

Adult Learning Inspectorate (2005) *Building the Future*. Adult Learning Inspectorate.

Builder Group (2003) Top 50 clients. *Building*, **8** (Suppl.).

Construction Clients' Forum (1998) *Constructing Improvement (Including the Pact)*. CCF.

Construction Clients' Forum (1999, 2000) *Surveys of Construction Clients' Satisfaction*. CCF.
Egan Sir J. (1998) *Rethinking Construction*. UK Government Department of the Environment, Transport and the Regions.
Egan Sir J. (2002) *Accelerating Change*. Strategic Forum for Construction.
Gallup (1995) Gallup survey of construction clients. *Building*, **29**.
Hughes W. (2004) Keynote address. In: *Proceedings of the 5th Annual Construction Marketing Conference*, London. Marketing Works Training and Consultancy Ltd.
Jones D. (2003) Interview. *RICS Business*, February.
Kings S. (2002) *Pricing Documentation for Constructors' Estimators: Establishing a More Effective Approach*. Unpublished PhD Thesis, Nottingham Trent University.
Latham Sir M. (1994) *Constructing the Team*. The Stationery Office.
RICS (1996) *The Procurement Guide*. Royal Institution of Chartered Surveyors.
RICS (2000) *Surveyors' Construction Handbook*. RICS Business Services Ltd.
Skitmore R.M. and Marsden D.E. (1998) Which procurement system? – towards a universal procurement selection technique. *Construction Management and Economics*, **6**, 71–89.

8 Project team selection

Introduction

Project procurement is a team-based activity. Much has been written about the appropriate selection and operation of teams to ensure suitable leadership and cohesiveness, but much of what has been written cannot be directly applied to construction, since the client is not the direct employer of the members of the team. The transient temporary nature of the project team and the organisational structure within which the team carries out its duties means that long-term relationships are rarely possible.

In this context team selection can only be based upon evidence provided by the potential team members, hearsay from others with whom they have done business or co-operated and price. Kashiwagi (2004) emphasises that by selecting team members who can prove previous success, the client is likely to improve the odds of a successful project. In the UK more emphasis has been given to processes able to support accountability.

However, selecting the right project team is likely to be the single activity that can potentially reduce most client risk. An effective team can collaborate in the client's interest to ensure the delivery of the right project at the right price at the right time. An ineffective team can fail on all these criteria and involve clients in expensive dispute.

It is therefore remarkable that much of the project work let to consultants and constructors is based primarily upon selecting team members on the basis of price rather than the quality or suitability of their bid. This is one area where government and representative bodies have consistently advised that the quality and suitability aspects of a bid should dominate choice. Inexperienced clients are well advised to emphasise provable previous performance as a key selection criterion which takes precedence over price. After all, the project will add value to the client and this added value is the key to project success. It is probably good business sense to take longer in selecting the best team on the basis of evidence provided than assuming capability or suitability on the basis of size or promotional literature.

Since all projects will go through phases involving design, costing, planning, construction and commissioning, they will involve appointing consultants and constructors. The selection of the most appropriate project team is a key process in terms of likely project success. Processes that aid the selection of both consultants and contractors and incorporate aspects of quality and capability have been developed, typically those produced in the late 1990s by the (then) Construction Industry Board. Processes such as these have been widely used, particularly by public sector clients, as they also enable accountability through a clearly established algorithm.

Whilst process cannot supplant the development over time of effective teams, in cases such as the construction industry there is some anecdotal evidence to suggest that significant improvement can be achieved.

Traditional business practice of selecting on the basis of lowest price is not likely to prove the most successful strategy when the risk of a badly designed project of poor quality delivered late can result. Advice from, for example, the client's independent advisor is available to assist in this process, to ensure sensible competition between experienced professionals. The consultants assisting the client through the project are key to achieving short-term objectives and long-term strategy. Good design adds value in both use and worth. Early appointments are likely to include designers and cost managers. The client sponsor will need to manage the process of communicating the parameters established and the functional and design needs for the project. This will enable a concept design to be produced and costed and then developed through an iterative process reflecting the development of the project brief.

Selecting consultants

The process of selecting and appointing the design team and the cost consultant is carried out by the client, but he or she may seek the advice of an independent adviser. The terms and conditions of these appointments are governed by the procurement strategy adopted for the project.

The following alternatives exist in selecting the design team:

- Single appointment – either of a multidiscipline firm which can itself provide the full range of architectural and engineering design services required, or of a lead consultant, normally the architect on building projects, who will subcontract design of other disciplines to independent professional firms and be responsible for their work and its co-ordination.
- Separate appointments – for each of the design disciplines required, with one firm being appointed design team leader with responsibility for co-ordinating the work carried out by the others.

The former has the benefit of administrative simplicity and of single-source responsibility for design. The latter offers the chance of selecting the best firm in each discipline but makes communication more difficult. The final selection will depend on the particular features of the project.

The selection of the design team and cost consultant (and other consultants as appropriate) will require the client making a balanced judgement on the following factors, listed in order of priority:

- Capability – the experience of the firm in projects of similar size and function and the availability within the firm of sufficient, uncommitted resources for it to meet the demands of the project; the demands of the project programme may be particularly important.

- Competence – the performance of the firm on past projects, to be ascertained by detailed, confidential references from past clients; efficient performance by design consultants cannot be taken for granted.
- Staff – the personal capability and experience of the key staff whom the firm proposes to employ on the project.
- The cost – quoted by the firm; unless large differences exist between offers from competing firms, this factor should not be critical.

Value for money, not lowest price, should be the aim in the selection of the design team members. The Construction Industry Board (1996) provided guidance on establishing a quality/price mechanism. Percentage weightings are given in this guidance to quality and price – the more complex the project the higher the ratio should emphasise quality.

Indicative ratios	**Quality/price**
Feasibility studies	85/15
Innovative projects	80/20
Complex projects	70/30
Straightforward projects	50/50
Repeat projects	20/80

Quality is assessed under four main headings and weighted; recommendations for the weightings are:

Practice or company	20–30%
Project organisation	15–25%
Key project personnel	30–40%
Project execution	20–30%

To enable a bid to be considered it must achieve a minimum quality threshold, e.g. 65 out of 100.

The resultant assessment is achieved by marking each of the quality criteria out of 100 and multiplying by the weighting percentage.

Consultants passing the quality threshold can then be interviewed and the lowest compliant bid scored at 100, multiplied by the quality/price ratio and added together to give a total score. The highest scorer wins (see Figure 8.1). Guidelines for the value assessment of competitive tenders in the procurement of professional services have also been published by the Construction Industry Council (1998).

Fees payable to professional consultants may be expressed in one of the following ways:

- As a percentage of the construction cost of the project
- As a lump sum
- On a time charge basis
- A permutation of all three

Example of a completed tender assessment sheet

Tender assessment sheet

Project quality weighting:	**65%**		**Project:**
Project price weighting:	**35%**		**Tenderer:**
Quality threshold:	**65** (to be compared with total quality mark)		**Assessor:**

Quality criteria	**Project weighting** (A)	**Marks awarded** (B out of 100)	**Weighted marks** (A × B)
Practice or company	25%	64	16
Project organisation	15%	80	12
Key project personnel	40%	65	26
Project execution	20%	75	15
	100%	**Total quality score**	69
Price criterion			
Tender price	260		
Price score	100	100	100
Overall assessment			
Quality weighting × quality score	= 65% × 69	= 44.9	
Price weighting × price score	= 35% × 100	= 35.0	
	Overall score	= 79.9	
Signed:		Date:	

Example of a tender assessment comparison sheet

Tender assessment comparison sheet

Project quality weighting:	**65%**		**Project:**
Project price weighting:	**35%**		
Quality threshold:	**65**		**Assessor:**

		Firm A		**Firm B**		**Firm C**		**Firm D**	
Quality criteria	**Project weighting**	**Marks awarded**	**Weighted marks**	**Marks awarded**	**Weighted marks**	**Marks awarded**	**Weighted marks**	**Marks awarded**	**Weighted marks**
Practice or company	25%	64	16	80	20	90	22.5	100	25
Project organisation	15%	80	12	70	10.5	80	12	80	12
Key project personnel	40%	65	26	80	32	80	32	90	36
Project execution	20%	75	15	70	14	70	21	80	16
Total quality score	100		**69**		**76.5**		**80.5**		**89**
Price criterion									
Tender price			260		320		280		425
Price score			100		76.9		92.3		36.5
Overall assessment									
Quality weighting × quality mark		= 65% × 69	= 44.9		× 76.5 = 49.7		× 80.5 = 52.3		× 89 = 57.9
Price weighting × price score		= 35% × 100	= 35		× 76.9 = 26.9		× 92.3 = 32.3		× 36.5 = 12.7
		Overall score	= 79.9		76.6		84.6		70.6
Signed:				Date:					

Figure 8.1 Example of consultant selection sheets. (Source: Construction Industry Board 1996.)

Percentage fees based on the out-turn construction costs are not generally recommended as they increase when cost overruns occur.

Lump sum fees may be the most satisfactory form of remuneration, provided only that the scope, value and timescale of the project can be established with reasonable accuracy before the appointments are made and that the services to be provided by the consultants can be accurately defined. Where lump sum fees are to be paid, the client will need to establish systems for monitoring the consultants' performance to ensure that they provide the full, specified service and do not skimp their services to save money.

Where time charge fees are to be paid, the final amount of fees payable is not fixed and there is a substantial risk that this amount may exceed initial estimates.

Selecting contractors

Selection of the most appropriate contractor is a fundamentally important part of the procurement process. Ensuring that the chosen contractor has the correct combination of technical skills, managerial expertise and financial resources has long been recognised as critical to project success, and has conventionally formed part of the process of tender list selection.

Contractors tender for projects in competition, seeking to obtain the project at the maximum level of profit with regard to the competition. They often tender for work at fixed prices, although generally their resource costs are not fixed. They make assumptions as to likely plant and labour outputs, which may or may not be achieved, and all of the work is generally subject to the vagaries of the weather.

It is generally recognised that in the modern construction marketplace there are many different types of client and contractor, and that the commercial relationship between them can take many forms, ranging from, at one extreme, a traditional, often confrontational, supplier/customer arrangement where the employer decides precisely what is required and the contractor simply provides what the client asks for, to, at the other extreme, close collaboration arrangements such as partnering, where employer and contractor generally work together to jointly develop the project and to resolve any problems that may arise.

In addition to those relational constraints the procurement of works and services in the public sector in Europe is constrained by law under the various European Union Procurement Directives (see Chapter 4). These regulations, which are incorporated into UK law through the Public Works Contracts Regulations, prescribe a methodology that must be followed when inviting tender for public sector works or services, which is designed to ensure maximum fairness and transparency in the tendering and contractor selection processes.

It is against this background that the client is going to make a choice as to which contractor should carry out the project.

There exists a long-held belief that for traditionally procured projects, if all of the contractors invited to tender meet the requisite criteria and all tender on the same set of basic information, then the firm submitting the lowest price must represent best value for money. This perception has now been discredited in many parts of the world as it became increasingly apparent that where this was the case the tendering contractors were sometimes prepared to take substantial commercial risks to obtain the work. Increasing project complexity (in terms of both the project itself and the context in which it is constructed) has led directly to more complex contractor selection methodologies based on a combination of price and other factors.

In the UK the selection of a main contractor was the subject of a code of practice published by the Construction Industry Board in 1997 (Construction Industry Board 1997). This code of practice set out an overall procedure that could be followed to aid selection, divided into three separate stages:

(1) The qualification and compilation of the tender list
(2) Tender invitation and submission
(3) Tender assessment and acceptance

Qualification

This process is intended to ensure that, before being included on a tender list, all contractors are assessed to make sure they are equally capable of providing the necessary technical and management expertise, as well as the financial resources, to complete the project successfully.

The qualification process may be carried out for each project where, depending on which method of procurement is being adopted, each applicant would be required to go through the selection process before having his or her name included on the tender list.

Alternatively, where a client or organisation has an ongoing programme of construction work, then they may wish to maintain a standing list of approved contractors.

The purpose of prequalification is to ensure that the firms chosen to tender for a project:

- Have the necessary technical and managerial skills to complete the work
- Have the necessary resources both technical and financial to complete the project
- Wish to submit a genuinely competitive tender

The process therefore consists of:

- Identifying those firms interested in tendering for the work
- Selecting the most appropriate of the interested contractors to form the final tender list

For a traditionally let project using either a bill of quantities or specifications and drawings the optimum number of contractors invited to tender is recommended generally to be between 5 and 8. Any less increases the risk that the tender process will not be genuinely competitive, whilst longer lists than this are likely to result in the best firms declining the opportunity to bid on the grounds that their chances of success will be too small.

A general list of factors can be considered when selecting contractors, a weighting being placed on these factors to reflect relative importance in terms of the project which can vary from project to project and client to client.

The following is a list of the general factors that may be used when prequalifying a contractor (in no particular order of importance):

- Financial stability and capability
- Current project commitments
- Geographical area of operation
- Areas of specialisation
- Management expertise
- Technical capability and resources
- Quality of site supervisory staff
- Evidence of standards of quality and workmanship
- Safety record and procedures
- Project performance record

The findings from each of these criteria should be based upon the best evidence available. It is better if these functions are carried out completely independently of each other and the results brought together only at the final stage so that the result of one is not affected by a knowledge of the result of the other investigation.

Production of the tender list

Once the factors to be considered have been decided, it is then necessary to apply a weighting, depending on the relative importance of each of the factors. Each member of a small selection team created from the consultants with representation from the client organisation acting independently would assess the scores for each contractor against the various factors being considered.

A number of rating systems are used, but the most common involve an assessment on a scale of 1–5, 1–10 or 1–100. A typical 1–10 grading represents the following:

Assessment	Rating
Unsatisfactory: falls short of acceptable level	1, 2
Below average: falls below level occasionally	3, 4
Average: acceptable level	5, 6
Above average: meets requirements generally	7, 8
Outstanding: often exceeds requirements	9, 10

An example matrix analysis is shown in Figure 8.2, and the contractors obtaining the highest scores would be included in the tender list.

All contractors now invited to tender should be capable of carrying out the project to the standard required and have the resources to complete within the contract period.

Frequently construction clients require contractors to additionally provide a performance bond, parent company or bank guarantee as a safety net for the client in the event of non-completion by the contractor.

Factors or attributes	Weight	Contractor A Rating award		Contractor B Rating award		Contractor C Rating award	
Financial stability	20	95	19.0	100	20.0	80	16.0
Management ability	20	100	20.0	95	19.0	95	19.0
On-site supervision	6	90	5.4	100	6.0	90	5.4
Technical expertise	20	80	16.0	100	20.0	90	18.0
Quality of material	6	95	5.7	95	5.7	100	6.0
Quality of workmanship	10	95	9.5	90	9.0	100	10.0
Quality of industrial relations	4	100	4.0	100	4.0	80	3.2
Safety procedures	4	100	4.0	90	3.6	80	3.2
Keeping to programme	10	100	10.0	100	10.0	95	9.5
Totals	100		93.6		97.3		90.1

Contractors are rated out of 100 for each of the factors or attributes, the best of the three being given 100 and the marks of the other two related to this.

$\frac{\text{Rating}}{100} \times \text{Weight} = \text{Award}$ e.g. $\frac{95}{100} \times 30 = 28.5$

Figure 8.2 Matrix analysis for selection of contractors.

Tender evaluation

Modern practice is that tenders are evaluated against a range of criteria, including financial issues, technical and commercial considerations. Each element can be separately weighted and scored, and drawn together at the last stage of the procedure. Sometimes the early part of the evaluation is carried out blind, i.e. not identifying the firm whose bid is being evaluated.

Evaluation will typically comprise: (1) a desk-top evaluation of the bid submission and (2) an interview process.

Price

Evaluation of contract price forms part of the desk-top process. Several techniques could be used, but the most popular involves comparing the various tender prices and scoring them with a preset formula. One popular and simple formula allocates 100 marks to the lowest tender and ranks each other tender according to the percentage difference between it and the lowest. If we assume that contractor A submits the lowest price then the score for any other contractor (x) is calculated as:

$$\text{Score (x)} = 100 - \left(\frac{\text{Price (x)} - \text{Price A}}{\text{Price A}}\right) \times 100$$

Technical evaluation

In evaluating tender bids for a large project, various headings may be used to assess the ability to carry through this particular project. These could include:

- Management team
- Personnel issues
- Proposed approach to the work
- Work control
  - — environmental practices and proposed solutions
  - — productivity
  - — safety
- Quality assurance
- Number of subcontractors: perhaps identifying specific firms if important
- Project materials

Commercial evaluation

This would involve checking the contractual and commercial terms of the tenders. Both technical and commercial evaluations will typically form part of the desk-top process, but may also include 'reality checks', involving a site visit process.

Interview

The contractor interview is an important element of the evaluation of the potential contractor. It should be used not only as a means of learning more about the contractor, but also as an opportunity for the client's team to meet the contractor and to assess how well the teams will be able to work together. It is also an opportunity for the contractor to assess the client's team, and this is especially important where a collaborative arrangement is envisaged.

Both client and contractor should field the key people who will be responsible for carrying out the project on a daily basis, and who will therefore be responsible for project success. The interview panel must be carefully briefed, and must agree beforehand on issues to be addressed.

The contractor should be asked to organise visits to suitable projects, both completed and in the course of construction, and to facilitate meetings with previous clients and end users. Such meetings will usually be private, and the contractor will not generally be allowed to participate.

Where a project is relatively large or the work is more complex a much more structured approach to the interview may be appropriate. When the Corporation of London was involved in selecting design and construction teams for city academies the interview consisted of a day of site visits, meetings with key players and an extensive, thorough and comprehensive interview in front of an expert panel.

Weightings

It is usual to weight price and non-price (the so-called 'qualitative' factors) in the assessment of the final tender evaluation in a similar way to that which can be adopted in the selection of consultants. Typical price/quality weightings will vary from 15% price/85% quality to 85% price/15% quality depending upon the complexity of the project and the environment, both physical and commercial, in which it is to be constructed.

Other factors that may also be evaluated, usually on a pass/fail basis, might involve health and safety record, financial position, etc.

The award

The technical, commercial and financial evaluations will be drawn together and a final decision made. Selection of tenderers and evaluation of submitted tenders may seem to be a long, laborious process, but the choice of contractor will generally be a critical decision in successful completion of a contract.

References

Construction Industry Board (1996) *Selecting Consultants for the Team: Balancing Quality and Price*. Thomas Telford.

Construction Industry Board (1997) *Code of Practice for the Selection of Main Contractors*. Thomas Telford.

Construction Industry Council (1998) *A Guide to Quality Based Selection of Consultants: A Key to Design Quality*. CIC.

Kashiwagi D.T. (2004) *Best Value Procurement*, 2nd edn. Performance Based Studies Research Group.

9 Managing the procurement process

Strategic aspects

Once the decision to construct a new building has been made by the client the process of initiation and management must be commenced. The definition of the project (the expectations and the boundaries) is a process that must be completed immediately. This process will include identifying the project function and an outline of the time, cost and performance constraints established in the business plan. This sets the parameters within which the project must be managed, and influences the resources that must be obtained, monitored and controlled and the quality that must be assured if the client's needs are to be satisfactorily met. Particular attention should also be given to the people who are and will be involved in the project; they will need to be motivated and managed.

The subsequent management of events is a complex and demanding process involving a wide range of professionals and specialists. There is no permanent management team and there is rarely, if ever, an opportunity to experiment with design or construction methods prior to the actual process of construction. Experience and prior proven competence are key qualities to enable a successful outcome. Walker (2002) refers to the matrix structure which construction professionals, constructors and even clients have become used to and made to work.

Relatively few clients are sufficiently experienced to manage the process and it is probably best achieved by the client appointing a project manager, who may be an employee of the client or external to the client organisation. Unless the client is confident in his or her own capability or those of his/her employees the project may become an unnecessary distraction to core business and it is often best to appoint an external individual or organisation to carry out the project management process and identify an employee to liaise between them and the internal stakeholders.

The project manager must be fully informed about the project, the stakeholders, the constraints and the resources and must be given appropriate and sufficient authority and accountability to mobilise these resources and make decisions regarding the project. The project manager will carry out and/or direct necessary functions, systems and controls in order to achieve project completion.

The client's role now becomes one of formal approval of and support for the project manager's actions and thus the client must ensure that the necessary resources, organisational structure and contractual arrangements are in place for a successful project outcome.

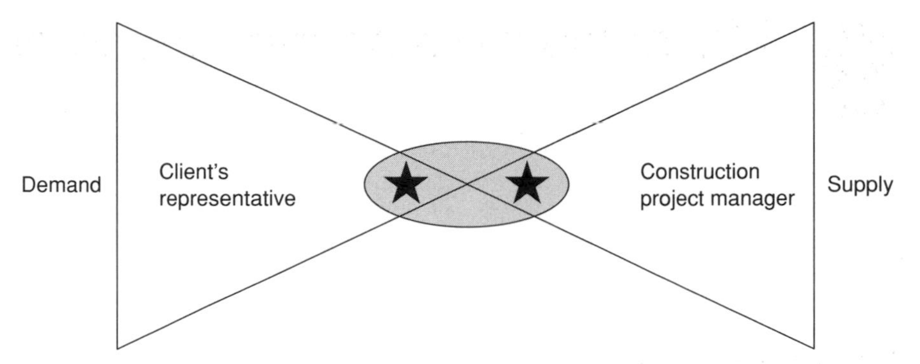

Figure 9.1 The bicameral approach to project management.

Some major client organisations, for example the Defence Estates Agency and the NHS, have taken this principle further by adopting a bicameral (the so-called 'bow tie') approach (see Figure 9.1).

This approach reflects the fact that where complex buildings and multiheaded clients are concerned it is rare to find a single individual who knows enough about both the client's business and the construction process to satisfactorily manage the demand and supply sides of the whole project single-handed. The approach would typically be adopted in conjunction with a collaborative method of procurement, and responsibility for overall project delivery is shared between the client representative (called in some organisations the project sponsor and in others, for example the NHS under the ProCure21 initiative, the project director) and the construction project manager. The client's representative is responsible for ensuing that all client (demand side) functions are carried out in a satisfactory manner, for developing the brief, for resolving any disputes between client side stakeholders, end users, etc. and for presenting a single point of contact with the construction team. The construction project manager is responsible for ensuring that the project is delivered in accordance with the client's expressed needs.

Under this approach the project manager and the client representative plainly need to work extremely closely together and to 'sing from the same hymn sheet', and therefore considerable resources need to be invested in developing and nurturing the relationship between them. This is usually done as part of the partnering process by the partnering facilitator.

Failure to ensure that this issue is properly dealt with can be clearly seen in the case of the new Scottish Parliament Building at Holyrood in Edinburgh, where the project was finally completed almost ten times over the original budget and three years late. During the course of this project no less than three project managers resigned in fairly quick succession, and whilst this issue was by no means the only cause of project failure, the lack of any kind of synergy between the project sponsor and any of the successive project managers was clearly an important issue. The project sponsor's attitude is clearly evident in her reported comment (Fraser 2004) that whilst the project manager was replaceable, the architect, in fact the alleged root cause of many of the project's problems, was not.

Regardless of the form of project management adopted, three major tasks require early attention:

- The development of a project execution plan acknowledging project objectives and constraints and recognising their prioritisation.
- The identification and appointment of a project team to carry out the plan to successful fruition.
- The development and implementation of an appropriate procurement strategy.

Developing the project execution plan

Earlier chapters have referred to the importance of the briefing process, project definition and procurement strategy. Developing the project execution plan from this information is the key primary role in the management of the project.

As a first stage it is advisable to identify all stakeholders (those parties with interest in the project's outcomes), noting their particular interests in and extent of influence on the project outcomes. Once they and their impact on the project have been established a view needs to be taken of the extent to which they can positively contribute to the project design or logistical planning or the extent to which they will be counter-productive.

It is now crucial to identify and disclose all constraints, including:

- The total funds at the client's disposal for the project and their availability or readiness as the project progresses and there are contractual obligations to pay.
- The required completion date, any important interim milestones when certain stages must be achieved and any flexibility between desired completion and absolute last delivery date.

The project execution plan will also be influenced by the project brief, including:

- The required functional performance of the final product and any indications of standards of quality.
- Whether a balance is to be struck between capital and running costs or whether initial capital cost is the primary factor.
- The likely impact to the client organisation of risks inherent in the project processes associated with time, cost and function.

A statement of the project execution plan will acknowledge the constraints and requirements of the project and enable initial consideration to be made as to which procurement strategy to adopt, bearing these factors in mind along with the extent of risk that the client is able to absorb.

Identifying the project team

Normally the project manager will need to be supported through the project by a team – a group of people with particular skills, capabilities, experience and knowledge who are specifically appointed to carry out the project plan to successful completion.

This team will assist in monitoring, controlling and generally managing the project. Levels of authority and accountability for team members must be defined. Identification of the individuals who will make up the whole team is not necessary at this stage but a 'feel' for the skill requirement and the size of the team is necessary. It may also be advisable for the project manager to appoint certain 'experts' with skills over and above his/her own (e.g. procurement specialist, specification writer, planner/programmer) at this early stage.

Implementing the procurement strategy

This is a key and early activity and warrants dedicated thought and review; it is referred to in Chapter 7. A strategy must be developed and implemented to ensure appropriate procurement in light of the project execution plan (see above). It must be clearly communicated to all the key players.

Special attention should be given to maintaining compliance with this strategy throughout all stages of the project. This will ensure the client has a greater chance of obtaining the right project at the right time at the right cost with the minimum of conflict.

Once this stage has been reached the next key stage is to procure the necessary resources and establish the methodology for their mobilisation.

Financial resources

Normally the funding function is regarded as a matter for the client. A specific individual, company or manager may be appointed by the client to take over this role, in which case boundaries to accountability and responsibility must be acknowledged.

Funding is a vital prerequisite to the project. By the time the procurement strategy is set, sufficient funding must be planned to be available at appropriate times in the pre-construction and construction process. The means of funds availability must also be planned.

The timing of funding requirements is a consequence of contractual agreements, whether they be agreements to purchase land, to design, to construct or to project manage, etc. Once the project plan is in place, it is possible to take account of each contractual agreement and to plan the timing of necessary expenditure in the form of a cash flow. In this context the project manager, perhaps with support from the project cost consultant, should take control to ensure that funding arrangements are appropriate.

Human resources

The human resource element required for the completion of the project may be viewed as three distinct groups: the client team, the project team and the construction team.

The client must be prepared to allocate appropriate in-house personnel to the project, or to appoint consultants for this purpose, in order to uphold the integrity of the brief within the constraints. The client may, as in most current construction projects, appoint consultants to design and cost a project, but the client should have his/her own staff member to provide liaison and a focus for decisions. If this is not possible it is desirable to explicitly hand these duties over to the appointed project manager.

The project team may be part of, or external to, the client and/or the project manager's organisation. The team should be appointed by and be accountable to the project manager, on behalf of the client. Each team member will have a specific task and/or role and his/her level of authority and accountability must be defined. Membership of the project team may vary throughout the project as the requirements for each stage differ. Key to the successful procurement of the project is the project manager's ability to coordinate and integrate the roles of each member of the team and to ensure that communication lines are maintained effectively.

The construction team (both the main and specialist contractors) will normally be led by a main contractor, construction manager or an equivalent. It is important that all decisions and directions are communicated to the construction team only through the project manager or other person designated for this purpose by the client. If this is not carried out and controlled there is likely to be misinformation, confusion and conflict which can potentially lead to project failure.

It is important to realise that many projects fail because of a lack of sufficient attention to people, both the individuals and groups. Understanding and communicating agendas and requirements help to build a level of trust and openness necessary to ensure the client receives a successful and complete project.

Physical resources

Other resources such as land, buildings, plant, machinery and materials are needed for project completion. At an early stage it is necessary to identify who is responsible for making these available and when and where. Some often-encountered problems are highlighted below:

- Where physical resources such as land, plant and machinery have to be provided by the client, the client must ensure their availability in a suitable form at the right time; failure to do so could be costly and will impact on success.
- Should the client require alternative accommodation, or where specific arrangements for decanting an office and staff are necessary, this is normally a matter for the client and/or his/her organisation. Alternatively these requirements may be part of the project but should be costed and contracted for accordingly.
- Design resources must also be appropriately selected and in place at the right time. Normally this is the responsibility of the client but may be part of the project manager's brief. The criteria for the selection of the design team should include factors of capability, competence, staff and cost. Value for

money, rather than cost, should be the major influence in the selection process.

- Construction resources will be contracted by the client or on behalf of the client by the project manager, and must be appropriately selected and in place at the right time. The criteria for the selection of constructors have been discussed elsewhere (see Chapter 8 'Selecting contractors').

Once the resources for the project are in place a temporary organisational structure must be established to implement the project execution plan. Additionally systems and controls need to be developed to ensure that the implementation process is controlled within the established objectives and constraints.

Temporary organisational structure

A 'temporary organisation' is created for the life of the project which enables effective communication and decision-making, management of client input, co-ordination of functional and administrative needs and the resolution of conflicts. This organisation may be headed by a key member of the client organisation or (more usually) by the project manager. This designated person must have sole authority to communicate decisions to the project design and/or construction team. If more than one individual is empowered to instruct, or to require changes, extreme confusion tends to occur.

Building users, specialists, facilities managers, maintenance staff, finance and account personnel, legal advisers and security personnel can all have input to the project from time to time throughout the existence of this temporary organisation, by invitation or by right. As appropriate, it can meet with designers or constructors to ensure effective communication. All decisions, however, should be exclusively reserved for the client's executive or the project manager.

Contractual arrangements

Having selected an appropriate procurement strategy it must be implemented. A range of contractual agreements are available, many of which are based upon 'standard forms of agreement' already in common use in the construction industry. The selection of the appropriate contractual forms is a consequence of the procurement strategy and the client's attitude to risk. The project manager will normally give advice upon the implication and selection of the appropriate form of contract.

Systems and controls

Once decisions have been made about the project execution plan, resources, the organisational structure and contractual arrangements, the project manager is responsible for implementing the procurement process and managing the project to its successful completion. Systems to enable the project manager to do this include:

- Financial systems that will ensure payments are made in accordance with contract agreements.
- Decision systems that will ensure decisions are communicated at the appropriate time and with the appropriate authority.
- Design change systems that will implement and monitor change as it becomes necessary.
- Cost and time monitoring systems that will chart real progress against the plan.

These systems will also enable the project manager to inform the client and other stakeholders of the current position at any point in time.

Controlling these systems to ensure that they are appropriately applied and not abused is a matter for the project manager. The diversity of these systems may mean delegating elements of the process to members of the project team (if appointed) or perhaps ensuring that existing client systems are specifically adapted.

Methods of control are most important to ensure that the costs remain within the budget, the pace is maintained within the programme, the quality or performance is within the standards set and all the systems are properly employed. This is of prime importance for project success.

The project manager will achieve cost control by ensuring appropriate pre-construction cost planning and an adequate system of regular cost reporting. Slippage will be immediately visible and hence easily manageable. A cost consultant can provide these services.

Pace control can be achieved by monitoring progress against the project programme or time plan. Progress should be assessed against imposed milestones, checkpoints or contract completion dates. Thus slippage can be immediately seen, communicated and action directed.

Quality and/or performance can be controlled through the provision of a sound contract which includes a detailed specification and by appointing constructors and consultants who have quality assurance systems in place. It may also be appropriate to appoint a dedicated quality control person, such as a clerk of works. His or her role would be to inspect the construction work during the progress of the project, observing and measuring quality and thereby ensuring that ongoing work is satisfactory and matches the specification.

These aspects of managing the project are addressed in more detail below.

Time management

The overall programme, or time plan, of activities that constitute the project must be developed at an early stage in the procurement cycle. It is common for IT-based programming systems to be adopted in carrying out this process and there are a number of effective and reliable systems of this type currently available. The programme forms the framework within which the designers and the constructors must start and complete their activities. It also serves as the framework within which other key stages, such as land purchase, funding and planning permission should be completed.

This programme of activities should be feasible and realistic. Insufficient design time may have an effect upon function, quality and cost, as will insufficient time allowed for construction. The programme must identify those tasks that are critical to completion on time and those tasks where some flexibility may be available. Time for approval processes – both statutory approval and client approval – should also be included as specific activities; this is invariably vital, as the need to obtain approval is likely to be a prerequisite to proceeding to the next stage of the project and further work.

The process of time control is in many ways analogous to that of cost control – there is only so much available. Thus, a time control system should embrace:

- A time budget – the overall project duration as fixed either by specific constraints or by the selected procurement strategy; this period becomes a key parameter for management of the project.
- A time plan – the division of total time into interlinked time allowances for readily identifiable activities with definable start and finish points.
- The overall project programme linking the above.
- A time checking system – the actual time spent on each activity must be monitored closely against its allowance, and divergence must be reported as soon as it is identified.

If an activity exceeds its time allowance there are essentially only two forms of corrective action available:

(1) The resequencing of later activities. This may involve abandoning low priority restraints and/or it may require phased transitions from earlier activities to later activities which logically follow them.
(2) Shortening the time allowance for future activities. This can usually be achieved by increasing specific relevant resources.

If neither is done, the overall time budget is likely to be exceeded and the project will finish late.

It should be noted that there is a risk that certain unavoidable events (for example flooding) may delay the overall completion of a project. Such possible events should be identified and contingency provision should be made for their effects to be taken into account as soon as possible, taking cognisance of contractual conditions. Other measures may also be considered, should the client so wish, in mitigation of serious delays and associated costs.

It is important to take into account at all times the fundamental relationship between time, quality and cost. It must be understood that:

- Any extension of the overall timescale for a project always generates additional costs to constructor and/or client. Every project contains time-related costs,

whether these are openly stated or not. The decision as to who carries such additional costs depends on the detailed contractual arrangements between the parties; it is likely that some of them will be borne by the client.
- Making up for lost time by resequencing later activities may be achievable but is often only at the risk of compromising quality or cost control.

Time control is as important during the design stages of the project as during the construction stage. Designers must work to a series of deadlines (milestones) at which different aspects of the design should be frozen in order that costs and the overall programme be kept under control.

At a very early stage in the project it is advisable for the client and/or the project manager to consider developing a time contingency (or reserve), with strict procedures for its allocation in specific events. This concept is essential on projects that are subject to external time constraints, for example where a building has to be available for occupation before the lease on another building expires.

Design management

Whilst the responsibility for achieving a successful design solution to the client's requirements lies chiefly with the architect, responsibility also rests with the client for ensuring both that his or her own needs are met and that the impact of his/her development on the local environment is acceptable.

The formulation of an accurate design brief and the development of design in strict accordance with that brief are key processes for the client or project manager to oversee and are addressed in Chapter 6.

In practical terms this means:

- Close collaboration between the client, the users and the design team during the development of the design.
- A procedure for formally checking that the developed design meets but does not exceed the requirements of the design brief.
- A formal presentation of the developed designs to the client and users.
- A formal 'sign-off' by the client that the design is correct and acceptable.

The client must indicate clearly the programme for project delivery and the design team should respond to this in terms of their strategy for meeting critical points in this programme.

Undoubtedly, through the project there will be incremental changes to the design, because, for example, of a change in the needs of the client or perhaps changes in technology or legislation. Formal processes for absorbing, reacting to and communicating these changes should be laid down and agreed by all parties.

It is unlikely that one person will have all the design skills for any other than the very simplest of projects. The collaboration of several designers will therefore be necessary, for example in specialist buildings or buildings that have sophisticated mechanical and/or electrical installations. Special attention needs to be paid to the integration of elements of specialist design and to effective communication and information flow.

Stakeholders, other than those involved in the project procurement process, are key to ensuring that the proposed project will be functional or will meet its spatial or other client value propositions. Many stakeholders are unlikely to be conversant with reading and interpretation of design drawings. It is therefore important to ensure that the design team present their proposals in a form that can be readily understood. Pictorial views are useful here and modern technology, in particular computer aided design (CAD), easily enables three-dimensional presentations.

Design risk

It is important for clients to know that when they appoint consultants or constructors to design projects the risk is safely transferred. Whilst at first sight this may be easily achieved, in practice it is somewhat complicated.

In most cases responsibility for the action follows the action, whether under contract or tort. In other words, designers can be held responsible for the consequences of the design failure, though not in all cases. Additionally the allocation of liability through contract does not always ensure the capability of the defaulting party to meet the financial consequences following the event.

In common law, and by statute, the provider of a product is liable for its suitability or fitness for purpose. Most clients will see their project as a product and expect such liability, particularly when they procure the project on a design and build basis. In the latter, and under common law, this is the case. It is not the case when a client purchases a service, such as the purchase of an architect's design service. Here the liability is not for suitability, or fitness for the purpose, but for care and skill in the pursuance of the architect's duty.

Standard forms of agreement for the appointment of consultants will reaffirm this limit of liability and in some cases attempt to restrict the extent of liability in express terms. The standard contracts of appointment of both the Royal Institute of British Architects and the Royal Institution of Chartered Surveyors indicate some evidence of this.

In the development of standard forms of contract, the industry and its professionals have benefited enormously from increased common understanding of the meaning of clauses. Case law and legal commentators have enabled a detailed understanding of the extent of liability and allowed clients to adopt standard contracts instead of having to write them.

Unfortunately there is a suggestion that in the contracts developed through the Joint Contracts Tribunal the influence of the client has not been as strong as it might be.

Some commentators, including Murdoch and Hughes (2000), refer particularly to the JCT 'with Contractor's Design' form of contract, prepared specifically for design and build contracts. In this case the fundamental common law principle that the design and build contractor is responsible for suitability or fitness for purpose of the project is reduced to that of an architect, or care and skill. In taking this approach, the client, in the event of design failure, is placed in a position of having the burden of proof to show that there has been negligence in design, a much more onerous position.

Paradoxically the only standard design and build contract prior to the publication of the JCT form was that prepared by the (then) National Federation of Building Trades Employers (NFBTE – the contractors). In this form, fitness for purpose was part of the contractor's liability. In the more recently developed New Engineering Contract forms of contract this liability is again possible.

The appointment of specialists to both design and construct parts of the project has produced a complicated legal position.

Where, in a traditionally let project, a specialist who has carried out a level of design for the specialist work becomes a subcontractor to the main contractor, a particularly complex situation is formed. This is the case under the most commonly adopted standard forms (the JCT forms) where the contractor does not owe any duty to the client for design. Whilst there may be duty from the specialist to the contractor, no such duty is owed to the client.

In cases like this, forms of collateral warranty or parallel contracts have been adopted in an effort to protect the client.

This book does not attempt to explore the complexities of contract law, or clauses in standard forms of contract, but seeks to highlight areas of concern, in which clients need to seek additional advice where design risk is high. One aspect of design risk that needs attention by the client, or the client's advisors, is design liability insurance. Such is the financial structure of the construction industry consequent upon the size of firms and the low level of assets needed, that insurance protection can be vital.

There is little consolation for a client knowing that there is clear contractual liability by a defaulting designer, if the designer has insufficient assets to meet the consequential costs.

Protection through design liability insurance is available providing the sum insured is adequate and providing the policy is active (the premium having been paid) when the claim is made.

There are difficulties in ensuring insurance continuity where either design and build contractors or specialist subcontractors are employed as it is difficult post completion for any client to demand proof.

Again, this is a matter of the extent of risk associated with the particular work. Sometimes the client will be best advised to seek insurance for his or her risk outside the contractual project framework. Examples of this include assignable latent defects insurance giving protection for up to 10 years but costing up to 1.5% of the construction cost.

In summary, design liability is as complicated as the contractual interrelationships associated with construction projects. However, treated as a key aspect of risk assessment, most project risks can be protected by the prudent client.

Cost management

It is essential that the client understands the difference between:

- An estimate of cost – the provision of an informed opinion at a particular time in the development of the design of what the final cost of the project is likely to be.
- Cost control – the management of the consequences of the design and the influences of external or site-related factors so as to achieve value for money and ensure that the final cost does not exceed the budget.

Estimates are unlikely to be accurate unless cost control is exercised. They provide a measure against which to control costs. They can also be used to assess project viability, obtain funding, manage cash flow, allocate resources, estimate duration and prepare tender prices.

The techniques used to produce estimates vary according to the type and level of data available at the time of preparation. The accuracy of the estimate will improve as the design of the project develops and more details become available. The range of accuracy is likely to be in the region of +/–30% at project proposal stage to +/–5% at final tender stage. Providing cost control is being exercised, the accuracy of early estimates can normally be of sufficient precision for them to be valid parameters for decision-making and for the management of the project.

For estimates to be effective, they should be supported by:

- A list of assumptions on which the estimate is based.
- A risk analysis – an identification of the potential risks together with an assessment of their probability, their likely impact (cost consequence) and the time at which they may occur.
- A sensitivity analysis – a statement of the comparative effects on the total estimate due to changes in principal data or external factors such as interest rates, inflation, etc.

An estimate should include an amount to cover the uncertain cost of risks. This is known as the *contingency*. It is important that contingencies are sufficient to cover risks and are not eroded to facilitate any lack of cost control.

Cost control procedures will be more effective the earlier they are instituted. By way of simplistic illustration:

- Cost varies with (but not in direct proportion to) size: once the size of a building is fixed, so is the general level of cost.
- Cost varies with the complexity of design; innovative design tends to be expensive.
- The selection of the most economical design for basic elements such as foundations, structural frame, external cladding and roofing is of far greater cost significance than the types of finishing.
- The overall cost of mechanical and electrical systems is largely governed by early decisions as to the type of system selected.

Methods used for cost control differ radically between the pre-construction and construction stages of a project. Cost control during pre-construction depends partly on the procurement strategy but more on control of the design process within that strategy. Pre-construction cost control, in simple terms, comprises:

- Preparing a cost plan – a balanced allocation of the project budget into readily identifiable elements of the building, with allowances for, say, contingencies, reserves for changes in market prices, etc.
- Checking the cost of each element as it is designed against its allowance in the cost plan. It is important to note that if this cost, as designed, exceeds its allowance in the cost plan, the excess can only be corrected by:
  - redesigning the element to reduce its cost or
  - transferring money into that element from contingencies or from another element yet to be designed.

  If neither is done, the control budget will be exceeded.

Successful cost control during the construction stage depends on avoidance of major change after commitment to build. Effective management here depends on:

- Design work being completed and fully co-ordinated before a commitment to construct is made (this applies in a progressive way when design and construction overlap).
- The contract being administered efficiently and promptly in accordance with its terms and conditions.
- Changes to the design being minimised after construction has started.

Regular cost reports should be produced throughout the construction stage of the project. From these, potential overspending can be identified before it occurs and corrective action can be taken. The client should, however, recognise that such corrective action is not always beneficial, since:

- It is likely that cost savings can be made only by reductions in standards or by omissions of part of the work remaining to be finalised, that is the visible finishing and fittings. This is likely to result in the requirements of the brief not being fully met.
- Late cost savings are inefficient as any amount saved will be reduced or may be negated by the costs of disruption inherent in making the changes needed to generate these savings.
- Where inflation is having an impact on the economy, estimates and cost control procedures should take this into account (i.e. the increase in construction cost between the date of the estimate and the date when the work is carried out).

The client should distinguish between inflation and the effect on construction prices of market conditions at the time of tender. Contractors tend to adjust their tender prices by reducing or increasing their target profit margins in accordance with commercial factors. It is unwise therefore to base early estimates on any assumptions as to market conditions at the time of tender as the construction industry is subject to wide and comparatively swift changes in workload in accordance with economic conditions.

It can be expected that the project will not proceed as originally perceived or planned – the unexpected will happen. Such unforeseen happenings should be covered in the cost control system by contingency sums and the client should have a policy for the management of such contingency funds to ensure that, at all times in the project, the remaining contingencies match the remaining risks. The client may also establish and control a client reserve, not disclosed to the consultants or the contractors, to cover, for example, late changes in user requirements or unforeseeable third party events.

A particular concern of the client's may be the running costs of the finished project for the life of the client's planned tenure or even whole-life costs – the costs of use over the whole life of the building – a stance that is becoming increasingly popular. Often higher initial building costs result in lower running or maintenance costs during the project's life.

Estimating future running or maintenance costs is difficult to achieve in practice. Pasquire and Swaffield (2002) identified a range of barriers to the adoption of life cycle and whole-life costing techniques, including a lack of suitable or usable data in a reliable form concerning component life and replacement costs. Potentially some data are available, but many organisations maintain separate sections which carry out day-to-day repair and replacement but rarely collect the costs and information in a recoverable way for individual buildings.

They also identified the emphasis by clients on lowest initial costs as a significant barrier, particularly when current fiscal policy allows commercial clients to obtain full tax relief on property running costs. Initial costs include design, construction and installation, purchase or leasing, fees and charges.

Future costs include all operating costs, such as rent, rates, cleaning, inspection, maintenance, repair, replacements/renewals, energy and utilities use, dismantling, disposal, security and management over the life of the built asset. Loss of revenue may also need to be taken into account, for example to reflect the non-availability of the revenue-generating building during maintenance work.

The timing of future costs must be taken into account in the comparison of options. Future cost flows are discounted by a rate that relates present and future money values – which may include an allowance for inflationary changes.

Opportunity costs represent the cost of not having the money available for alternative investments (which would earn money) or the interest payable on loans to finance work.

There is growing awareness that unplanned and unexpected maintenance and refurbishment costs may amount to half of all money spent on existing buildings. Estimates of the value of the unplanned portion in UK construction output range from £8 billion to £20 billion per annum.

Of course life cycle costs or whole-life costs become more important where the client organisation carries the risks associated with these through the project. The

Private Finance Initiative (PFI) and public private partnership (PPP) projects present this scenario and are referred to in Chapters 13 and 14.

Quality control overview

The quality (or performance) of the project, the completed building, is governed, progressively, by:

- The project brief – a clear statement of the standards of quality required. This statement of standards must be clear and unambiguous: vague phrases such as 'the highest quality attainable within the budget' should not be used.
- The original design, which must indicate:
  - the components selected for the building and their integration in the whole design;
  - how components and systems interface;
  - how mechanical and electrical systems are integrated into the overall design.
- The completeness of design before construction starts.
- The specification in terms of both:
  - the conversion of the quality standard demanded by the project brief and design drawings into precise written statements of what standards must be met and what extent of work must be done for the contract to be completed satisfactorily; and
  - the setting out of criteria against which the standard of the finished work will be judged by reference to standards, codes of practice or similar where available.
- The quality control system – control mechanisms that can be applied to the execution of the work on site; for example the detailed on-going supervision by the contractor, the programmes for testing and the procedures for rectifying any defective work.
- The inspection systems – the independent inspection and verification of the contractor's work by the design team. This includes procedures for witnessing tests, for commissioning plant and systems, for precompletion inspection for defects lists and defect rectification and for final hand-over.

The client may choose to appoint a quality inspector (or a clerk of works) whose sole function is to inspect completed work and verify it against the imposed quality standards. However, such an inspection and verification regime should be recognised as a last line of defence. The key to quality control on site is to specify clearly and to monitor closely the quality control activities carried out by the contractor while work is being done.

Many construction companies and firms of consultants within the building industry have adopted, or are in the process of adopting, formal quality assurance systems for their own work or services in accordance with BS 5750, ISO 9000 series or equivalent standard. These organisations are better placed to understand and to deliver required quality products and performance.

Change control overview

With designed projects – particularly those based upon lump sum or fixed price contracts – avoidance of change during the work is the best strategy. Clients should avoid making changes to 'signed-off' and agreed designs as these can be the cause of significant extra costs and delay.

However, circumstances (such as the economic environment or the client's needs) do change and so change to a project may be unavoidable. Consistency of approach to the implementation of such changes is vital; it helps to ensure transparency and effective communications and contains cost and time within understood parameters.

Client changes (as distinct from design development) are changes that are made either by: (1) the design team, with the approval of the client, to a design feature which it has been decided should be altered in order to improve the initial design or to remove inconsistencies; or (2) the client, after the design brief has been agreed. The most significant of these are changes to those elements of the project that are critical to timely and economic completion.

Client changes can be relatively minor, such as adding a few extra power points, or they can be major, with major cost implications, such as the addition of an extra storey. The cost of client changes depends largely on when they are made, for example:

- Before the construction contract has been let, where the cost can be contained to that of the changed feature itself.
- After the work has commenced, when the cost will probably be disproportionate to the value of the change. It can disrupt the contractor's work and invariably gives rise to a higher cost than if the change had been included in the contract as let.
- Cumulative minor changes, which can have a serious effect: disruption claims caused by a large number of small changes are common.

Some client changes are unavoidable. Examples of such changes are:

- Compliance with changes in legislation
- Requirements of the health and safety or fire prevention authorities
- Those required by unforeseen ground conditions
- Previously unforeseen users' requirements

The contingency in the project budget should be sufficient to take account of the likelihood of such changes, based on risk assessments.

Changes proposed prior to construction may be either unavoidable or optional. If they are unavoidable, the client should authorise a transfer from the contingency in the budget to cover them. If they are optional, the client is advised to approve them if it can be demonstrated that they offer good value for money (or a saving) and that there are sufficient funds available to pay for them.

Changes proposed after the construction contract has been let can have major time and cost implications and should be avoided if at all possible. If they are not essential, it is recommended that they should be deferred until the project is complete and then reviewed to see if they are necessary and economically justified.

Commissioning

Once the building work is complete, the systems that will support user comfort (e.g. heating, ventilation and lifts) must be commissioned to ensure they are working effectively and reliably. In relatively simple buildings, the client can insist that this is a function the contractor must perform. Where buildings have sophisticated systems controlling the internal environment or facilitating staff movement or safety, commissioning can be established as an independent activity carried out by specialists. However commissioning is facilitated, it must be complete before any building can function effectively and be handed over.

Occupation and take-over

The client is responsible for addressing the issues of occupation, staffing and subsequent operation and maintenance of the building. This activity is separate from the design and construction process, although it will affect it, and will have its own time, resource and cost implications which should be incorporated into the overall project plan.

For large projects, the client may wish to arrange for the nomination of a member of part of his or her organisation to act as occupation manager to manage this activity or may appoint an external facilities manager.

Occupation plans should be established during the design stages of the project and should cover:

- The operation of the building on a regular on-going basis
- The hand-over and acceptance of the building from the contractor(s)
- The progressive final fitting-out (if any)
- The physical occupation of the building with minimum disruption to the client's operations

Conclusion

This chapter has discussed the management of the project and has frequently referred to the project manager. Whilst there is a growing appreciation of the need for a member of the project team to manage the process and the appointment of a project manager is increasingly common, it is, of course, possible for the management role to be achieved without this appointment. The fragmentation of the design and production processes, however, makes this task very difficult for a member of the design or construction team to carry out as well as their primary role. It is not to be recommended. Managing the realisation of the product is a key role in achieving the client's aims and a separate appointment is advantageous, particularly to inexperienced construction clients.

References

Fraser Rt. Hon. Lord of Carmyllie (2004) *Official Report of the Holyrood Enquiry*. www.scottishparliament.uk/vli/holyrood/inquiry/sp205-00.htm.

Murdoch J. and Hughes W. (2000) *Construction Contracts: Law and Management*. E. and F.N. Spon.

Pasquire C. and Swaffield L. (2002) *Best Value in Construction*. Blackwell Publishing.

Walker A. (2002) *Project Management in Construction*, 4th edn. Blackwell Publishing.

10 The value of design

Value context

Good design can be appreciated by the corporate client, the individual user or by the public. In the case of the public it is often a matter of taste whether the new building is admired, accepted or disliked. Churchill once asserted:

> 'There is no doubt whatever about the influence of architecture and structure upon human character and action. We make our buildings and afterwards they make us. They regulate the course of our lives.'

However, the public see only the influence of the project in its physical context. A case can be made that there is a responsibility for architecture to be socially appropriate as well as serving the needs of the individual client, even though it is the individual client who will meet the cost. Buildings can contribute positively to their location, whilst some equally can become eyesores.

Construction is the physical response to a specific client need or value proposition as expressed through design. The design responds to a business case carried out in a micro and macro economic context, and within a period of particular social attitude. As Saxon (2003) argued, 'Construction is a means, not an end'.

The value of the designed and constructed project flows from the use of it over its physical, functional and economic life, and a client's view of whether a design is good may, in some cases, be more to do with its usefulness, its efficiency or its real estate value. For example:

- Owner-occupier clients will be largely concerned with the value or worth of the building in use, how functional it is, its costs in use and perhaps its real estate value.
- Developer clients, particularly those who develop speculatively, will be concerned with achieving an outcome that will meet the demands of the current property market and will return highest profit.
- Investor clients will be concerned with achieving extended physical, functional and economic building life and real estate value – buildings attractive in the marketplace for lease or rental.
- Image-conscious clients may wish to compromise economy or even functionality in order to achieve a building that they feel is important to their public presence or market.

For each of these examples value outcome must be greater than cost. Sometimes marginal benefits are acceptable (as in the case of image-value buildings), but in most cases maximum benefits are desirable since client organisations will inevitably wish to gain the maximum benefit for their investment. The predominance of inexperienced clients often means that they focus on the maximum volume they can obtain for initial capital cost. The process that follows can result in a project that fails to focus upon the primary aim of the value proposition as it refocuses on short-term objectives.

The construction industry has traditionally not perceived value in the context of the client's business case or expression of need and has focused upon cost. Clearly this is the consequence of segregating the design and construction functions and failing to drive the importance of value through the supply chain.

Oscar Wilde defined a cynic as 'one who knows the cost of everything and the value of nothing'. On that basis construction is a cynical industry. It is also a victim of that cynicism. As Saxon (2003) puts it, 'Construction usually sees itself as a cost to its customers, not as a value adder. This leads to lowest-cost tendering, commoditization, adversariality, and all the other ills'.

Duffy, in his Introduction to the *Design and Economics of Building* by Morton and Jagger (1995), takes this idea further:

> 'It is no longer enough to consider the costs and value of construction independent of the way clients look at buildings. Ultimately clients are interested not just in the productivity of the building process but in the occupancy costs in relation to their own economic objectives. Clients are now becoming interested in a new and most important concept: measuring the productivity of building use through time.'

The Royal Academy of Engineering identified the generic ratio between capital cost, lifetime cost and occupier staff costs over 20 years as 1:5:200 (Royal Academy of Engineering 1998). To define potential benefit Saxon (2003) argued for a five-unit ratio, 0.1:1:5:200:250+, which included the consultancy cost (0.1) to define and design and a minimum figure for value added by the occupier. He argued that this could also easily be 2500 and that the ratio between consultancy input and occupier value added is thus between 1:2500 and 1:25 000.

Better building design leads to more valuable buildings. Judgement as to what constitutes good design can be affected by the perspective of those who judge. Users, traders, owners and the general public will all have a view – mostly different to each other: users will evaluate its function and comfort, traders its marketability and owners its real estate value or efficiency. The role of the architect is to serve all these criteria in a balanced way.

Boyd (1965) argued that architecture is an art which commences with a puzzle – the unique set of problems set by the client and a host of other factors including site conditions, function, budget and time constraints. Within these limits the architect must use the 'sternly practical business of providing bodily shelter as a medium for artistic expression'.

Present-day architecture commences with the concept of one or two practitioners but over time relies upon the coordination of the contribution of many participants including specialist designers, specialist suppliers and specialist constructors. Frequently coordination results in compromise which can affect the integrity of design and consequently value. Architecture can become a process fragmented in nature; however, collaboration with an integrated team can enable improved building design, thus satisfying all stakeholders.

Best and De Valance (1999) highlight the concept described by Lovins and Browning (1992) where the architect draws a building and then throws the drawings over the wall to the mechanical engineer with the request to 'cool this (building)'. This lack of collaboration or coordination at an early stage can result in complex bespoke solutions which are expensive.

The value of good design

Design is an important factor in ensuring good working conditions for staff and convenience for members of the public who need to visit the building. A well-designed building is a good investment; it can contribute to both the economy and the efficiency of the building through its life. Good design can be realised by not only focusing on aesthetics but also paying attention to:

- Internal and external layout, with particular attention to efficient and economical use of space.
- Efficient use of energy, for example in the heating system (e.g. ensuring good insulation) and in the mechanical services.
- Building maintenance costs.
- Building and space flexibility, to allow for changing requirements.

The blame for a poor building is just as likely to be ascribed to the client as to the designer. Equally, a well-designed building reflects credit on those who commissioned it as well as on the designers who conceived it. The setting of the design (i.e. relationships with the neighbourhood) must also be considered, not just the design in isolation. Furthermore, the client should expect to consider alternatives before committing to the 'signed off' design.

The important value proposition which is developed for any project and which is underpinned by a specific business case will enable the balance between the primary criteria of cost, time and function to be established and a procurement strategy to be developed in the context of the client's attitude to the risk generated by carrying out the project.

The segregation of design and construction inherent in the procurement strategies adopted in most cases does not, however, enable value to be enhanced. Constructors or specialists best able to contribute are not involved until most design and all cost and time parameters have been fixed.

There is, therefore, a lack of continuity in the process, which results in a fracture between the focus by the client on the value of the output and the focus by the construction professionals on the cost and timing issues associated with project delivery. The potential for innovation and information to flow from producer to designer or from specialist to customer is vastly reduced or eliminated.

The significance of the balance between project value propositions and cost and time factors is considered below.

Sometimes the value proposition may require, for example, the mechanical installation to drive design. Take cold stores, where much of today's food is stored before distribution to our supermarkets. In design terms they are simple structures, the predominant purpose being to cover the cooling equipment. To achieve greatest value the equipment should dominate the design, though aesthetically this is likely to produce an unimpressive result.

Best design is a combination of inspiration, understanding and application. Concept design is probably better in the hands of the architect and application in the hands of the specialists and constructors. The key is understanding the client's value proposition and functional needs. Collaborative procurement systems enabling constructors and suppliers to contribute to design will enable better design.

Best design will help to deliver best value. Most clients will be advised to focus on value outcomes rather than cost parameters. Techniques such as value management can significantly contribute to this concept and will enable a fragmented team to 'buy in' to the project. Where collaborative procurement processes such as partnering are appropriate these too can contribute to success.

Managing value: value management and value engineering

Value management is not a new concept and is increasingly being requested by experienced construction clients as a way of achieving best value for their investment, particularly where projects need to satisfy client needs in terms of functionality. Many construction clients see value management as essential to help them to maximise value for money. Less experienced clients tend to be unaware of the benefits that can flow from its use and are less likely to require this service. They are very dependent upon their consultants for strategic advice and appear only infrequently to be recommended to adopt value management measures.

Ashworth and Hogg (2000) refer to claims of success in financial terms of 10–15% improved value against a fee of 1% but also suggest that less than 10% of projects adopt the techniques due to lack of appreciation by the client and lack of marketing by practitioners.

Commercial and retail clients are facing increased pressures in their attempts to retain their competitive edge and they must pay greater attention to quality of both product and service. There is no doubt about the increased need for techniques that offer better value. Satisfying the needs of clients has become crucial, and is one of the UK construction industry's key performance indicators (see the Constructing Excellence website, www.constructingexcellence.org).

Consultants are being required to broaden their roles and responsibilities in order to achieve better integration and compatibility with corporate client objectives, and expectations have generally risen within the whole project team. Justification of expenditure is increasingly important.

Government is insisting that 'best value' be demonstrably achieved on all publicly funded projects. In the procurement guidance of the Office of Government Commerce, the responsibility for optimising and achieving value for money is explained.

Value management is a structured approach to the examination and development of a project, which will increase the likelihood of achieving a range of user's requirements at optimum value for money. It is a continuous process from inception to completion.

Value management

Value management is used as a generic term to indicate the broad process involved in managing value. It is a structured process of dialogue and debate among a multidisciplined team of designers and decision-makers during an intense short-term process. The primary objectives of value management include the development of a common understanding of the design problem and design objectives and the achievement of a consensus about the comparative merits of alternative courses of action.

It is early in the design phase when the bigger decisions are made and when the greatest potential for making better design decisions therefore exists (see Figure 10.1).

It is inappropriate to carry out a functional analysis with the aim of eliminating unnecessary cost in the very earliest stages of design. To be able to discuss elements and components of buildings assumes that the decision that a building provides the solution to a problem has already been made. It also assumes that a design solution exists. It cannot be taken for granted that the nature of the problem is fully or indeed widely understood or consistently understood by all participants or even within the client organisation.

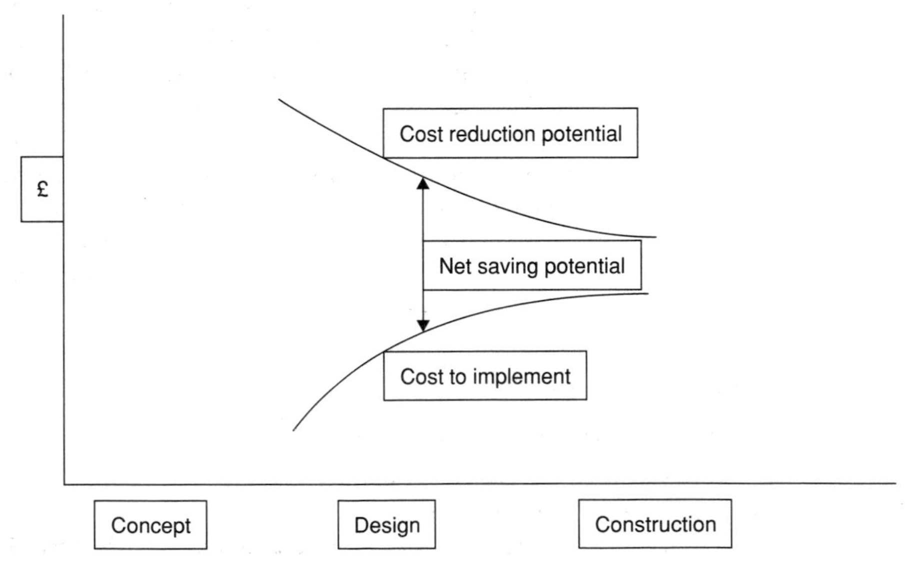

Figure 10.1 Cost benefit.

This lack of clarity and/or agreement in the very early stages of the conception of a project is common and can lead to buildings that fail to satisfy a client's needs. Value may mean different things to different members of the client organisation. Some way of making explicit what value means to the client is therefore needed. As this has to be done before any designs are produced, discussing functions at this early stage is not possible. Thus, the methodology for value engineering as described below may also be viewed as inappropriate at this stage.

In practice it is often the case that a value engineering study is initiated when things start to go wrong or in response to a projected cost overspend, rather than being built into the design programme in advance. The potential for making better and/or more cost-effective design solutions is then even more limited. It is better to ensure that the right decisions are made the first time than to correct them once they have gone wrong.

'Value management' is the term adopted to describe the process at this earlier phase, with 'value engineering', as described below, coming later. Sometimes 'value management' is used as a generic term to incorporate both sets of procedures where value engineering is seen as a sub-process of value management.

Value management workshops

There are varied views about how value management should be applied to construction. Seminal work by Kelly and Male (1993) comprehensively discussed how this can be achieved through a process of workshops. For example the first workshop may take place at the end of the concept stage when the building of a new project is first suggested as a possible solution to a perceived problem (RIBA Stage A) (Figure 10.2). This workshop is closely linked to the decision whether to proceed with the project. The primary purpose of the workshop is to confirm the need to build before the client becomes committed to expenditure. The secondary objective is to ensure that there are clear project objectives which are understood by all the parties.

The next workshop could occur at the end of the feasibility stage (RIBA Stage C) when the outline brief has been completed and the design team has formulated a selection of schemes for which they have produced estimates of cost. During this second workshop issues such as ensuring that the outline design scheme is selected in accordance with the appropriate performance criteria are addressed and any changes needed to be made to the design in the light of changes to the objectives originally formulated are considered. These may be due to changed factors or where new information has come to light or circumstances have changed so that it would be sensible to amend the objectives in line with these.

The benefits of value management

The adoption of value management may achieve a range of benefits, including:

- Confirmation of the need for a new project
- The open discussion of objectives
- The development of a structured, rigorous and rational framework for evaluation and re-evaluation

- Decisions supported by data and made on the basis of defined performance criteria
- Improved accountability with an established audit trail
- Consideration of alternative solutions
- Decisions made with greater confidence
- A greater potential for increasing value for money in the project
- Better understanding within the team and better communication throughout the organisation
- Increased satisfaction with the end product due to participation by users
- Optimisation of the overall efficiency of the project

Integration and application of both the rational problem-solving skills and the 'softer' interpersonal skills associated with effective people management are key to successful value management. Use of the various structures and procedures proposed for the workshops will facilitate the rational processes and provide a climate for creative thinking, 'ownership' and commitment.

The potential benefits of the adoption of value management are significant. The process is always to some extent dependent upon the skill of the value management facilitator.

Value engineering

Value engineering is a systematic procedure aimed at achieving the required functions at least cost. Value engineering is based on the assumptions that all parties understand

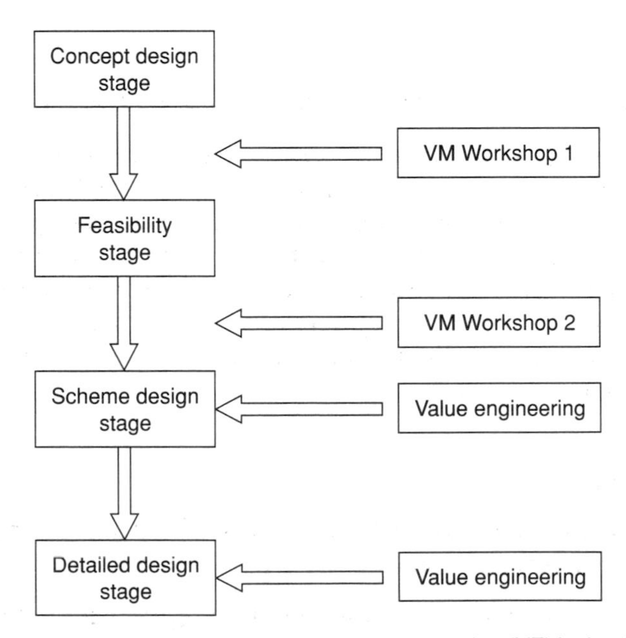

Figure 10.2 Value management (VM) and value engineering (VE) in the design process.

the functions required and all design alternatives provide the same level of functional performance. Each can then be assessed on the basis of cost alone.

Within this frame of reference, an increase in value can be directly related to a reduction in cost. The original concept of value engineering can be traced back at least to the development of techniques and procedures developed by George Jackson Churchward, Chief Locomotive Superintendent of the Great Western Railway in the early years of the twentieth century, to guide the design of a new range of railway locomotives. In an attempt to achieve maximum value for money and efficient operation the designers introduced disciplines to ensure that every component delivered maximum functionality at minimum cost.

The concept was applied in the USA when the General Electric Company (GEC) was faced with a significant increased demand for a number of key components. Components were in short supply, so a task force was set up to address the problem. Their task was identified by Lawrence Miles, a purchase engineer with GEC, as being:

> 'If we can't get the part, then we must get the function.'

This approach resulted in a method that identified the function of a scarce component and then sought its replacement with an alternative component that would serve the same purpose. It was noticed, as a side effect of this process, that many substitutions resulted in a reduced cost. The basic philosophy of value engineering was therefore established as being *to eliminate the cost that did not contribute to the performance of the required function.*

The value engineering concept was further developed by GEC over the following 10 years and gradually became more refined and started to spread throughout the US manufacturing industry.

Application of value engineering in the construction industry

Owing to the complexity and fragmentation of the construction process it is inevitable that there is unnecessary cost in the design of most buildings. Value engineering works by bringing together in a team a range of professionals, including architects, engineers and quantity surveyors, and introducing a systematic approach during the design process to identify and overcome potential problems at an earlier stage in the project and produce better value for money. By pooling their expertise under the guidance of a value engineering team leader the project gains through their collected experience.

Method of approach

The set of procedures for the conduct of a value engineering study comprises five distinct stages or phases:

(1) *Information phase*: an information-gathering process which focuses attention upon the component being studied. Particular importance is given to the use of functional analysis.

(2) *Speculation phase*: creative thinking techniques are used to generate alternative ways to provide the functions identified in the previous stage.
(3) *Evaluation phase*: the solutions generated are evaluated in terms of their feasibility and cost. Ideas are combined and consolidated to produce a list of, say, five or six ideas that are worthy of further consideration.
(4) *Development phase*: the surviving ideas are developed in detail, ensuring that all of the interfaces with adjacent components are fully accounted for.
(5) *Presentation phase*: the best solution is identified and recommendation is made. This is often done by means of a formal presentation followed up by a written report.

Value engineering workshops

Value engineering is achieved though a framework of workshops, including a series of mini-workshops conducted at key stages of the project, or a two-day workshop.

Mini-workshops

A value engineer joins the design team at the briefing stage and acts as facilitator throughout. Two mini-workshops, following the job plan format, are then held to review the proposals at about 15% and 35% design completion.

The two-day workshop

The two-day workshop appears to be the most popular approach, where all members of the project team are able to focus on the issues in an intense and focused way.

As with value management there are significant benefits to the client where the project adopts an approach to value engineering at an appropriate stage in the project. This may be particularly so with complex projects or where the project team is fragmented.

References

Ashworth A. and Hogg K. (2000) *Added Value in Design and Construction.* Pearson Educational.

Best R. and De Valance D. (1999) *Building in Value: Pre-design Issues.* Arnold.

Boyd R. (1965) *The Puzzle of Architecture.* Melbourne University Press.

Duffy F. (1991) *The Changing Workplace.* Architectural Design and Technology Press.

Kelly J. and Male S. (1993) *Value Management in Design and Construction: The Economic Management of Projects.* E. and F.N. Spon.

Lovins A. and Browning W. (1992) Green architecture: vaulting the barriers. *Architectural Record*, December.

Morton R. and Jagger D.E. (1995) *Design and Economics of Building.* E. and F.N. Spon.

Royal Academy of Engineering (1998) *Study of the Long Term Costs of Owning and Using Buildings.* RAEng.

Saxon R. (2003) *Povey Lecture.* Joint Contracts Tribunal.

11 Risk

The need for risk management

Construction projects are typically complex. They have time, cost and quality targets which the client requires to be met, and compromise is usually needed to attain that unique combination of targets which will generally represent the client's definition of value for money for the project. Quality, for example, may be sacrificed in order to achieve a time target or to comply with a cost limit, or time may be compromised in order to achieve high quality. Since the major purpose of the project procurement strategy is to ensure that the client is given the best possible chance of fulfilling his or her objectives as defined by 'value for money', it is clear that the development of the most appropriate strategy must at the very least be informed by the likelihood that the targets might not be achieved. Some analysis of the commercial, technical, social and political risks that might tend to knock the project off course must therefore be considered in developing the procurement strategy. It is important to note that risk management cannot eliminate risk, but risk management techniques can be used to reduce the impact of events that may cause failure to reach the desired targets. It is also important to remember that there is considerable research evidence to suggest that significant cost savings can be achieved by paying attention to contract strategy based on systematic consideration of risk. The proposals for funding a project should therefore include recommendations on contract strategy.

Risk management may, depending on the structure and organisation of the project team, be undertaken by the project manager or by an independent risk manager, but the success of the chosen risk management strategy will largely depend on the stage of the project at which it is introduced. It is widely recognised that deficiencies in the project brief lead to uncertainty and greater risk during the contract. For many projects in the past, risk management has been a subconscious and somewhat haphazard process, but if it is to be effective it needs to be formalised and embedded into the procurement process.

This chapter therefore considers some issues to do with risk analysis and management. It is not suggested that a detailed risk assessment would be required for every project. Plainly the extent of the risk assessment needs to be appropriate to the type and size of the project, the prevailing economic and political circumstances and the degree of complexity inherent in the project and the commercial environment in which it is to be procured. Nonetheless some consideration of the likely risks must always form part of the project procurement strategy.

Risk and uncertainty

It is important to distinguish between events that are 'risky' and events that are 'uncertain'.

Risk events are events the occurrence of which can be predicted to at least some degree, generally based upon historical data or experience and making a decision according to the probability of a particular event occurring. In other words, a forecast is made on the basis of past data and therefore within some degree of certainty.

Uncertain events are random events which defy prediction. Risk events can be managed; uncertain events cannot. It should therefore be a fundamental rule of risk management to reduce uncertainties to a minimum. This is typically done by converting them into events that can be predicted, generally by gathering more data, or by engineering out of the process the events that are likely to give rise to them.

It will generally be impossible to 'manage out' all of the uncertainties, and the usual way to manage those that remain is to include some form of float or contingency allowance in programmes and budgets.

Risk management strategies

A number of formal risk management strategies have been developed by various organisations, and widely used models in construction include the Association for Project Management 'Project Risk Analysis and Management' (PRAM) (APM 1997) and the Institution of Civil Engineers 'Risk Analysis and Management for Projects' (RAMP) (ICE 1998). Most models, however, involve the following four stages:

(1) Risk identification
(2) Analysis, involving quantification of the risk effects and prioritisation
(3) Choice of an acceptable risk management strategy
(4) Risk monitoring and control

Each of the above is considered in detail in the following sections.

Risk identification

Once the project and its objectives have been well defined, the process of risk identification can take place. There are a number of approaches that can be used. Some or all of the following may be employed:

- Assumption analysis
- Lists
- Brainstorming
- Delphi technique

Assumption analysis

As its name suggests, this is where risks are assumed based on previous experience and knowledge. The identified risks are assessed using a description. For most business decisions there are four main categories into which risks can be placed:

(1) High probability – high impact
(2) Low probability – low impact
(3) High probability – low impact
(4) Low probability – high impact

Those with a high probability of occurrence and a high impact on the project would be those typically considered for risk management.

Lists

Checklists permit rapid risk identification and avoid problems being overlooked. Prompt lists act in a similar way. A typical checklist is shown in Figure 11.1.

Brainstorming

Brainstorming is perhaps best know as a problem-solving technique, but it works equally well for risk identification. To ensure that the best outcome is achieved, a facilitator should set the ground rules and explain the procedures for all members of the team. The proceedings start with a framework for the session, taken from prompt lists and checklists, looking first at whole project issues and then focusing on specific phases. No criticism is allowed.

- Participants are encouraged to suggest any idea
- Quantity is required (the more ideas the better)
- Combination and improvement are sought (this encourages the generation of better ideas by building on others)

Delphi technique

The Delphi technique is a structured way of obtaining a group consensus on the risks and their impact. It is better suited to assessing risks, but can still be effective for risk identification.

The technique requires a chairman or co-ordinator to act as the central hub. The chairman distributes questionnaires or asks participants to generate individual lists of potential risks. These are then given to all members, each of whom individually considers the probability of the risks occurring. These independent opinions are collated and redistributed to gain consensus. No participant discusses his or her opinions with anyone else.

Once the process has gone through sufficient iterations and the chairman feels that no further benefit will be gained from continuing, the results are reported.

(1) The following check list gives both generic and specific risk issues likely to be encountered by most types of project. It is based on a number of more specific check lists from a variety of sources, and includes lessons learned from particular projects.

(2) Each question should be considered in turn by the project team and/or risk consultant, and should be answered by one of **NOT APPLICABLE**, **YES**, **NO** or **UNKNOWN**. Every question where the answer is NO or UNKNOWN requires a risk issue to be raised and risk mitigation or contingency actions to be identified.

(3) Respondents should consider both those aspects of the project for which they are responsible, and the complete project in the broader sense.

PROJECT:	**RESPONDENT:**	**RISK CONSULTANT:**	**DATE:**	
RISK TYPE	**RISK AREA**	**UNCERTAINTY**	**NA/YES/NO/ UNKNOWN**	**ACTION**
1. Requirement	1.1 Clarity	Is the requirement well understood?		
	1.2 Volatility	Is the requirement stable?		
	1.3 Specification	Are all required specifications available and adequate?		
	1.4 Interfaces	Are all interfaces well defined and acceptable to us?		
	1.5 User	Is the required user interface clearly defined?		
2. COMPLEXITY	2.1 Project	Is the complexity of the project acceptable, i.e. not expected to cause problems?		
	2.2 Size	Is the size of the project manageable, i.e. not expected to cause problems?		
	2.3 Integration	Has sufficient time/effort been allocated to system integration?		
	2.4 Subsystems	Are all subsystem interactions defined and acceptable?		
	2.5 CM	Is configuration management adequately controlled?		
. . . etc.				

Figure 11.1 Example check list. (Source: Association for Project Management in their 'Project Risk Analysis and Management' (PRAM).)

One of the benefits of the Delphi technique lies in its flexibility. Because it is an individual activity carried out in isolation, there is no need for all of the participants to be together either geographically or in time. It therefore presents a totally different approach from brainstorming.

Cause and effect

Risks do not generally occur in isolation – they have one or more causes and one or more effects, and it is important that these should be considered as part of the risk identification process. A single risk may have only minor consequences, but a combination of seemingly 'minor' risks could have serious consequences for the project. Hence the importance of a comprehensive analysis.

A further factor that needs to be considered is risk dependency – that is, the extent to which the probability of risk events occurring is modified by the occurrence of other events either individually or in combination, for example:

- If event A occurs then event B will not occur
- If event A and event B occur then event C is inevitable
- Etc.

It should also be borne in mind that risk events may only occur at certain stages in the process, for example during the design stage, or perhaps during various stages of the construction process. Risk management therefore needs to be a dynamic process.

Risk examples

Examples of some of the major risks that may require attention during the design of the procurement strategy might include:

- External risks such as those that arise from economic, legal or political environments
- Financial risks:
  - currency conversion risk
  - funding risks
- Site risks:
  - restricted or occupied site
  - planning difficulties
  - access
  - environmental risks
- Client risk:
  - lack of client knowledge and/or experience
  - complex and multi-headed clients, leading to lack of clarity, misunderstanding or disagreement concerning project objectives
  - likelihood of post-contract changes
- Design risks:
  - selection of an inappropriate consultant team
  - poor brief

  - incomplete pre-contract design
  - complex building works, novel design or lack of buildability
  - inadequate co-ordination of design or production documentation
  - inadequate design and co-ordination of construction and services
- Selection of an inappropriate contractor:
  - inadequate selection process
  - lack of clarity in terms of the trading relationships between employer and contractor and between contractor and supply chain
- Construction and delivery risks:
  - weather
  - constructability
  - health and safety
  - availability of key resources

Analysis, involving quantification of the risk effects and prioritisation

Once the risks have been identified, then their effects must be analysed before attempts can be made to manage them. Risk analysis therefore involves deciding the following:

- Risk probability: how likely is it that the event will occur?
- Amount at stake: how much is the possible loss?

Risks are typically addressed through a combination of *qualitative* and *quantitative* assessment. Qualitative assessment is typically a relatively simplistic technique used to describe risks in linguistic terms. Quantitative assessment on the other hand describes risks in terms of their mathematical probability of occurrence and the numerical consequences of their occurrence.

Qualitative assessment

Qualitative assessment seeks to describe and understand each risk in order to assess the likelihood of occurrence and the likely consequences in terms of impact on project performance. Various linguistic scales are used ranging from the simplistic high/low probability, high/low impact to more complex and subtle subdivisions where the variables may be measured on more complex scales such as:

Probability	**Impact**
Very high	Catastrophic
High	Very serious
Moderate	Moderately serious
Low	Inconvenient
Very low	Insignificant

If a finer distinction than this is called for then a quantitative measure is probably the best approach.

Quantitative assessment

Here the likelihood of an event occurring is given a numerical probability. Probability is measured on a scale of 0 to 1. An event with a probability of 0 will not occur, and an event with a probability of 1 will definitely occur. An event with a probability of 0.5 has an even chance of occurring or not. The possible consequences of a risk could be quantified in terms of:

- *Performance*: the extent to which the project would fail to meet the user requirements for standards and performance. Benchmarks and key performance indicators (KPIs) need to be included here.
- *Time*: the additional time required beyond the original completion date anticipated for the project.
- *Cost*: the additional cost over and above the original estimate for the project out-turn. This includes not only the contract sum but also all associated fees.

This method of analysis requires: (1) probabilistic combination of individual uncertainties; and (2) estimates of uncertainty in predicting the cost and duration of activities.

Quantitative techniques are less subjective than the qualitative approach, although in many cases the raw data have had considerable subjective input.

Quantitative analysis often involves more sophisticated analysis techniques which usually require computer application. Computer-generated models and analytical techniques can be useful indicators of trends and problems for attention, but they should not be relied on as the sole guide to decisions. Their accuracy depends on the realism of assumptions made, the skill of the model builder and the accuracy of the data used.

An alternative method for considering project risks is to analyse any risk independently of others, with no attempt to estimate the probability of occurrence of that risk. The estimated effects of each risk can then be accumulated to provide maximum and minimum project outcome values. In other words, neither a subjective nor an analytical value is given for the probability of occurrence of the risk event. Instead, each risk event is compounded to determine the possible effect upon the project, and then, by applying a range of maximum (critical) to minimum (minor) project consequences, the full extent of the particular risk can be seen.

Sensitivity analysis

This technique is used to consider the effect of changes with regard to those events that are deemed to be a potential risk to the project. It calculates the effects on the project using a range of values for the event changes. For example, there may be a risk that the cost of steel will increase by 2.5, 5, 10 or 15%. The project outcome is evaluated for each of these potential cost changes and the result can be plotted on a graph to show the percentage variation (risk v cost change).

The results of a sensitivity analysis can be shown graphically on a 'spider diagram'. A sensitivity analysis is very useful, because often the effect of a small change in one variable (for example cost or duration) produces a marked difference in the project outcome. When several risks are being assessed in this way, a 'spider diagram' provides an effective way of demonstrating those risks that are most critical and sensitive. These are the ones the project manager must act upon.

A sensitivity analysis can be performed for all the risks and uncertainties that may affect a project, in order to identify those that have a large impact on the cost, time or other objectives. It may also be used to identify the variables to be considered when carrying out a probability analysis (see below).

One problem with sensitivity analysis, however, is that each risk is considered independently, with no attempt made to quantify their probabilities of occurrence. Another limitation is that in reality a variable would not change without other project factors changing, and this is not shown in the sensitivity analysis. Eventually, however, given sufficient practice, users can more easily identify those risks that have a large impact on the project, and so the number of risks in need of consideration can be reduced.

Probability analysis

This technique can extend beyond the limitations of a sensitivity analysis by specifying a probability distribution for each risk and then assessing the effects on the risk events in total. However, careful interpretation of the results is essential.

One important stage in this type of risk analysis is assessing the range of probabilities that could result. 'Monte Carlo' simulation (random sampling) can be used where calculation of data inserted into an equation would be difficult or impossible. This procedure may be used in a probability analysis as follows:

(1) The variations to the risks being considered are assessed and a suitable probability distribution of each risk is selected.
(2) For each risk a value within its specified range is selected. The value should be randomly chosen and within the estimated probability distribution.
(3) To establish the outcome for the project, a calculation is made based on the combined values for each risk.
(4) The process in (3) is repeated several times in order to produce the probability distribution of the project outcome.

Decision trees

For all major projects, there are almost certain to be a number of routes through which the objectives may be achieved, and selection of the most appropriate procurement route may not therefore be straightforward. Decision trees are a useful way of assessing the likely impact of risk on managerial decision-making. They provide a

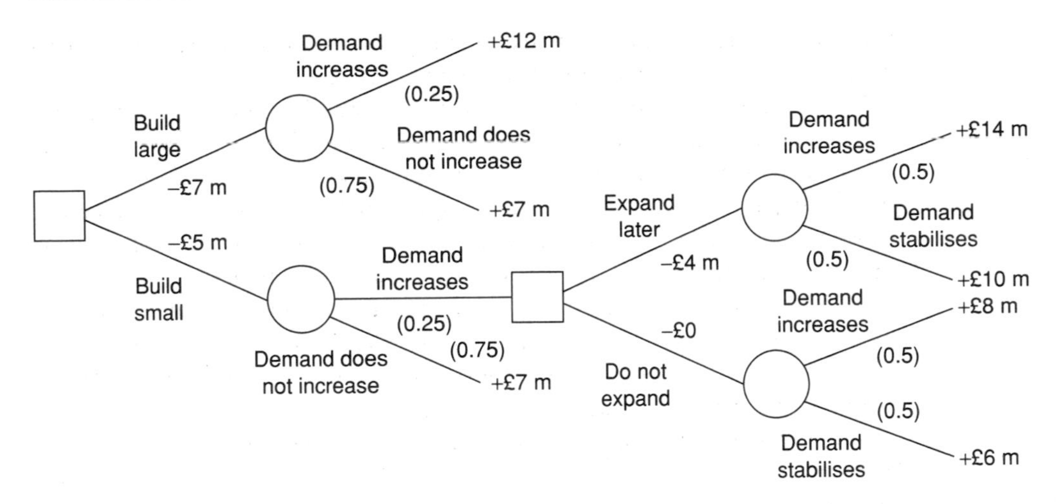

Note: by convention, a square node depicts a choice to be made, while circular nodes correspond to chance events.

Figure 11.2 Example decision tree.

graphic display of the range of possible outcomes and can be used to help make a particular choice of action(s), or provide justification for it. An example decision tree is shown in Figure 11.2.

The most popular approach is to evaluate the expected outcome at each node in some numerical form, and the accumulated 'value' at that point is then given by multiplying the net value by the probability that the outcome will occur. This criterion says that the expected value, or worth, of a profit of £x that materialises with probability p is £(px). Thus the expected value at each circular node is the sum of all the £(px) quantities. Note that summing together just the probabilities out of any one chance node has to equal 1, since precisely one branch must occur. (Remember that the probability of a certain event occurring is, by definition, 1.)

To find expected values at each square node, it is necessary to subtract the costs associated with the particular branch. By working backwards exhaustively through the entire decision tree, the most appropriate decision at each node can be found simply by choosing the branch that maximises expected profit at every step. This enables the best overall strategy to be implemented – or at least gives a more solid basis on which to make decisions than purely by intuition alone.

Figure 11.3 shows the expected values inserted into the decision tree. The uppermost circular node has an expected value £8 250 000, derived from £12 million × 0.25 + £7 million × 0.75. The second chronological decision node (should warehouse be expanded later or not?) has an expected value of £8 000 000, being the larger of £8 million × 0.5 + £6 million × 0.5 = £7 million (if not expanding), and £14 m × 0.5 + £10 million × 0.5 + (–£4 million) = £8 million (if choosing to expand). The first chronological decision can be seen to be between expected profits of £1 250 000 for building the large warehouse immediately, against £2 250 000 for the more cautious 'wait and see' option.

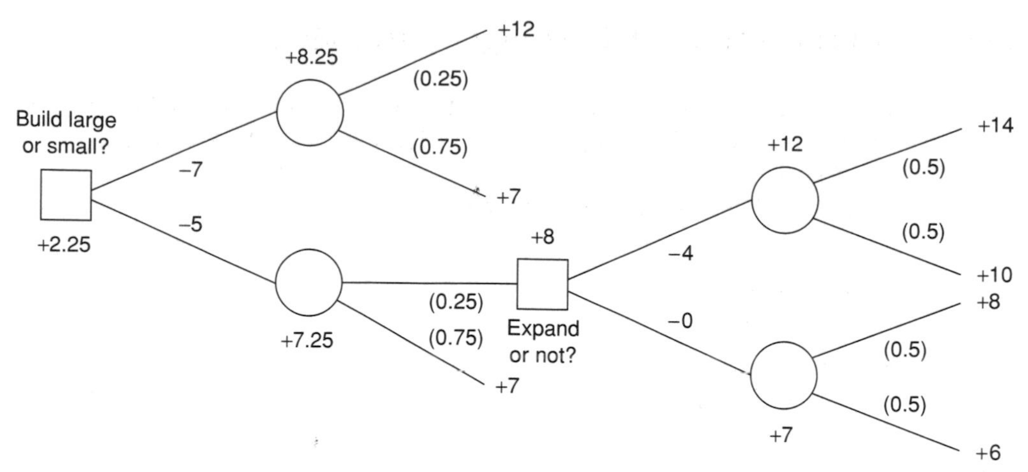

Figure 11.3 Example decision tree with values inserted.

In this example, therefore, application of the decision tree method strongly supports building the smaller warehouse for now.

There are some obvious limitations to this technique. Used naively, the user puts faith in the accuracy of all the probability figures and cost estimates. If any of these happen to be seriously wrong, so too might be the recommended decision.

The usual way to overcome this problem is to conduct a sensitivity analysis, similar to that described above, especially if there are grounds for reasonable doubt as to the figures given in the decision tree – perhaps because not all values are based on well-researched, historical data. Clearly, if costs and probabilities are no more than wild guesses, the method, used simplistically, has little more to offer than blind intuition. However, it comes into its own when large and complex decisions are broken into a series of smaller ones, any of which can be subjected to minor changes in associated numeric values to assess its overall impact.

In summary, decision trees:

- Can provide a powerful technique for convincing oneself and others of the most suitable course of action. Their graphical representation sometimes makes the solution obvious. By the addition of estimated costs, values of outcomes and probabilities, they provide a basis for analysing complex problems. This system can therefore help to clarify and communicate the option available to the project manager.
- Can be used when choosing the appropriate procurement route, the method of construction or even whether to proceed with a project claim. Their more obvious use is by clients in choosing between alternative development projects.

While this approach to risk and decision-making is not without limitations (in common with most tools in management science), a key advantage is that it encourages the client, project manager and/or decision-maker to assess some degree of probability of an outcome occurring and to take rational and logical action at each decision point.

Choice of an acceptable risk management strategy

This again comprises two main activities: (1) an initial response once a risk is known, and a strategy devised to mitigate the risk; and (2) allocation of the risk to the person best able to control it.

Typical risk management strategies are:

- Risk avoidance
- Risk reduction
- Risk transfer
- Risk sharing
- Risk retention

Risk avoidance

Risk avoidance requires some action to be taken to ensure that the event giving rise to the risk does not occur. Typically one would remove the cause or consider an alternative solution.

Risk reduction

Risk reduction requires some action to ensure that, if the event occurs, the impact will be reduced as much as possible. The most popular option would include consideration of alternative solutions and a more detailed examination of the risks involved.

Risk transfer

Risk transfer requires someone else to take the risk. Options here might include insurance or subletting. In the event that major risks are involved, the 'empty chair' theory of risk management says that if no-one in the team has the capability to manage the risk then an additional team member should be introduced and paid to take the risk. This technique has been used effectively in some major privatised infrastructure projects in circumstances where either local institutions have no track record in the international market place or currencies are not directly convertible in the international markets. In these cases additional organisations have been invited to join the team simply to act as guarantors against default.

A typical example of the transfer of risk in the selection of procurement route might be a decision to use design and build rather than a traditional or management-based approach. In this case the whole of the design risk is placed with the contractor rather than with the employer.

Risk sharing

Risk sharing is generally achieved through the mechanism of the construction contract. For example in many contracts the risk of adverse weather is shared in that the contractor would be given an extension of time but no additional monies. The client would therefore not be able to deduct liquidated and ascertained damages, but neither would he or she have to pay any additional preliminary costs.

Risk retention

Risk retention means that the risk is noted and remains. In effect someone 'takes the risk'. Taking the risk is a perfectly acceptable risk management strategy provided that the decision is based upon sensible and carefully considered criteria.

Care should be taken when considering the available risk management options to ensure that the potential impact of each risk is not outweighed by the direct costs of risk management. A risk management plan will help to keep a check on the overall strategy chosen. An awareness of these problems or sources of risk is essential in order to give focus to specific aspects of the procurement process. How these factors are to be dealt with will heavily influence the choice of contract strategy.

Risk allocation

Risk should be allocated on a solid basis of responsibility and control to those parties best able to manage it, and in a manner likely to optimise project performance. The responsibility for allocating risks should rest with the risk manager. Ideally, the final allocation of risk should be done through the contract documents. Traditionally in construction, clients have been extremely shy of taking any responsibility, but more enlightened clients have recently come to the view that value for money in the finished project can be significantly enhanced through a proper and responsible allocation of risk.

It is also worth remembering that incentives and risk go together in that a party responsible for carrying a risk has the incentive to minimise its impact, but assuming responsibility for a risk often also has a significant cost attached to it which must be paid eventually by the client. The analogy here is the example of an insurance premium.

The traditional method for allocating risks in the construction and real estate industry is as follows:

- From client to designer and contractor(s)
- From either client, designer or contractor (main or sub) to insurer
- From client, funder, developer, purchaser or tenant to professionals (architect, engineer) or contractor by collateral warranty
- From main contractor to subcontractor (domestic and nominated)
- From contractor (main or sub) to guarantors or sureties

Note that historically this process has been driven from the top downwards. That is, clients have unilaterally assigned their risks to designer and contractor regardless of who is best able to control them. This process, commonly known as 'risk dumping', is now widely considered to be counterproductive and a more collaborative method of risk allocation by mutual consent is recommended.

In a sensible risk management regime, risk should be equated to the expertise, role and reward of the party to whom it is ascribed. When ascribing risk, therefore, it is reasonable for a contracting party to bear risk in any of the following cases:

- If the party can cover a risk by insurance, and it is reasonable for the risk to be dealt with in this way
- If the risk is of loss due to the party's own misconduct or lack of reasonable care
- If it is in the interests of efficiency to place the risk on the party
- If the economic benefit of running the risk accrues to the party

An important consideration when allocating project risks is the willingness of the respective parties to take on the risks. The allocation of project risks, contractually, has historically been in the hands of the client, but modern strategic procurement techniques now hold that the most cost-effective allocation of risk is derived from consensus, not decree.

Risk monitoring and control

The main evidence that risk management is taking place on a project is the existence of a risk-reporting procedure. This means that the risk management can be audited, be measured and provide a background of experience for future projects. The procurement strategy must therefore include provision for a risk register in which risk management techniques are formally recorded, and for regular updating of the risk management strategy as the project proceeds.

References

APM (1997) *PRAM: Project Risk Analysis and Management Guide.* Association for Project Management.

ICE (1998) *Risk Analysis and Management for Projects.* Thomas Telford.

Further reading

Edwards L.J. (1995) *Practical Risk Management in the Construction Industry.* Thomas Telford.

Smith N.J. (1999) *Managing Risk in Construction Projects.* Blackwell Science.

12 Partnering culture and the management of relationships

Introduction

The aim of this chapter is to introduce some of the major practical issues that influence the successful use of partnering and other collaborative procurement techniques. The discussion also includes an introduction to some of the skills required to successfully implement a partnering arrangement.

Collaborative procurement: an overview

During the past two decades the British construction industry has undergone, and is continuing to undergo, massive cultural shifts, largely driven by an increasingly discontented customer base. As long ago as 1983, the British Property Federation (BPF), an organisation representing some of the largest private sector property employers in the UK, drew attention to the problems faced by employers in ensuring that projects were delivered on time, within budget and to the expected quality standards. Their response to the perceived failings of the construction industry was to develop and unilaterally introduce their own procurement methodology. The proposed methodology (known as the BPF system) was intended to reduce the cost and time taken to procure construction projects, and included such advanced concepts as an employer's project manager, tenders produced on the basis of priced activity schedules which were to be used as 'milestones' to trigger payments to the contractor, and financial incentives to the contractor to improve project performance. At the time, these proposals were considered to be too radical, and the BPF system failed to achieve general acceptance. It did, however, throw the problems faced by employers into sharp focus, and began a long running debate about the future role of the employer in the construction procurement process.

The Latham review and subsequent developments

In 1994 Sir Michael Latham published *Constructing the Team*, his government-commissioned root and branch review of the UK construction industry. Amongst his major findings were that:

- Profit margins in construction were roughly half those that were accepted as the norm in other comparable industries.
- The construction industry had an exceptionally poor record of financial investment in the future.
- Construction industry customer complaints far exceeded those in any comparable industry.

Amongst his major recommendations, all of which were subsequently accepted and endorsed by Government, were:

- The introduction of improved dispute resolution mechanisms in construction.
- Construction contracts should not be placed on the basis of lowest price alone.
- A collaborative style of construction procurement should be adopted as general practice, where employers and contractors work together rather than in the confrontational way that had become accepted as the norm.

The Latham review was followed by a further review of construction, *Rethinking Construction*, carried out by Sir John Egan in 1998 (Egan 1998), and by a review of UK public sector procurement practices by Peter Gershon in 1999 (Gershon 1999). Following the Latham review, Government began to introduce the idea of collaborative construction procurement in the public sector through a series of Procurement Guidance Notes published by the Treasury between 1995 and 1999. This information is now publicly available through the Office of Government Commerce website. Additional government advice has also been provided in a range of publications, amongst which the most notable are probably the UK National Audit Office publications *Modernising Procurement* (National Audit Office 1999) and *Modernising Construction* (National Audit Office 2001).

The rise of collaborative approaches to procurement

Although a collaborative approach to public sector construction procurement perhaps first came to prominence in the UK as a result of the Latham review, in fact the concept is much older. In building, a good British example of a successful early strategic alliance was that operated for a number of years in the 1960s and 1970s between Marks and Spencer and Bovis for the construction and refurbishment of department stores. There are also illustrations of collaborative approaches in allied industries, for example the concept of 'alliancing' has been used in oil exploration for more than a decade.

Overseas, 'partnering', arguably the most popular term used to describe the collaborative approach, has been used extensively in the USA, where one of the best known examples is that of a substantial number of large road schemes beginning in the late 1980s. In Australia the use of partnering has had a lengthy and somewhat chequered history, principally due to a number of contractors attempting to exploit the concept in a rather cynical way, but has now developed into the concept of relational contracting. The Australian National Museum Project in Canberra is, at the time of writing, probably the world's largest and most complex collaborative scheme. Collaborative procurement approaches are being increasingly widely used around the world for large, complex and risky projects.

All of the above approaches differ slightly in the mechanisms used and in the legal significance of the 'partner' status, but all are essentially methods of harnessing the combined skills, expertise and efforts of all involved in the project to ensure successful project completion. The technique therefore requires a significant shift in the relationship between employer and contractor away from the conventional 'customer/supplier' approach, where each of the parties fights for his or her own goals, and towards a 'shared risk/shared reward' approach, where all parties involved commit themselves to work for the good of the project.

The successful implementation of this method of procurement also requires the use of specialised supporting tools such as formal risk and value management strategies and the implementation of formal quality management procedures. Note, however, that simply putting these tools in place will not, of itself, guarantee success. The tools must be used in a sensitive way as part of a cohesive and carefully designed strategic procurement methodology where the emphasis is firmly rooted in shared goals and a mutual commitment to project success. Experience particularly with the management of quality has, for example, shown that the simple adoption of formal quality assurance systems such as the ISO 9000 series did little to improve the quality of the finished work. What is required is a cultural shift such that a 'right first time every time' attitude is accepted as commonplace. Whilst, therefore, implementation of the techniques is important, this must be accompanied by an attitudinal change by all concerned in the project if the shift away from confrontational methods and towards collaborative methods of construction procurement is to be effective.

One major advantage of the collaborative approach is that it aims to improve value for money by placing the employer and, in many cases more importantly, the end user once more at the centre of the project development and construction process. Indeed the success of the methodology requires the employer and the end users to be full and active members of the project team.

Collaborative procurement techniques have played a significant part in the majority of the projects highlighted by the UK Movement for Innovation (M4I) and the Construction Best Practice Programme (CBPP). Industrial experience over the past several years in the UK and elsewhere also clearly shows that, provided the parties to the project are carefully chosen and fully committed to the relevant principles and philosophy, and that the contractual basis supports these, the approach works well in the overwhelming majority of cases.

Collaborative approaches to construction work

Some general issues

Collaborative approaches recognise that project success is more likely to occur if all of the parties involved work together for the good of the project rather than independently and for the good of themselves. The process of bringing the parties together is commonly facilitated by a third party in an intensive one or two day workshop. During the workshop, considerable emphasis is placed upon breaking down organisational barriers and traditional management systems, and on deriving an agreed set of common goals to which all concerned are willing to subscribe. These approaches are well founded in established organisational and management theory.

Collaborative approaches such as these demand that complete trust exists between the parties, and that all involved must have both the willingness to maximise value for money for the project as a whole, and the opportunity to earn an enhanced reward should they manage to do so. Of these, trust is probably the most important as a prerequisite for successful collaborative working.

Contractual approaches

Collaborative techniques must be underpinned by contractual forms which encourage the parties to collaborate rather than to work in competition. A short review of each of the major forms of contract commonly used in the UK is given below.

The JCT forms of contract including the Standard Form of Building Contract

The JCT range of contracts is very popular in the UK, not least because of their broadly risk-neutral drafting. However, although the JCT produced a Practice Note explaining the use of the JCT 98 contracts in a partnering environment (Practice Note 4 – *Partnering*), there was a commonly held view that this range of contracts does not in fact support the basic principles of the partnering philosophy. In short, the JCT 98 range of contracts were perceived to be confrontational in that they specifically required that parties to the contract serve formal notices in order to protect and preserve their contractual rights, and thus encouraged the parties to take up entrenched positions.

The JCT has plainly recognised that this is a problem, and plans to introduce a new form of contract for project partnering, to be called the JCT/BE Partnering Contract in late 2005/early 2006.

The GC Works range

The GC Works range is the range of contracts produced specifically for government works, of which the most widely used is the GC/Works/1 Contract for Building and Civil Engineering Major Works. The GC/Works range is generally regarded to be less risk-neutral than the JCT range, with a higher proportion of the risk being perceived to be passed to the contractor, and although the forms are primarily intended for government projects, the GC/Works/1 form is occasionally used by clients looking for a less risk-neutral approach than is provided by the JCT range. Although some of the procedures in the GC/Works range still require the parties to serve formal notices if particular events arise, the form does at least appear to recognise the fact that co-operation is required. GC/Works/1 Clause 1A states that:

(1) The employer and the contractor shall deal fairly, in good faith and in mutual co-operation, with one another; and the contractor shall deal fairly, in good faith and mutual co-operation, with all his or her subcontractors and suppliers.
(2) Both parties accept that a co-operative and open relationship is needed for success, and that teamwork will achieve this. The project team for this purpose shall include, but shall not be limited to, the PM and his or her representatives; the quantity surveyor; the contractor's agent; and major subcontractors and suppliers engaged in the works from time to time.

The New Engineering Contract (NEC) Engineering and Construction Contract (ECC) Edition 3

The NEC ECC form is a popular choice for contracts let on a collaborative basis. Its flexible structure together with clear drafting, in simple English, is seen as a significant advantage, and the contract includes a similar requirement to that included in GC/Works/1 for the parties to work together in a spirit of mutual trust and co-operation. A partnering agreement (option X12) is available for incorporation in the contract. The partnering agreement, which does not create a legal partnership, does nevertheless establish a framework for the partnering arrangement, which includes the formation of a core group, key performance indicators and incentive payments.

It is interesting to note that NEC ECC Edition 3, published in 2005, has been endorsed by the Office of Government Commerce, and one must therefore question whether this form will eventually replace GC/Works as Government's preferred contract.

The ACA Project Partnering Contract (PPC 2000)

The PPC 2000 contract, published by the Association of Consultant Architects (ACA), was the first generally available standard form of construction contract that is specifically built around project partnering. Partnering is therefore at the very core of the contract, and the partnering agreement is thus legally binding. The contract incorporates all of the features recommended in *Constructing the Team* (Latham 1994) and *Rethinking Construction* (Egan 1998). The contract also incorporates a number of key features including:

- A multiparty approach – all project partners become signatories to the contract.
- An integrated approach to design and construction – the contract provides for the early selection of a partnering team before design commences, thus encouraging the contribution of all parties during the design process.
- Supply chain management on an open-book basis.

- A partnering timetable to govern the activities of all parties in the partnering process.
- A core group responsible for management of the partnering arrangement.
- Incentives, continuous improvement and performance measurement.
- Formal risk management procedures.
- Non-adversarial problem resolution processes.
- A partnering advisor to assist the project team to effectively manage the partnering arrangement.

Notwithstanding the fact that this is the first contract specifically designed for partnering, initial response has been mixed. Some project teams do not want their partnering arrangements to become legally enforceable, and some purists have argued that to make them enforceable is a direct denial of the benefits of partnering itself.

Perform21 Public Sector Partnering Contract

This new form of contract, published in 2005 on behalf of the Federation of Property Societies, aims to provide a simple form of contract consistent with the principles of partnering and best value for use in the public sector. The form therefore sets out the essential risk distribution between the parties in very simple terms, whilst leaving procedural matters to be agreed between the parties. The suite includes a separate partnering agreement and professional services agreement, and provides options for:

- Term maintenance on a measure and value or target costs reimbursable basis.
- Authority design and lump sum or target cost/reimbursable basis.
- Contractor design on a lump sum or target cost/cost reimbursable basis.
- Subcontract on a lump sum or target cost/cost reimbursable basis.

Target cost contracts

Whilst it is perfectly possible to arrange a working partnering agreement within a conventional lump sum contract, the desirability of incorporating incentives renders this arrangement less than totally satisfactory. Many partnering contracts are therefore arranged on the basis of a 'cost plus' contract incorporating a target cost and a guaranteed maximum price. The target cost is thus predefined as a 'fair price' for the work rather than the lowest price obtainable in a tender competition, and is commonly derived from the employer's pre-contract budget and cost plan, which normally acts as the 'control price' for the work. The price-competitive element is usually derived from competition between contractors on the basis of the percentage addition to be applied to the net cost in respect of profit and overheads.

The aim of the project team is then to secure maximum value for money by completing the contract as far below the target cost as possible but without loss of function. Contractors or subcontractors bringing forward alternative approaches or other innovations that aid this process are awarded a pre-agreed percentage of any

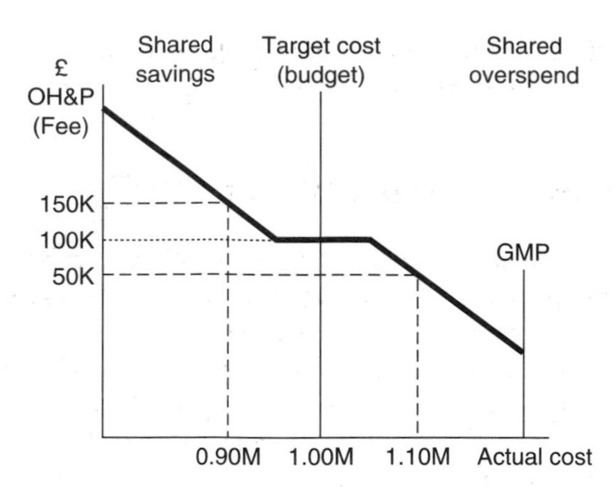

Figure 12.1 Target cost/guaranteed maximum price contracts. (Reproduced with permission of *Knowles Management*.)

cost savings as a reward for their innovation (so-called 'gain share'). The approach also commonly recognises the fact that events may occur that are the fault of neither party, but which nonetheless might increase the project cost. In this case the additional costs are often shared between the contractor and the employer (the so-called 'pain share'). The employer's liability in this respect is commonly limited by the introduction of a 'guaranteed maximum price', which represents the absolute limit of the employer's liability unless, of course, the employer him- or herself causes the project to overspend by, for example, ordering variations to the work. The process is shown diagrammatically in Figure 12.1.

Here it is assumed that the initial target cost is set at £1 million, the contractor's agreed level of overheads and profit (OH&P) is set at 10% and the pain/gain share percentage is set at 50%. If the target is achieved the contractor's return will therefore amount to £100 000. If, however, the contractor is able to bring forward innovations that reduce the actual cost to £900 000, then his or her return rises by 50% of the value of the savings to £150 000. Conversely, if unforeseen events drive the project cost up to £1 100 000, then the additional cost is shared between the contractor and the employer, and the contractor's recovery is reduced to £50 000. Where gain share provisions are used, it is also common to cap the employer's contribution through the agreement of a guaranteed maximum price (GMP).

Note that the commonly recognised deficiencies of cost plus contracts are largely negated by allowing the contractor to improve his or her return as a reward for improving value for money. This approach is usually reinforced by an 'open-book' approach to project accounting, where the contractor's records are made fully available to the employer's quantity surveyor, and the contractor is then paid the agreed value of resources supplied. The approach has a number of advantages, and greatly reduces the potential for disagreement over the value of the work carried out and consequent disputes.

There is, however, a major concern over the setting of the initial target cost, in that it is plainly in the contractor's interest to set the target as high as possible. There is therefore an important role here for the employer's cost consultant in ensuring that the initial target cost and guaranteed maximum price are realistic.

What advantages does partnering bring?

Experience across a wide range of public and private organisations shows that partnering and similar techniques can, if properly applied, significantly enhance the chances of successful project completion in accordance with the employer's time, cost and quality targets. At its best, partnering will result in significant savings in time and cost and in improved quality, thus leading to better value for money. In addition, by adopting a joint approach to the solution of problems as they arise, partnering can significantly reduce the number and severity of contractual disputes, and should therefore ensure that settlement of the final account is much less problematic than would otherwise be the case.

Managing a successful partnering arrangement

It is strongly recommended that all partnering arrangements should be facilitated by a suitably experienced external facilitator. The facilitator's role will include working with the parties to establish the initial partnering relationship, and helping them to reinforce the relationship as the project proceeds. The process will include the identification of a set of overall project objectives applicable to all parties, which all of the parties then undertake to work actively to accomplish. The agreed mutual objectives then become the 'code of practice' for the project, and form the basis for the project key performance indicators against which eventual project success will be measured. Successful partnering depends upon many factors, but perhaps the most important are that all the parties involved in the project are committed both to the agreed project objectives and to the partnering philosophy. Working in this way is often very different from the more confrontational approach adopted in conventional 'low bid' tendering situations.

The role of the workshop facilitator

In general terms, the aim in the partnering workshops is to achieve a shared understanding or definition of a problem and thence to generate better quality solution(s) through the processes of synergy and consensus decision-making. This is not an easy task, and much depends on the skills of the facilitator.

The individual who has responsibility for this role not only needs to have a thorough understanding of the techniques discussed above, but also must have an extremely high level of interpersonal skills. The facilitation role is often misunderstood, and it should be emphasised that this person's job is to enable and guide rather than to play a central part in the substantive aspects of the discussion or impose solutions. The facilitator should not be seen as the 'visiting expert'.

Not only is the task difficult because of its nature, but also it is all to do with orchestrating professionals with often widely divergent backgrounds, interests and objectives. Time limits and pressures also mean that the group of perhaps up to ten or twelve individuals (ideally, no more than this) has to begin working very quickly as an effective team. The time pressures involved in the workshop process therefore often result in less security with the unfamiliar tasks and roles.

Compare the characteristics below:

- *In co-operative groups*:
  - people work together
  - process issues are worked on covertly
  - people negotiate
  - information passes on a need-to-know basis
  - conflict is accommodated
  - politics are rife
  - people co-operate to get the job done
  - feelings are not part of the work
  - trust and openness are measured
- *In effective teams*:
  - people trust each other
  - feelings are expressed freely
  - process issues are part of the work
  - commitment is high
  - objectives are common to all
  - listening is maintained
  - conflict is worked through
  - decisions are by consensus
  - people are open

The greater the ambiguity and uncertainty in a particular situation, the greater is the need for the group to exhibit the characteristics of the effective team earlier in order to achieve its task, but the more difficult it is likely to be to do so. There are, however, a number of measures that have been found useful in assisting the facilitator to manage the process and in increasing the chances of a successful outcome. The guidelines are presented briefly below. They have much in common with those frequently advocated for the effective management of meetings. It should be emphasised that they are of even greater importance where the climate is felt to be characterised by uncertainty and potential conflict, as is frequently the case when new teams come together for the first time.

Facilitation guidelines

Agenda

It is essential that the agenda is drawn up and circulated in advance. Giving participants an idea of what to expect and what will be expected of them will help them to prepare themselves. It may also go some way towards dispelling doubts, criticisms and scepticism, as well as helping to structure the workshop. It will form an *aide-mémoire* and time-keeper and controller. A constant and obvious reminder of purpose, tasks and timing in the form of a large clear agenda displayed in a prominent position sounds so obvious that it is rarely done!

Structure and climate

It is imperative that the facilitator goes through the agenda at the start of the workshop, together with the objectives and the procedures to be adopted. The right climate must be created, and the expectations of the group managed regarding the attention they need to pay to process issues: that is, how they are working together.

Regular summaries

Regular summaries of progress serve much the same purpose as the agenda. They help to clarify communication and maintain control. They can be used as a psychological booster to focus team members' attention on what has been achieved so far and/or as a way of getting things moving forward and keeping them on track should time be running out.

Control of phases

Clear separation between stages is necessary, but this does not mean that the facilitator should show no flexibility. It is important that he or she should be able to read the mood of the group. Often a short break will help to get things moving again if participants are flagging and/or ideas are lacking. However, obviously it would hardly be wise to interrupt a brainstorming session in full swing with ideas flowing thick and fast unless it appeared that the whole programme would be jeopardised as a result if not. Obviously judgement is required: too much control will destroy the flow, too little will open the floodgates.

Gatekeeping

Needless to say, it is the facilitator's job to 'gatekeep'. This is where his or her interpersonal skills are really stretched. He/she has to ensure that each individual is contributing and that vociferous individuals are not allowed to dominate the discussion.

Team composition

It is plainly important to ensure that the partnering team includes all of the major stakeholders. In the case of public authorities it has been found particularly important to involve both audit and legal representatives at least in the initial sessions. It is important that the facilitator is seen as an impartial 'third party' who is not directly involved in the project. The perceived 'neutrality' of this person is useful in helping to guide discussion in an unbiased manner.

Some organisations hire external 'consultants' to lead workshops, on the basis that such people have a greater degree of expertise in facilitation skills and a greater degree of objectivity.

The management of relationships

The successful implementation of collaborative procurement strategies such as partnering and strategic alliances has been held to be largely dependent upon the formation and development of trusting relationships. The aim of this section is therefore to introduce some of the major concepts and techniques involved in relationship management, and to provide a brief overview of the problems of implementing these in a construction context.

Important issues to be considered include:

- Concepts of relationship management – why is it important in the modern construction industry?
- The importance of 'cultural fit' in the context of organisational relationships, and an understanding of why alignment of cultural values and philosophies is an important determinant of project success.
- The nature of corporate culture, how it may be measured and the problems inherent in achieving cultural correspondence in the construction supply chain.
- The nature of trust and its importance in relationship management.
- The need for team building as a component of modern collaborative strategies for construction work.
- The importance of mutually agreed project objectives and role. Function and relevance of role of documents such as the project charter.
- The need for reinforcement of the relationship throughout the currency of the contract and how this may be achieved.

Background

Historically the construction industry across the world has been seen to be increasingly confrontational in nature, with relationships between the various members of the supply chain being characterised by a culture of mistrust, greed and opportunism. In recent years this prevailing culture has been challenged by a number of influential construction commentators and clients anxious to improve on the industry's perceived shortcomings in terms of quality and predictability of time and cost.

Following an intensive period of introspection in the UK, 'best practice' in the management of the construction industry supply chain has now been substantially reviewed and redefined by authorities such as Sir Michael Latham, Sir John Egan and others. The outcome of these influential reviews, supported by the UK Government in its advice to public sector construction clients, is that in most cases (the exceptions may be highly experienced employers dealing with well-defined and clearly documented projects) improved project performance is likely to be dependent upon re-engineering supply chain relationships such that mistrust, greed and opportunism are replaced wherever possible by trust and a shared sense of purpose.

This need for a substantial redefinition of the way the construction supply chain works has also been noted in a number of other countries across the world. A considerable body of experience in a construction context exists particularly in Australia and the USA, and in oriental cultures, especially Japan, the existence of mutual dependency networks based upon shared obligations has long been a feature of business in general. Similar issues have also been addressed in other industries dependent upon significant outsourcing of stock and components, such as retailing, motor and aircraft manufacture, or where the risks are particularly high and the rewards, although potentially great, are unpredictable, for example oil exploration.

The successful implementation of integrated supply chains of this kind requires a fundamental rethink in terms of the way relationships between the various parties involved are developed and managed. Significant issues in this process include:

- Establishing the 'cultural fit' between the partners
- A clear understanding of each partner's needs and desires, followed by the identification of, and commitment to, a set of mutually agreed objectives
- Team development
- The nature of trust
- Maintenance of the relationship
- Assessment of the risks involved in dependent relationships

Establishing the 'cultural fit'

'Cultural fit'

Experience shows that organisations are more likely to form successful partnerships if the cultural and ethical norms governing their respective business behaviours are both similar and clearly understood by all involved. The procedures governing the selection of a business partner (for example employer and contractor or main contractor and subcontractors) must therefore be structured so as to pay close attention to ensuring that the organisations involved share, as closely as possible, the same ethical and cultural values and beliefs.

In the past, in construction, this issue has, in many cases, been clouded by emphasis on a simple 'supplier/customer' relationship between the various members of the supply chain. Here suppliers are told explicitly what is required, and their sole responsibility is limited to supplying what is asked for. The relationship is analogous to the purchase of consumer goods, and the motivation of those involved in the transaction is largely to maximise their own advantage, thus leading all too often to a commercial culture based upon opportunism and greed, claims and litigation.

Best practice in the modern world, however, recognises that construction is largely a bespoke industry, where much (perhaps most) of the added value contributed by the various different members of the supply chain arises from a joint understanding of, and approach to, the problems to be solved. Here the aim is to maximise the synergy arising from the combination of talents and experience of all members of the supply chain, such that the whole becomes greater than the sum of the parts. The relationship is therefore much more that of partnership in the joint achievement of maximum value for money for all concerned, rather than simply the provision of some predetermined service or component. The importance of shared values and a joint approach to problem solving is therefore self-evident, and in order to achieve this we need some understanding of the determinants of 'culture' itself.

What is culture?

Culture has been defined as a learned, shared, compelling, interrelated set of symbols whose meanings provide a set of orientations for members of a society. These orientations, taken together, provide solutions to problems that all societies must solve if they are to remain viable. The 'culture' of an organisation therefore defines the rules by which that organisation exists and develops. Culture, by its very nature, is dynamic (that is, it is constantly evolving, constantly growing), but the underlying cultural values of an organisation generally remain relatively stable, at least in the short and medium term, and are the benchmark against which the operation of the organisation itself is judged.

All companies will operate within a hierarchy of cultural value systems, including the national culture, the prevailing business culture and the company's own corporate culture.

Business culture is generally considered to be a subset of the overall national culture, and comprises the rules by which business as a whole is conventionally conducted. The prevailing business culture therefore provides an overall definition of ethical boundaries (for example the limits of competitive behaviour) and defines acceptable local priorities, hierarchies and codes of conduct in business dealings.

Corporate culture, or the culture of the individual organisation, establishes 'the way things are done' within the organisation, and therefore reflects both the underlying norms of the host society and the business environment in which the organisation normally operates. Corporate cultures therefore tend to reflect the cultural preferences of those in positions of power, and will include things like formal and informal codes of conduct, formal and informal hierarchies and power structures, role modelling, and special language and jargon. An understanding of the cultural 'baggage' of the environment in which an organisation operates (or wishes to operate) is therefore a vitally important ingredient for successful relationship management.

Note, however, that, in all except the smallest organisations, corporate cultures are rarely monolithic, and that many subcultures may co-exist under the main corporate umbrella. Remember also that personal attitudes and patterns of behaviour are often largely formed from past experience and may, in some circumstances, be significantly affected by personal hidden agendas. The cultural attitudes and behaviours expressed by individuals may not therefore always completely match the cultural norms promoted explicitly by the corporate image.

Understanding culture

In the light of the above comments, it follows that the culture of an organisation is not always easy to understand. At one level it is shaped and influenced by the environment in which the organisation typically operates. It is therefore intimately influenced not only by societal norms in terms of the predominant language, education system, social hierarchy and religion, but also by past corporate experience and the normal business practices predominant in the organisation's usual marketplace. At another level, however, corporate culture may be seen as the sum of the component parts of the organisation – that is, the aggregate of the various subcultural group beliefs and personal attitudes and behaviours of those with influence within the organisation. The resultant blend is also likely to be heavily dependent upon the way the organisation is structured and managed.

Cultures may therefore not always be explicit, and in addition to the 'public face' promoted by corporate management may include unspoken 'hidden agendas' which, whilst they may not be expressly articulated, may be even more powerful drivers of business behaviour than the explicit factors.

Hofstede (1980) is perhaps the best known management theorist to propose a cultural measurement system. He proposed four measures of culture within the context of project management, and theorised that all cultures could be categorised according to how far each of the measures was evident:

(1) *Power distance*: indicates the importance of hierarchical control. A greater power distance indicates the importance of hierarchical structures, whereas a smaller power distance equates to a more participative management style.
(2) *Individualism–collectivism*: the extent to which people are concerned with their own goals or with the goals of the group.
(3) *Uncertainty avoidance*: the extent to which the society tolerates ambiguity. Cultures with a high uncertainty avoidance score try to limit conflict and avoid confrontation. Those with a low score encourage conflict and risk taking.
(4) *Masculinity–femininity*: the rational achievement orientation versus the emotional/affiliation orientation.

For Asian cultures, Hofstede subsequently added a further dimension of *Confucian dynamism*, which emphasises the importance of persistence, status, thrift and a sense of shame. He has more recently added the concept of short-termism/long-termism to indicate the cultural propensity to either the short-term or long-term view.

Achieving cultural fit

The problems of achieving 'cultural fit' between two or more organisations are therefore considerable, and are in large part similar to those involved in personnel selection. Whilst many believe that it is possible for cultural convergence to be engineered, generally through a process analogous to mediation, the process will plainly be more difficult and take longer the further apart the organisations are to begin with. Where time is at a premium there is therefore a pressing argument for choosing to collaborate with organisations that already have a close cultural correspondence.

Preliminary assessment of commercial supply chain partners may therefore comprise an initial assessment of price, coupled with non-price factors such as technical and managerial competence, financial standing, etc. Following this initial trawl a number of potential partners will emerge, and the final assessment must then be substantially based upon some form of cultural assessment. This final assessment will often comprise both a review of the corporate culture and an assessment of the actual members of staff (particularly managers) that it is proposed to assign to the project. In this respect it should be noted that on some large collaboratively procured schemes (for example the Australian National Museum project in Canberra) human resources professionals have been employed as an integral part of the assessment team.

Mutual objectives

The development of a set of measurable mutual objectives, which all parties are prepared to commit themselves to achieving, is a central part of a truly collaborative procurement approach. The process of developing the mutual objectives is conventionally carried out in a workshop environment at the beginning of the project, and will form a major plank in any ensuing commercial relationship. Successful development of a satisfactory set of mutual objectives will firstly require each of the organisations involved to critically review and articulate their own objectives in a way that can be understood by all of the other organisations involved. The process will then require the various partners to jointly review all of the proposed objectives, seeking areas in which modification or compromise of their own objectives can lead to improved mutual benefits. This process also enables all concerned to begin to identify and understand the risks to which they will be jointly and individually exposed, and to develop agreed mutually dependent strategies to manage them.

This process is conventionally managed in a workshop environment, using a process of facilitated discussion and negotiation. It is plain that this process requires that each of the partners is open and honest about what they want to achieve, and that there is no room here for hidden agendas or the seeking of opportunistic advantage.

In the case of a construction project it is likely that the employer's declared objectives will initially centre around the achievement of acceptable performance criteria in terms of time, cost and quality, but experience shows that this may not always represent the whole picture. One case in point concerned a local authority project to redevelop and extend an adult retraining centre in a town suffering from high unemployment caused by the collapse of the local coal mining industry. The project was initiated by the local authority technical services department who, in the initial partnering workshop, expressed their project objectives in terms of the conventional time, cost and quality criteria. Contributions from the training centre management, however, showed that one of their primary objectives from the project was to maximise the employment opportunities arising from the construction work for local unemployed workers who had been retrained in construction skills. This issue was not articulated in the tender documents, and therefore came as a complete surprise to all of the other stakeholders. An improved briefing and project inception process would help to remove this type of confusion.

From the contractor's point of view, the major objective will usually be to achieve a certain level of profit together with an acceptable limitation of risk. A subsidiary objective in some cases may be for the contractor to develop his or her relationship with the employer, thus leading to future work. A clear understanding by the employer of the commercial realities of the scheme from the contractor's point of view is an essential element in establishing the required mutual trust and commitment. It is therefore important that contractors are open and honest about their commercial objectives, and equally that the employer agrees that the required profit levels are acceptable.

The agreed set of mutual objectives then forms the basis for the project key performance indicators, and a set of business principles derived from these normally forms the basis for the project charter.

Team development

Most construction projects allow little time for formal team-building in advance of the start of construction work. On the contrary, in practice the traditional procurement route relies heavily upon project participants simply taking up predetermined standardised roles (employer, contractor, architect, quantity surveyor, etc.) as a strategy to enable the construction team to move from team formation to some minimum acceptable level of performance in the shortest possible time. Any innovative ideas that occur as the result of improved team interaction happen largely by accident as the project proceeds, and this approach therefore runs totally counter to the fact that the maximum scope for innovative change occurs at the beginning of the project.

We have already seen that the successful use of modern collaborative approaches demands the early development of a team approach, freed as far as possible from the constraints imposed by rigid adherence to traditional roles. The careful choice of supply chain partners, coupled with the use of a workshop approach to partnering, risk and value management as early as possible in the process, helps the project team to move quickly through the well-known forming/storming/norming/performing cycle very quickly. This then permits maximum advantage to be taken of any innovative ideas that arise.

Some of the main characteristics of effective teams have been identified as follows:

- Clear focus of goals
- Operate in a creative way
- Strong focus on results
- Clear definition of roles and responsibilities
- Well organised and managed
- Good utilisation of individual skills and talents
- Mutual support
- Positive 'can do' attitude
- Resolution of disagreements without confrontation
- Open communication
- Objective decision-making processes
- Regular evaluation of their own effectiveness and effective action on the results

There is a considerable body of theory and practice relating to personal motivation and team development, personal interaction and conflict resolution. The origins of the discipline lie in the work of pioneering management theorists such as Drucker, Argyris, Likert, Hetrzberg, McGregor and others during the 1950s and 1960s, and active research in this area continues to the present day. Central to the whole debate, however, is recognition of the fact that simply bringing highly skilled, competent and motivated people together, even when they belong to the same organisation and may have agreed to the same goals, does not necessarily make them behave as a team. Teamwork depends upon a recognition of mutual dependence arising out of an acceptance of one's personal and technical shortcomings and a recognition that the skills and knowledge possessed by other team members supplement one's own. Tacit recognition of these issues by all team members will then allow the team to move forward and will help to ensure that the whole performance of the team exceeds the sum of the parts.

There are many issues to be resolved here, perhaps the most important of which is the management and resolution of interpersonal conflict. There are many ways in which this can be handled, but it is certain that conflict avoidance (simply sweeping the problem under the carpet) is not the best approach. Experience seems to show that compromise can be useful in some cases, but the best approach appears to be to create a team atmosphere where members feel confident that they can raise issues in an environment where their concerns will be taken seriously, confronted and addressed in a positive way by the other team members. Meredith and Mantel (1995), for example, quoting from a study by Hill and White (1979), report how a construction project manager handled a difficult conflict:

> 'The project manager did not flinch in the face of negative interpersonal feelings when listening to differences between people. "You have to learn to listen, keep your mouth shut, and let the guy get it off his chest." In short the manager:
>
> - Encouraged openness and emotional expression
> - Set a role model for reacting to personality clashes. It was observed that a peer would often intercede and act out a third party conciliation role much like the manager

- The manager seemed to exhibit the attitude that conflict could be harnessed for productive ends
- Although the manager usually confronted conflicts, he also avoided face to face meetings when outside pressure was too high.'

Some organisations have developed specific team-building exercises designed to encourage teams to work together to meet some shared intellectual or practical challenge. Many of these programmes are based out of doors, and they often rely in whole or in part upon some shared fear or perception of danger. Typical activities would include night-time orienteering or navigation tasks, rock climbing, abseiling, potholing, etc., where although the activity is carried out under the supervision of an appropriately qualified expert there is still a sufficient personal challenge to encourage both individuals and the team as a whole to work together to accomplish the task in the face of various obstacles. There is significant evidence that activities such as these not only develop team unity but also may engender greatly enhanced levels of personal satisfaction among team members.

Considerable interest has also been expressed in the management of teams, and here we can, I believe, learn much from other disciplines that rely for their success on team rather than individual performance. Perhaps some of the best examples lie in the sports and leisure industry, particularly perhaps in the UK in sports such as professional football or Formula 1 motor sport. Whilst it is of course accepted that high performing teams need 'star' players, the difference between the best and worst seems to be largely due to the quality of the team's management, and we may be able to enhance construction performance by observing and learning from the acknowledged high performers in these fields.

The nature of trust

The concept of trust between the parties involved is usually considered to be a fundamental prerequisite for successful partnering, but relatively few commentators have attempted to analyse the nature of trust, or how it can be developed.

Trust between two parties may be viewed in two basic ways. Optimists tend to believe from the beginning that their partner is trustworthy, and work on that basis until they are proven wrong. Pessimists on the other hand take the view that trust needs to be earned, and they will therefore behave with considerable caution until they are sure their partner 'can be trusted'. Plainly, in an environment such as a construction project where time is usually critical, a trusting relationship both between the parties and, perhaps more importantly, at a personal level between the members of the project team needs to be forged as quickly as possible if the consequent benefits are to be maximised. In essence, what is required is confidence between the partners and the team members that they can rely on each other. Here again, establishing cultural fit between the partners and between the individual members of the team is of paramount importance, and will be substantially reinforced by the team-building activity discussed above.

Although trust as a concept is quite hard to define, the following characteristics of a trusting relationship within the context of a construction project have been identified:

- Full and open communication
- A consistent view from all personnel
- Long-term commitment
- Full cost transparency
- Price is not the overriding factor
- Honesty and openness
- Early input into the project design
- Mutual advantage
- Words backed up by actions
- Receptiveness to ideas from partners
- Good attitude and loyalty
- Honouring price commitments
- Providing help to other team members with no strings attached

Many of these characteristics are, as would be expected, similar to those displayed by effective teams.

There is also clear evidence to show that trust is enhanced through the use of appropriate contractual mechanisms. Where contracts encourage the partners to view each other in a confrontational supplier/customer way through the use of strict contractual procedures and strongly defined roles and hierarchies (sometimes termed arms-length contractual approaches as, for example, in the JCT Standard Form of Building Contract) there is a natural tendency for the parties to rely on the contract to resolve their problems. On the other hand, where contracts are written in a much less formal way and where much greater importance is placed upon mutual obligations and a shared approach to problem solving (an approach sometimes termed obligational or relational contracting) there is much less pressure for the partners to resort to acting in a confrontational way.

This distinction between short-term (i.e. supplier/customer) transactional contracts and longer-term relational contracts is one that has occupied the minds of legal theorists for over a quarter of a century. Basically the discussion centres upon the extent to which project risks can be foreseen and quantified, sometimes called the degree of 'presensiation'. In the case of simple supplier/customer transactional contracts, such as the purchase of a bus ticket or a can of food, the transaction is completed within a short period of time and the risks are very transparent. Such contracts can therefore be very prescriptive, and have a high chance of being able to provide a contractual remedy for virtually any event that is likely to occur. In construction, such remedies frequently involve the development of complex and prescriptive sets of rules such as, for example, those included in the UK JCT 98 Standard Form of Building Contract for the valuation of valuations.

In the case of longer-term relational contracts, however, future events that might occur during the currency of the contract are much more difficult to foresee with any degree of clarity. There is therefore an argument for structuring contracts, and contractual terms and conditions, which facilitate the parties finding mutually acceptable, and possibly even extra-contractual, solutions to problems through discussion and negotiation. Such discussion may of course in some cases involve the services of a third party (dispute resolution advisor, mediator, facilitator, conciliator, adjudicator, etc.), rather than the simple application of a prescribed set of rules. Some indication of the modern development of such systems is seen clearly in the development of adjudication as a mandatory form of alternative dispute resolution (ADR) in many British construction contracts, and in the role of the partnering facilitator in the Association of Consultant Architects Standard Form of Contract for Project Partnering (Association of Consultant Architects and Trowers and Hamlins 2000).

It is important, however, not to lose sight of the fact that the development of mutually trustworthy relationships may be heavily influenced by the relative power of the organisations involved. Significant power factors would include:

- Differential commercial strength, sometimes called 'leverage' (for example where one organisation is much larger and financially more powerful than the other or, on the other hand, where one organisation has a monopoly of supply over something that is indispensable to the other).
- The degree of mutual dependence (that is the extent to which mutual success is critical to both of the organisations involved).

It is also important to recognise that a mutually supportive relationship will only continue to thrive whilst both parties continue to gain some advantage from it.

Maintenance of the relationship

Once established, collaborative relationships need to be reinforced on a regular basis if the benefits are to continue to accrue. Maintenance of the relationship will involve consideration of a number of distinct factors, for example:

- The preservation of personal and organisational ties.
- Periodic re-evaluation of the mutual benefits to be gained from team membership.
- The need for an 'induction mechanism' to ensure that incoming new team members accept the goals, culture and behaviour patterns established by the team, and are appropriately incorporated into the team structure.
- The need to ensure continuous improvement and to prevent the relationship from becoming too 'cosy'.

It is held by many commentators that the preservation of personal and organisational ties depends largely upon positive social interaction between team members (i.e. people are more likely to work more harmoniously and to perform better if they are working with people they like rather than those they do not). In this context team maintenance begins to reach outside the work environment, and will often include social events in addition to directly work-related activity. On the other hand there is the need to guard against the team becoming simply a social organisation, and to ensure that team members remain focused upon the need for continuous practical improvement in team performance. Again there are many similarities here between commercial teams and other more overtly competitive organisations such as sports teams.

There is also the need for each of the partners in the collaborative relationship to periodically and objectively review both the cost of their contribution to the overall team goals, and the benefits accruing to them from team membership. This process should also include a periodic reassessment of the risks arising from continued membership. A good example here would be the risks faced by a relatively small company from becoming too dependent upon one much larger partner. Here the risks faced by the small company might be much greater than those faced by the larger one, particularly where there are alternative suppliers in the market for the products or services provided by the smaller organisation.

Long-term collaborative relationships must also recognise that their membership will change during the currency of the project, and that these transitional events need to be carefully planned. There are a number of documented examples in areas such as vehicle manufacture where long-term relationships which have been forged between organisations have been driven by a few key individuals, and when one of these key individuals is removed (for example by promotion or transfer to another project) the relationship collapses. There is therefore the need to plan for these transitional events if the organisational relationship is to continue to function effectively. In essence what is required is an induction mechanism to ensure that both incoming or replacement team members are in tune with the goals and ideals of the team and their initial interaction with other team members is carefully managed until they are fully absorbed into the team structure and totally comfortable with the team ethos. Many of the problems faced by new team members (mutual goals and objectives, shared culture and trust) are analogous to those facing the initial team members upon the original formation of their partnership. They are, however, made more difficult because the team has already settled into an operational state. New members will therefore have less of an opportunity (and will consequently find it more difficult) to influence the pre-existing team goals to match their own personal aspirations. There are obvious parallels here with the human resources processes commonly used for staff induction, and the use of human resources techniques may be very useful here.

The project supply chain

Historically, in construction projects let using either traditional or design and build arrangements, the project supply chain has been generally considered to comprise basically the main contractor and those organisations upon which he or she relies in order to complete the works (subcontractors of various kinds, materials suppliers, etc.). The supply chain was balanced in most cases by a 'demand chain' comprising all of those elements upstream of the main contractor and including bodies such as the end user(s), project funders and various other parts of the client organisation who would benefit from the proposed project. The main dividing point between the demand chain and the supply chain was therefore perceived to be the main construction contract, thus mirroring the traditional supplier/customer arrangement. Management of the supply chain thus rested virtually completely with the contractor, and management of the demand chain rested with the employer.

Collaborative procurement techniques have, however, tended to change this position. Under a collaborative arrangement the relationship between employer and contractor changes from a supplier/customer arrangement to more of a partnership of equals. Under these arrangements the demand and supply chains have tended to become concatenated into a single supply chain dedicated to satisfying the needs of the eventual consumer (the end user), and the rationale for treating the construction contract as some sort of natural division between demand and supply chains therefore becomes much less clear. Under collaborative-type arrangements, the whole of the project team (including substantial elements of what used to be considered as the contractor's supply chain) are expected to contribute to project success through the use of innovation and continuous improvement, and in return to participate in sharing any consequent rewards. This is a very substantive issue for the many main contractors who act primarily as management organisations subletting most if not all of the construction work.

There is also the issue that conventionally the project's temporary organisation has consisted of representatives of the relevant participant organisations, and has tended to exist outside the constituent firms. It has therefore developed a large degree of autonomy. Whilst this is in many ways a good thing, it may mean that none of the specialist organisations involved in the supply chain actually assumes ownership of the scheme, and the project team is thus in some respects left to fend for itself.

This model does not fit well with modern collaborative procurement methodologies, where ownership of the scheme by all participating organisations is often seen as a prerequisite to mutual project success. This is not to say, of course, that the project management team should not enjoy a significant degree of autonomy, merely that for the project to be successful all parties must be prepared to provide wholehearted support.

The challenge then is to ensure true ownership of the project by all of the parties in the extended supply chain (thus ensuring a high degree of 'buy in' to the project goals) whilst still allowing the project management team sufficient freedom to manage the way in which they think best in order to ensure project success.

Initial partnering exercises tended to be very unfair to subcontractors and suppliers in that although they brought forward many innovations, the traditional supply chain management processes of many main contractors effectively prevented any sharing of the consequent rewards. This issue was of great concern to employers, many of whom realised that the majority of innovation was in fact being driven not by the main contractors themselves but by their supply chain. Employers therefore began to take an active interest in the way main contractors managed their supply chains in an attempt to ensure that the contractual and business arrangements and culture set out in the main contract were cascaded down the supply chain. This then poses many problems as to: (1) how the contractor's supply chain should be managed to maximise value for money and incentivise innovation; and (2) how the merged demand and supply chain should be managed in order to maximise project performance.

These problems tend to become more significant as projects and their supply chains become larger and more complex, and much more work is required to build the necessary theoretical models to support practical improvements in project success.

A crude analysis would seem to indicate at least four generic alternative approaches:

(1) Management of the extended supply chain by a specialist independent project manager (the *holistic* approach).
(2) Separate management of the demand chain by a project sponsor or project director and of the supply chain by a construction project manager, with the project sponsor or director and construction project manager working closely together with joint overall responsibility for project delivery but each having virtual autonomy over their own areas of responsibility (the *bicameral* approach).
(3) Management of the extended supply chain, including the employer, primarily by the main contractor (the *unilateral* approach).
(4) Management of the supply chain as a partnership of equals (the *collaborative* approach).

The holistic approach

Where project managers have been employed, overall management of the project has historically been considered to be the project manager's responsibility. This obligation is made clearly evident in the UK Government GC/Works/5 Framework Agreement for the Appointment of Consultants, where the duties of the project manager as set out in Annex 1 provide that the project manager is required to:

> '... carry out the services and obligations necessary to achieve the satisfactory completion of the project at or below the approved cost limit ...'

In the past, however, detailed management of the contractor's supply chain has been left to the main contractor, but increasingly the employer's desire to influence this process has led to the project manager becoming involved. This process has apparently been used with some success by experienced clients working in long-term strategic relationships with contractors and specialist subcontractors, but it is unlikely that satisfactory results will be obtained from a one-off arrangement.

The bicameral approach

It has already been pointed out that merging of the demand and supply chains to include 'upstream' stakeholders such as the client, funder and end user has led to problems, and these become especially problematic in the case of complex projects with complex multiheaded clients.

In these cases, some organisations have opted to divide the conventional project management role into two by separating the functions of demand chain and supply chain management. Here the demand side of the project is typically managed by a 'project sponsor' or 'project director', with the supply side being managed by a conventional construction project manager. The project director is often drawn from within the employer's organisation, and generally takes responsibility for the major strategic decisions, whilst the management of the supply side is controlled by the construction project manager.

This model has been adopted with varying degrees of success by several UK public sector organisations. The major problem of course is that overall project management is only as good as the relationship between the project sponsor and the project manager, and great care and considerable effort therefore needs to be taken to ensure that this relationship works effectively.

The unilateral approach

In this approach the contractor takes the lead. A good example is the development of prime contracting within Defence Estates, where the role of the prime contractor is to integrate all the activities of a pre-assembled supply chain. The process is claimed to replace short-term contractually driven adversarial relationships with long-term strategic alliances based on trust and co-operation. Central to the arrangement is the construction of the pre-assembled supply chain based upon clusters of specialists (e.g. mechanical and electrical (M&E) services, groundworks, frame and envelope or internal finishes) working together to provide specific elements under a cluster leader. The job of each cluster is then to design and deliver an integrated part or element of the building. The responsibility of the prime contractor is therefore to bring together the entire supply chain, from clusters to the end user, to design for best value.

An evaluation of the prime contracting pilots conducted by the Tavistock Institute found that the most fundamental lessons learned concerned the key competencies that clients should look for when appointing the prime contractors. These were established as: (1) the ability to manage and integrate a number of different kinds of design inputs in a way that goes well beyond what is generally meant by design management in construction; and (2) the ability to work with the specialist contractors in order to manage costs with a new sense of rigour.

A further factor in assessing the competence of the prime contractor is his or her ability to identify the key interfaces at an early stage in design development, and it was found that, in both of the pilot projects, key gains were secured when the prime contractors brought together consultant designers to work with the project sponsor and user representatives.

Other areas where this approach has considerable potential include the development of industrial units and the refurbishment of social housing. In this latter category the prime constituents generally comprise activities such as:

- Refurbishment of kitchens and bathrooms
- Rewiring

- Upgrading of gas and plumbing installations
- Wall and ceiling insulation
- Repair and refurbishment of precast reinforced concrete dwellings
- Reroofing
- Etc.

The amount of work available in this sector in the immediate future is considerable, and a significant number of both main contractors and specialist subcontractors are seeking to specialise in this type of work. This situation is leading to specialist main contractors being able to offer prospective employers a package consisting of specialist social housing expertise which is sensitive to and experienced in the wider issues associated with working in this sector, coupled with a ready-made tried and tested supply chain which is prepared to guarantee delivery on time and on budget. The next logical step in this process is for contractors to develop a process to induct employers into their pre-assembled supply chain and thus to integrate the demand chain into their existing processes.

The collaborative approach

The collaborative approach relies upon all of the key participants acting together within a framework of a partnership of equals. Such an approach is incorporated into the PPC 2000 Standard Form of Contract for Project Partnering published by the Association of Consultant Architects. Under this form, all of the key participants of the projects are parties to the contract, and the idea of a contractual hierarchy based around subcontractors to a main contractor is therefore avoided.

Under this contract all of the major participants, including the employer, owe mutual obligations to each other to '. . . work in mutual cooperation to fulfil their agreed roles and responsibilities and apply their agreed expertise in relation to the project . . .'.

The parties elect members to a core group, which cumulatively has the objective of delivering the project to meet the agreed objectives.

Conclusion

This chapter has reviewed the factors underlying the development of collaborative methods of construction procurement, together with a range of contextual issues. It must be remembered that the successful application of partnering (and other collaborative techniques) is much more than simply and mechanically applying the tools. Partnering requires a substantial cultural shift, such that the parties to the contract work together as partners rather than as opponents each working for their own individual goals. The development of trust and good working relationships is, above all, the major critical success factor in achieving a successful partnering relationship.

Traditional approaches to construction work have historically been based upon a competitive tender process largely based on lowest price. The relationship between the parties has thus been grounded generally upon commercial considerations, with increasingly more complex forms of contract providing a framework for a culture of claims and blame allocation.

This approach has now been widely discredited, and there is an increasing realisation that successful projects are those where all of the supply chain members work together for the good of all concerned. This philosophy is reflected in modern collaborative approaches to construction procurement, and increasing attention is now being paid to the nature of the relationships that tie supply chain members together. These concerns are reflected in other industries such as retailing and motor manufacture; however, the techniques of relationship management pioneering in these industries need to be modified to fit the unique nature of the modern international construction industry.

References

Association of Consultant Architects and Trowers and Hamlins (2000) *PPC2000 ACA Standard Form of Contract for Project Partnering*. Association of Consultant Architects.

Egan Sir J. (1998) *Rethinking Construction*. UK Government Department of the Environment, Transport and the Regions.

Gershon P. (1999) *Review of Civil Procurement in Central Government*. HM Treasury, http://www.hm-treasury.gov.uk/documents/enterprise_and_productivity/ent_pep_gershon.cfm.

Hill R. and White B.J. (1979) *Matrix Organisation and Project Management*. Michigan Business Paper no. 64, University of Michigan.

Hofstede G. (1980) *Culture's Consequences: International Differences in Work Related Values*. Sage.

Latham Sir M. (1994) *Constructing the Team*. The Stationery Office.

Meredith J.R. and Mantel S.J. (1995) *Project Management: A Managerial Approach*. John Wiley.

National Audit Office (1999) *Modernising Procurement*. The Stationery Office.

National Audit Office (2001) *Modernising Construction*. Report by the Comptroller and Auditor General, HC 87 Session 2000–2001, 11 January. The Stationery Office.

13 Privately financed public sector projects (PFI and PPP)

Introduction

The concept of governments using private finance to procure public sector infrastructure has a long history, and many alternative forms have been used, from outright privatisation (usually with varying degrees of regulation and control) through a wide range of collaborative models. The purpose of this chapter is to explore the historical use of the Private Finance Initiative (PFI) and public private partnerships (PPP) techniques, and to review how successful they have been both as an investment vehicle and as a method of providing public services which governments have been unable to fund from capital resources.

Historical development

The structures we know today as the Private Finance Initiative (PFI) and public private partnerships (PPP) have their roots in Europe, in the beginnings of a demand for mass travel and long-distance commerce in the second half of the seventeenth century. Smith (1999) provides an extensive analysis of the early development of the role of government in public/private sector partnerships.

In Britain the movement began in the 1660s with the creation of the first turnpike trusts for the development of roads; it flourished in the form of canals and railways, nurtured by the explosion of British manufacturing industry which followed the industrial revolution, with private investment reaching peaks in the 1790s, the 1840s and the 1860s, each of which was followed by a dramatic financial collapse which ruined many investors both large and small. This development pattern of turnpike roads, followed by canals and waterways and later by railroads, was subsequently repeated in the USA between about the end of the eighteenth and the end of the nineteenth centuries.

In continental Europe the involvement of the private sector was facilitated through the use of concessions, where entrepreneurs were effectively given a licence to provide public services for a specified period of time, and to charge a fee. McCarthy and Perry (1989) report the granting of the first water supply concession, to provide a distribution system to parts of Paris, to the Perier (sic) brothers in 1782, although this was subsequently revoked following the French revolution. They also report that the use of concessions for water supply spread rapidly in France, Spain, Italy, Belgium and Germany from about 1850 onwards, although their use became less frequent as the primary infrastructure was completed.

Perhaps one of the best known early examples of international private sector involvement in major infrastructure work using the concession approach, and also a good early example of a public/private sector partnership, is the 90 mile long Suez Canal, completed in 1869 with a 99-year concession period. The concession was initially granted to a semi-retired French diplomat, Ferdinand de Lesseps, in 1854 by Said Pasha, Turkish Viceroy of Egypt. De Lesseps formed a company to carry out the development, but was apparently unable to raise all of the estimated £8 million construction cost from the private sector, largely due to strong British opposition led by the Prime Minister Lord Palmerston, who declared that the enterprise was not only impractical but also commercially unsound and would never make a profit. A substantial proportion of the finance was eventually provided by Said Pasha, who also provided much of the necessary labour under a *corvée* system. The concession was eventually terminated when the canal was nationalised by the Egyptian government in 1956.

This early project also illustrates many of the issues that have been identified as critical factors for success in more modern projects, in that:

- It shows the need for a politically skilled champion prepared to fight for the project against strong opposition.
- It displays the extended gestation period typical of many present-day large infrastructure projects.
- The events leading up to the repudiation of the concession agreement, the serious effects of the action on international trade and political stability and the frantic attempts to find a diplomatic solution to the crisis provide a vivid illustration of the vulnerability of large-scale long-term international privately financed infrastructure projects to shifts in the political landscape.

By the time the Suez Canal was opened the use of private sector capital for infrastructural development in Britain, mainly by that time in railways, had begun to gradually decline. This decline intensified during the closing years of the nineteenth century, when a wide range of factors including overcompetition, public concern about fares, charges and levels of service, increasing governmental control and the growing bargaining power of an increasingly well organised labour force meant that the returns paid to shareholders by the operating companies were no longer competitive with returns from alternative forms of investment. This in turn led directly to a shortage of the necessary investment funds, and thus to an inability to maintain and renew the facilities as they began to reach the end of their physical life. Significant private sector involvement in British transport infrastructure was effectively brought to an end as Europe lurched towards the First World War.

Early schemes were not, however, in the UK at least, an outstanding financial success. Fay (1947) reported that none of the early turnpike trusts ever achieved their target of paying off all of their debt within the initial 21-year term. Some early canal schemes did, however, prove extremely profitable; Deane (1984) for example reports that the Oxford canal paid annual dividends of 30% for more than 30 years, and that shares in the Birmingham canal which cost £140 each when the canal was constructed in the 1780s were selling for £2840 each by 1825. They were therefore, in general, very popular with investors, although there were occasional difficulties with funding. Pannell (1977) for example reports that construction of the Forth-Clyde canal in Scotland in the latter half of the eighteenth century was halted owing to difficulties in raising the necessary funds, and construction was only able to continue after the Government agreed to a £50 000 loan, approximately one third of the total cost of the works.

Deane credits the canals with producing '. . . a new class of investor, the canal shareholder, a non-participant investor who was readily transformed into a railway shareholder . . .'. Some authors claim that the need to protect the interests of these new private investors gave rise to a further expansion of Government's role as regulator. Burton (1972) explains:

> 'Canals required very large capital sums indeed. . . . Allowing promoters to encourage the public to invest in such projects with the promise of large profits was an obvious solution. . . . Although canals were largely built by joint-stock companies, before anything could be done a scheme had to be approved by Act of Parliament. The company and its plans were subject to parliamentary scrutiny. It was not a fool-proof system, but it did offer the investors some guarantee that a disinterested third party had given the plans consideration and approved them as practical.'

Despite Burton's assertions, however, such approvals often did not offer much protection. Burton himself goes on to admit that:

> 'In fact companies regularly underestimated the construction costs by an alarming amount, and overestimated the profits similarly. [Nonetheless] the canals were built, and the best of them provided handsome dividends.'

The point is confirmed by Pannell (1977) who, whilst discussing the construction of the Forth-Clyde canal in the 1760s, reports the final cost of the works as £150 000, which, he comments, was extremely close to John Smeaton's original estimate of £147 337, 'a very unusual circumstance when most engineers' estimates were, to say the least of it on the optimistic side'.

By the beginning of the 1790s canal construction was in full swing and shares were booming, but in 1792–93 the economic 'boom–bust' cycle generated a general financial collapse in which 'as usual, the loudest wails were raised on behalf of the poor city clerk, the widow and the orphan' (Fay 1947).

Notwithstanding the financial woes of the 1790s, by the mid-nineteenth century, and following on from the tremendous success of the world's first inter-city railway between Liverpool and Manchester which opened in 1830, the railway boom was in full flood. Spurred on by the average annual dividends of 9.5% paid by the Liverpool and Manchester company, much higher than the returns paid by alternative forms of investment available at equivalent risk, investors queued to pour their money into the new railway mania. Many companies followed the practice previously established by the canal promoters of issuing part-paid shares, whereby investors could obtain shares with a nominal value of, say, £100 often for as little as £5, with the remainder only being called for as and when required during construction. It was therefore possible, for a very small sum, for a person to be issued with shares of an apparently much higher value. These factors acted as a considerable incentive to those willing to speculate that the value of the shares would rise before future payments were called for, thus allowing them to trade the part-paid shares at a substantial profit.

Numerous companies were floated between 1830 and 1840, many of them deliberately fraudulent and most offering impossibly optimistic projected returns, and disaster was ultimately inevitable. Williams (1883) described how easy it was to part the general public from its money, and the 'dirty tricks' that some promoters engaged in:

> 'The capital required was small. A few knaves engaged an office, bought a map, struck out a railway in what appeared to be a suitable direction, gave it a plausible title, and with a sheet of foolscap and a "Court Guide" made a prospectus, on which they placed the names of a few noble lords, Right Honourables, ex-M.P.s, and merchants, to which an engineer, banker, and lawyer were added, and the whole was "served up", with imaginary advantages, and with the assurance of at least a ten per cent dividend. The excitement of the times prevented much chance of the detection of the fraud, and the inexperienced who also wished to speculate were taken in. . . . The names of gentlemen wholly unconnected with railways, and who would have utterly repudiated association with men who advocated them were unhesitatingly employed for the purpose of lending a supposititious countenance to bubble companies. One line was declared to enjoy the patronage of four gentlemen who had been dead for several months; and ten others had no knowledge of the existence of the scheme till they saw it paraded before the public avowedly under their own sanction. In another case, the three leading projectors of a very costly railway were notoriously "living by their wits", and could not have raised a hundred pounds among them except by fraud.'

The total amount of money actually invested in these schemes was staggering; Deane (1984) reports that, at the height of the construction in 1847, total railway expenditure '. . . was then running at a level which was more than the declared value of British exports and roughly a tenth of the total national income'. The result was predictable. Financial collapses in the 1840s and 1860s ruined both rich and poor.

As with the canal schemes of the previous century, a common problem was the tendency to dramatically underestimate the costs involved. Pollins (1971), for example, reports that, in 1836, the promoters of the London and Birmingham Railway announced that in order to complete construction work they would need to double their capital base from the planned £2.25 million to £4.5 million.

There are uncanny precursors here of twentieth century schemes [illegible] Channel Tunnel.

Modern models for privatisation

Many different arrangements have been developed for private sector participation, from outright and complete privatisation, through various degrees of partnership and shared control, and some of the best known are considered below.

Outright privatisation

Outright privatisation, whilst originally the major model for private sector participation in infrastructure provision, is now much less common. Perhaps the most well-known modern examples are seen in the areas of water supply, power generation and telecommunications, and the industries are usually subject to substantial degrees of government regulation and control. Nonetheless, from a financial standpoint, outright privatisation, with its total lack of reliance on any public sector investment, is plainly the most desirable option for those governments embracing a high degree of free market economics. The problem of course comes in ensuring that the service to consumers can be protected by an adequate regulatory framework, and that strategic public services can be maintained in times of emergency without the risk of government being subject to any form of blackmail by the service providers.

Outright privatisation is perhaps most successful in industries like telecommunications, where the service is provided at a comparatively low unit cost to a large number of consumers, and where the market can be easily opened up to a large number of potential competitors (that is, where a natural monopoly is unlikely to arise). The telecommunications industry is also typically characterised by rapid technological change and very high research and development costs, and a regulated and competitive private sector is therefore often seen as the most efficient way to promote improved consumer services. The only investment required by government is the establishment of industrial regulatory and consumer protection systems, and this may easily be recovered from the private sector operators through some form of licensing system.

The commercial nature of the outright privatisation model may, however, limit government's freedom to influence both the way in which the industry itself develops and the extent to which government can ensure that services are made available to those sections of society where, although there is a social need, the demand for the service is insufficient to ensure that it can be supplied at an economic price. A typical example commonly concerns the provision of services such as electricity supply to outlying or remote consumers, for example rural farmers or small isolated communities, where the full cost of providing the normal level of service may be prohibitive. Whilst it is possible for government to undertake a certain amount of 'social engineering' and to provide some support for such consumers through its regulatory frameworks, such moves are likely to be actively resisted by service providers unless government provides some form of subsidy.

Participative privately financed techniques

Models include corporatisation and the use of 'mixed' companies, management contracts, leasing, prefinancing and a variety of concession-based methods using some private sector resources to provide some combination of design, construction, financing, operation and subsequent transfer to the host government. Whilst all of these techniques allow government to provide large and costly infrastructure without the budgetary constraints normally imposed by substantial capital expenditure, it has been argued that in overall economic terms the use of private finance in this way is more expensive over time than the use of public capital, essentially because government is in a position to borrow money at more competitive rates of interest than the private sector. This is perfectly true, but this disadvantage has usually been considered to be outweighed by the short-term budgetary advantages. Some of these techniques may, in some countries, pose particular problems regarding law and taxation, and governments may need to give special consideration to this issue on a case-by-case basis.

Corporatisation and the use of semi-private (i.e. mixed) companies

Publicly owned corporations

A number of governments have created special-purpose corporations specifically to construct and operate infrastructure facilities. A wide range of options exists, ranging from companies that are totally publicly owned to semi-private (i.e. mixed) companies having both public and private sector shareholders.

One example of the totally publicly owned corporation is provided by the development of the French motorway network during the mid-1950s (Vornetti 1994). Here a number of free-standing concession holding companies were formed, but the whole of their equity was held by local authorities and by a public credit institution, the *Caisse des Dépôts et Consignations* (CDC). Although the French government had no direct stake in the companies, it was heavily involved in both the financial and technical aspects. In so far as the technical aspects were concerned the State was responsible for the initial design of the motorway, and also took the role of both main contractor for the construction work and maintenance contractor. In terms of finance, loans for the construction work, guaranteed by the State, were issued through the CDC. The total absence of any private sector involvement meant that the concession companies were simply a 'front' for the State; a device that enabled government to circumvent its own budget restrictions.

Other examples of publicly owned corporations that operate more like truly private companies exist elsewhere. In Hong Kong for example the Mass Transit Railway Corporation (MTRC) and the Kowloon-Canton Railway Corporation (KCRC) operate in this way. They do, however, operate very differently from conventional government departments in that:

- They operate their core business as private companies, and both are [illegible] the provisions of the Companies Ordinance.

- They may, within the context of their core business, plan new projects on their own based solely on financial viability.
- They are authorised to raise the necessary development funds from the capital markets.
- Their staff are not civil servants.

Neither operation receives a government subsidy, and both are generally considered to be very successful commercial entities. The MTRC particularly is one of very few underground railways in the world to run at a profit.

Semi-public corporations

In the case of semi-public corporations the company typically operates outwardly as a commercial concern, with private sector shareholders but raising its own capital funds on the open market. The company is not subject to direct governmental control, although it is common for government to retain a controlling interest, either by retaining a minimum of 51% of the share capital or by holding a so-called 'golden share'.

Cavanagh (1995) reports that the Hungarian government considered the development of a possible joint venture project structure for the development of new motorways. The structure envisaged the use of private sector finance and management forming a joint venture company with government. The company would then let construction contracts on a phased basis through a construction management route, and would let operations contracts by competitive tender. Government would limit tolls to provide a minimum rate of return, and would reimburse the operator through either a shadow toll or a minimum traffic flow guarantee.

A further example is the Japanese 'Third Sector' approach introduced in the mid-1980s. The declared objective of this initiative was to bring together both public (the 'first sector') and private (the 'second sector') industry expertise in individual project based companies (the 'third sector') with a view to:

- 'Tapping the vitality' of private sector management in order to promote large-scale projects with greater efficiency than might be expected from government acting on its own.
- Stimulating additional private sector activity in ways that would benefit regional development.
- Developing synergy between the public and private sectors in helping to solve local problems.

Projects would be financed partly through the private sector and partly through government subsidy, and the project companies would be staffed by secondment from the public and private sector partners' organisations. Special legislation was enacted, and a large number of such companies were established during the late 1980s and early 1990s. Examples of typical projects that were planned to be procured in this way included a large number of resort developments, the 15 km Trans-Tokyo Bay Highway between Kawasaki and Chiba prefectures, and the new 58 km Joban

Railway line linking Tokyo's Central Business District with the Tsukuba Science Park in the Ibaragi prefecture.

The approach does not appear to have been very successful, and a number of the proposed project companies had difficulty in attracting sufficient private sector partners. In addition, when the Japanese economy collapsed in the mid-1990s, many companies were left in chaos, with projects that were no longer financially viable and bitter complaints from the private sector partners about their bureaucratic public sector colleagues.

Management contracts

Existing services and utilities are commonly privatised through the medium of management contracts, where the private sector takes over responsibility for the operation and maintenance of some existing facility which has previously been operated by the public sector. This form of arrangement may be especially useful when the public sector has a facility that is in need of some upgrading or renovation, but does not have the necessary funds available to carry out the work. An alternative to a conventional management contract is a fixed term concession based upon a lease/renovate/operate and transfer (LROT) contract, as described below.

The leasing model or build-lease-transfer (BLT)

In this model the facility (for example a motorway) is typically designed, financed and constructed by the private sector, and is then leased back to government for some predetermined period of time at a pre-agreed rental. Rental costs are commonly based upon the costs of construction and finance and, according to Trosa and Schreiner (1994), will usually represent about 9 or 10% of the investment costs. During the period of the lease legal ownership of the facility rests with the private sector partner, and at the end of the leasing period government typically has the option to renew the lease, to buy out the private sector partner for a lump sum or to simply walk away from the deal, leaving the facility in private sector hands. Operation and maintenance of the utility during the lease period will typically be the government's responsibility and, in the case of a motorway for example, government may choose to either operate the utility effectively free of charge to the user (usually recouping the costs from general taxation or from some form of specific vehicle road tax) or charge a toll. It may also contract out the operation and maintenance, either as part of the original contract or to a separate contractor under an operation and maintenance contract, or retain the responsibility for operations and maintenance itself. Where operation and maintenance forms part of the original deal the project may be described as *build-lease-transfer-maintain (BLTM)*.

This form of procurement thus effectively provides government with a way of financing large-scale infrastructure out of ongoing revenue rather than out of capital expenditure, but the primary disadvantage is that the legal ownership remains with the private sector. It is also common with this type of project for government to contribute to the initial construction, perhaps by assembling and providing the necessary land and external utilities (for example drainage disposal, access roads, etc.), and/or provide assistance with initial route alignment, obtaining the necessary planning approvals, etc.

With this form of arrangement the private sector partner has the advantage of knowing that the operational risks he or she carries are largely mitigated by the length of the lease and the frequency and amount of the periodic rent review. There remains, of course, the political risk that government might renege on the terms of the lease. The developer, however carries no risk relating to the extent of usage of the facility, this being assumed by government. Risk transfer is therefore seen as minimal.

Trosa and Schreiner (1994) report that this model has been considered for use in Germany in the construction of motorways, Havens (1994) reports that a number of variations of the lease and lease-purchase models have been used for the construction of prisons in the USA, and Hollihan (1994) reports that power stations in Mexico have been procured using this approach.

Prefinancing

The prefinancing model is similar in many respects to the leasing model, but the major difference between them is that with the prefinancing model the private sector will initially finance and construct the project, with government paying off the full cost, including the financing charges, by a series of pre-agreed annual lump sum payments over some agreed period of time (typically around 15 years). With this model it is again usual for government to provide the land free of charge, and also to carry out preparatory work such as preparing initial route alignments and obtaining the necessary planning permissions. The final detailed design may be prepared by government, with tenders being invited for the financing and construction only, or the final design may be entrusted to the contractor on a design, finance and construct basis.

Tenders for the work will therefore essentially comprise an offer to finance and construct the works to some agreed specification in exchange for a laid down series of lump sum payments over an agreed time period. As in the case of the leasing model, government may either operate the facility itself or contract out operation and maintenance to the private sector.

With this approach it is common for ownership of the completed facility to pass to government and for the payments to begin immediately upon completion of the construction work, and there is therefore considerable incentive for the work to be completed in the shortest possible time. As with the leasing model, this approach has advantages for the developer in that all market risks are born by government, and the developer's risk is normally limited to cost and time overruns on the construction contract and fluctuations in the costs of providing the necessary capital.

It has been reported (PFI 1995) that the German government had begun a programme of 12 major motorway contracts using this model, totalling some 151 km, at an estimated cost of some DM 3.89 billion.

Mixed models

A development of the leasing and prefinancing models has been used for the redevelopment of the A2 motorway between Helmstedt and Berlin in Germany (Trosa and Schreiner 1994). On this project, ownership of the scheme passed to government immediately after construction, and the annual payments from government depend upon the traffic flows. Payments are calculated based upon the number of different types of vehicles using the road, with different types of vehicles generating different fees. The road is not operated as a toll road, and the government payments are based on traffic counts. The method can therefore be seen as a precursor of the British 'shadow tolling' approach incorporated into the design-build-finance-operate (DBFO) strategy, which is discussed in more detail below.

The risks for the developer and for government are thus significantly changed when using this approach. Since payments are based upon traffic flows, if traffic flows are lighter than predicted then the developer will suffer, but if traffic flows exceed expectations then government will have to pay more. The government payments are therefore much more difficult to anticipate and to plan for.

Concession-based methods

A wide variety of techniques have been developed based upon the concept of a fixed term concession. Where the concession makes provision for ownership of the facility to be vested in the host government, this may occur either upon completion of the construction work or at the end of the concession period depending upon the terms of the concession agreement. Note, however, that, in some cases, even if legal ownership is transferred upon completion of the construction phase, the operator is frequently required to be *deemed* to be the owner for taxation purposes. Upon completion of the concession period the host government generally takes over responsibility for operation and maintenance of the facility. At this stage the host government may, if it wishes, let a further concession for operation and maintenance, often on a *lease/renovate/operate/transfer (LROT)* basis (see 'Management contracts' above).

It is common for the concession period to include the design and construction period. As discussed earlier, concession-based approaches are among the oldest forms of public/private sector partnership. The terminology and acronyms used to describe concession-based projects are not, however, always used consistently, and different projects that apparently use the same terms may vary significantly in the actual contractual arrangements. On the other hand a number of terms are virtually synonymous. The principal variants in most common use are described below.

Build-operate-transfer (BOT)

BOT is often used as a generic term for concession-based agreements where the facility is designed, financed, operated and maintained by the concession company for the period of the concession, typically between 10 and 30 years. The United Nations Industrial Development Organisation (UNIDO) has produced detailed guidelines which discuss all aspects of the BOT family of approaches (UNIDO 1996).

Under the 'standard' BOT route the concessionaire's involvement with the project terminates at the end of the concession period, and at this point all operating rights and maintenance responsibilities revert to the host government. Legal ownership of the utility may or may not rest with the concession company. In many projects legal ownership of the constructed facility is required to pass free of charge to the host government immediately upon completion of construction, but the concessionaire retains the right to operate the utility for the full concession period effectively as a licensee.

Walker and Smith (1995) contend that the Hong Kong Cross Harbour Tunnel, first mooted in 1958 and eventually opened in 1972, is probably the world's first modern-day BOT project, but there are other possible contenders. Messent (1997) for example writes of the construction of Australia's longest bridge, the Hornibrook Highway across Moreton Bay near Brisbane, built in the 1930s by building contractor Manuel Hornibrook as a way of keeping his company going during the Great Depression. Messent reports that:

> '. . . Hornibrook personally took on the extraordinarily difficult task of privately raising the finance for the project during the black years of the Depression. The bridge was opened on the 4th of October 1935 at a toll of a shilling a car. It proved of national importance during the war years carrying military convoys, none of which paid any tolls.'

He goes on to report that the toll remained at one shilling per car (A$0.10 after decimalisation) until the bridge was handed over to government debt-free in 1975.

More recently British contractor John Laing entered into a joint venture in the late 1960s with the Spanish company Ferrovial for the construction and operation of a 109 km motorway between Bilbao and San Sebastian in Spain under the provisions of the Spanish government's Toll Motorway Construction Plan. Armstrong (1994) reports that the 35-year concession was awarded in 1968, and the road opened to traffic in 1976. According to Armstrong, support provided by the Spanish government included the following:

- A guarantee of 75% of the funds borrowed in international capital markets.
- Subject to a premium of 0.2% per annum, foreign currency would be supplied for the settlement of principal and interest at the same exchange rate as that prevailing on drawdown in respect of funds borrowed in the international capital markets.
- An interest-free advance was made, equal to the difference in revenue between actual receipts and those for the target traffic density, subject to repayment only when traffic density exceeds the target by 5%. In consideration of this support net dividends were capped at 10%.

Build-own-operate-transfer (BOOT)

This phrase describes the earliest concession model, although the late Turkish Prime Minister, Targut Ozal, is popularly cited as originating the phrase 'build, own, operate and transfer' in the 1980s in connection with proposals for the construction of power plants in Turkey. In this model, ownership of the utility rests with the concessionaires until the end of the concession period, at which point both ownership and operating rights are transferred free of charge to the host government. Host governments often consider legal ownership of the concession by the concessionaires to be undesirable, particularly where the utility is of strategic importance, and the use of this variant is now much less popular.

Build-transfer-operate (BTO) and build-transfer-service-maintain (BTSM)

Technically these terms could be applied to any project where ownership of the facility passes to the host government upon completion of the construction phase. In practice the terms tend to be used for projects where the host government takes possession of, and pays for, the facility once construction is complete, with the developer being responsible for maintenance and operation of the plant for the period of the concession. This approach has been used for projects such as chemical waste treatment and waste recycling plants, where very specific expertise is usually required in the design and operation, but the revenue stream is difficult to predict with any degree of accuracy. Payment for operation of the facility is often a combination of a fixed and predetermined 'availability charge' covering the fixed costs of the facility, supplemented by a further variable charge depending upon the extent of use.

Build-own-operate (BOO)

The limit of the build-own-operate approach is, of course, outright privatisation, but BOO projects are sometimes let on a concession basis for a fixed period of time with no provision for transfer of ownership to the host government. In projects of this type the developer is responsible for design, funding, construction, operation and maintenance of the facility during the concession period, at the end of which several options might apply. Typically the project agreement might provide for one or more of the following options:

- Renegotiation of the original agreement for a further concession period.
- Negotiation of a new agreement on a renovate and operate basis.
- Purchase of the facility by the host government (who may then of course let a separate lease, renovate, operate and transfer concession).
- Termination of the facility, in which case the developer may be responsible for decommissioning, demolition and reinstatement of the site. Here the project may be defined as *build-own-operate-remove* (*BOOR*). An example of a project of this type is the A$25 million Yan Yean Water Treatment Plant constructed outside Melbourne by a Transfield–North West Water joint venture, where the plant is required to be removed and the site reinstated at the end of the 25-year concession period (Young and Sidwell 1996).

Lease-renovate-operate-transfer (LROT)

LROT is an approach that may be used where a host government already owns a facility which requires modernisation and improvement, or that may be used as a 'follow-on' to an existing concession once it reaches the end of the concession period. In this case the private sector operator pays a rental to government and undertakes to renovate the utility to some pre-agreed standard, and in exchange is granted a concession to operate the facility for a fixed period of time and to charge a fee for the service. The private sector partner is responsible for financing and carrying out any required refurbishment work in addition to regular maintenance during the concession period.

This approach has been widely used in some parts of the world, for example for government-owned car parks and vehicular tunnels in Hong Kong, and for water and sewage treatment plants in Malaysia.

Design-build-finance-operate (DBFO)

DBFO is a term coined by the UK Highways Agency to describe their concession-based road schemes let under the UK Government Private Finance Initiative. The basis of the 'standard' DBFO contract was defined by the Highways Agency (1995) in the following terms – the private sector partner (the DBFO concessionaire) will be:

- Responsible for the design, construction operation, maintenance (i.e. capital, routine and winter maintenance) and operation of the project road.
- Responsible for financing the project.
- Granted a long-term right of access (probably 30 years) to the project road by the Secretary of State for those purposes.
- Paid by the Department for delivery of the specified service in the form of payments over the life of the contract.

In addition to the free provision of any necessary land, the public sector contribution is likely to include some or all of the following, depending upon the detailed requirements of particular projects:

- A right of access to the project road without charge.
- DBFO payments which are expected to include traffic-related payments, usually based upon 'shadow tolls', lane closure charges and additional payments for initiatives designed to improve safety.

Significant features of the DBFO contracts are as follows:

- No rights of ownership are conferred on the developer, neither does the developer at any point acquire any interest in the land. The Secretary of State remains the highway authority throughout the contract period.
- The DBFO contractor is merely given a right of access to the road, and effectively a 'licence' to operate it, normally for a period of 30 years.
- Five years before the end of the operation period, a joint inspection will be held to identify any work that the operator is required to carry out before the road is handed back to the Highways Agency.

- Payment is made on the basis of traffic flows at predetermined 'shadow tolls'. The level of the shadow tolls forms an element of the DBFO tender.
- Some protection will be offered in the form of increased payments in the event that the public sector partner changes the conditions under which the road operates (for example if other competing roads are upgraded during the contract period thus reducing traffic flows).

Design-construct-manage-finance (DCMF) and design-build-finance-operate-manage (DBFOM)

DCMF and DBFOM are terms used in connection with projects let under the British Private Finance Initiative (PFI). These projects are very similar in many ways to build-own-operate concessions.

References

Armstrong J. (1994) Transport: Europistas and Severn crossing. In: *Proceedings of Financial Times Conferences, International Infrastructure Finance*, London.

Burton A. (1972) *The Canal Builders*. Eyre Methuen.

Cavanagh A. (1995) Eastern Europe reviews the private finance option for roads. *Project Finance International Europe Market Report*, Autumn (Suppl.).

Deane P. (1984) *The First Industrial Revolution*. Cambridge University Press.

Fay C.R. (1947) *Great Britain from Adam Smith to the Present Day*, 4th edn. Longmans, Green and Company.

Havens H.S. (1994) Private sector ownership and operation of prisons: an overview of United States experience. In: *New Ways of Managing Infrastructure Provision*, Public Management Occasional Papers 1994, no. 6, Market-Type Mechanisms Series no. 8. Organisation for Economic Co-operation and Development.

Highways Agency (1995) *Roads DBFO Projects; Information and Prequalification Requirements*. UK Government Highways Agency.

Hollihan J.P. (1994) The evolution and future of BOT and alternative structures; advantages and disadvantages of BOT as a project finance mechanism. In: *Proceedings of Financial Times Conferences, International Infrastructure Finance; Build-Operate-Transfer (BOT)*, London.

McCarthy S.C. and Perry J.G. (1989) *BOT Contracts for Water Supply*. World Water.

Messent D. (1997) *Opera House Act One*. David Messent Photography.

Pannell J.P.M. (1977) *Man the Builder*. Thames and Hudson.

PFI (1995) *Federal Government Trunk Road Progress*. Project Finance International.

Pollins H. (1971) Britain's Railways, an Industrial History. David and Charles.

Smith A.J. (1999) *Privatized Infrastructure: The Role of Government*. Thomas Telford.

Trosa S. and Schreiner M. (1994) Private financing of public infrastructure in Germany. In: *New Ways of Managing Infrastructure Provision*, Public Management Occasional Papers 1994, no. 6, Market-Type Mechanisms Series no. 8. Organisation for Economic Co-operation and Development.

UNIDO (1996) *UNIDO BOT Guidelines*. United Nations Industrial Development Organisation.

Vornetti P. (1994) The French experience of partnership between the public and private sector in the construction and operation of motorways: theory and practice. In: *New Ways of Managing Infrastructure Provision*, Public Management Occasional Papers 1994, no. 6, Market-Type Mechanisms Series no. 8. Organisation for Economic Co-operation and Development.

Walker C. and Smith A.J. (1995) *Privatised Infrastructure: The BOT Approach*. Thomas Telford.

Williams F.S. (1883) *Our Iron Roads: Their History, Construction and Administration*. Bemrose and Son.

Young D.M. and Sidwell A.C. (1996) *Case Studies in Infrastructure Investment*. Construction Industry Institute Australia.

14 Public private partnerships (PPPs) and the UK Private Finance Initiative (PFI)

Introduction

The previous chapter introduced some general concepts to do with the procurement of public sector projects using private capital. This chapter continues that examination, but concentrates on the UK Government's vision of public private partnerships (PPPs) and the Private Finance Initiative (PFI). These arrangements build upon the earlier experience, but have moved the techniques on from utilities and transport infrastructure to the provision of schools, hospitals and a wide range of other public building types.

Following considerable confusion over the terminology, the UK Government, in 2000, clarified its view in *Public Private Partnerships – the Government's Approach* (HM Treasury 2000). This document defined three categories of PPP as follows:

(1) The transfer of private sector ownership into state-owned businesses, using the full range of possible structures (whether by floatation or the introduction of a strategic partner), with sales of either a majority or a minority stake.

(2) The PFI and other arrangements where the public sector contracts to purchase quality services on a long-term basis so as to take advantage of private sector management skills incentivised by having private finance at risk. This includes concessions and franchises, where a private sector partner takes on the responsibility for providing a public service, including maintaining, enhancing or constructing the necessary infrastructure.

(3) Selling government services into wider markets and other partnership arrangements where private sector expertise and finance are used to exploit the commercial potential of government assets.

The Private Finance Initiative (PFI)

The British Government's Private Finance Initiative (PFI) was introduced by the then Chancellor of the Exchequer, Norman Lamont, in his autumn budget statement in November 1992. The initiative marked a further step along the privatisation road which had been pioneered by Margaret Thatcher and continued by John Major, and was intended to encourage a major shift in the way government fulfilled its basic obligation to provide economically efficient infrastructure and services.

Many so-called 'public services' in Britain, including electricity, gas, water supply, telecommunications and rail services, had already been privatised with varying degrees of success, but the principles underlying the PFI envisaged much more. PFI was intended to go far beyond a simple sale of public sector assets. It was much more radical, and was based upon the premise that, rather than government committing capital investment to owning, operating and/or managing the means of providing the necessary services, substantially greater economic efficiency and lower costs might be attained by 'contracting out' the services themselves to the private sector. In other words, rather than building and owning a school or a prison or a hospital, government would simply buy the *service* it required from the private sector, for example custodial services for a given number of prisoners, education for a specified number of children or medical facilities to a defined standard. It would then leave it to the private sector partner to develop and put in place whatever mechanisms it considered to be necessary for the required level of service to be provided.

The initiative therefore offered the opportunity to shift a substantial balance of infrastructure expenditure from capital to revenue, or alternatively to encourage private sector capital investment in public services that would be additional to that already available from public funds.

It was, of course, recognised that privately funded and operated schemes might be more expensive than similar publicly funded projects, but an important principle was that PFI offered the opportunity to shift a substantial part of the risk burden inherent in the development of new projects from the public to the private sector. By carefully specifying the *service* that it required the private sector to provide, thus forcing the private sector to make the necessary investment in new construction, development of new technology, etc., government would no longer be responsible for cost and time overruns, or systems that were inefficient or that failed to work at all. All of the responsibility for these issues would ultimately rest with the private sector, for which government would in theory be happy to pay a higher price.

Many of these things have been done before. It has already been shown that the private sector has a long history of involvement in capital projects, and that the public sector only took control either for political reasons or when the private sector began to fail to deliver reliably. Several countries, for example the USA, Australia and New Zealand, have, in recent years, also embraced public/private sector partnerships, resulting in the 'semi-privatisation' of prisons, hospitals and educational facilities.

What was different about PFI, at least in modern history, is that, for the first time, private sector procurement, management and financing of key projects had been placed at the very heart of government's procurement philosophy. All major projects were required to be assessed prior to any government investment to see whether the private sector could offer a better deal.

The fundamentals underpinning the scheme were explained by the Treasury as follows:

- The private sector must genuinely assume risk.
- There should be competition where government facilitates a project or seeks a private sector partner or purchases services as a customer.

- The initiative would apply not only to infrastructure projects but also to any other capital investment providing services to the public sector.

Genesis

Prior to 1977 all British public sector capital expenditure was measured against the Public Sector Borrowing Requirement (PSBR), even when wholly financed from internal resources derived from user charges. From 1977 onwards, however, the focus for controlling the PSBR switched to external financing limits (EFLs). During the early 1980s there was considerable debate about whether the use of EFLs was prolonging the recession by effectively preventing investment in those profitable areas of government that could have a positive effect upon the economy. In addition it is often thought advisable to invest in public sector infrastructure in times of recession due to the more competitive state of the construction market. Expenditure capping on departments such as the Department of Transport and upon the nationalised industries was seen by some to be exacerbating the recession.

In response to these concerns, a National Economic Development Committee (NEDC) working group was established under the chairmanship of Sir William Ryrie to report upon the potential for using private sector investment in the nationalised industries. The ensuing report (NEDC 1981) proposed two conditions, known as the 'Ryrie Rules', which should be applied when considering the use of private sector funds in public sector capital projects:

(1) Privately funded projects must be tested against publicly funded alternatives to prove that they are cost effective.
(2) Privately funded projects should not be additional to public expenditure provision and public expenditure should be reduced by the amount of the private sector funds obtained.

Although the Ryrie Rules were specifically developed for the nationalised industries, they were subsequently assumed to represent Treasury policy on the use of private sector funds in the public sector. The Ryrie Committee's findings attempted to redefine the boundaries between the private and public sectors. Private sector business was, and is, directed by market forces, legislation, financial constraints, etc., but public sector decisions were supposedly made by government on the basis of long-term economic and social needs. The use of private sector funding by the public sector was therefore viewed by some as a relinquishment or an evasion of control. Indeed it has been claimed that the Treasury's 'hidden agenda' in the 1980s was actually to prevent the use of private finance for public schemes. David Willets, then a junior Treasury official, maintains that the Ryrie Rules 'were not intended to be met in practice' (Willets 1993).

Throughout the 1980s the Conservative government proceeded with its privatisation of the nationalised industries, a programme that was ridiculed by some as 'the great sell-off' and hailed by others as one of their greatest policy successes. Lord MacAlpine for example described the floatation of British Telecom as 'an outstanding success'. During this same period the government also began to contract out services to the private sector, and housing associations were virtually compelled to use private finance to build houses (Hancock 1995). In 1989 John Major, then Chief Secretary to the Treasury, announced that the Ryrie Rules, and in particular the rule regarding non-additionality, were hampering the use of private sector finance in infrastructure projects and the Rules were therefore abolished.

The final element in the genesis of PFI came with the 'Black Wednesday' foreign exchange market crash in September 1992. Major reductions in government capital spending were an inevitable consequence, and as a result John Major's 'Strategy for Growth' announced in October 1992 included plans to offset the reductions by encouraging private sector involvement in large infrastructure projects.

Contractual arrangements

A variety of contractual models have been used in the past. The prisons at Bridgend and Fazakerley were procured on a fairly conventional build-operate-transfer contract, with a concession period of 25 years and transfer of the building to government at the end of that time (Allen 1997). For road schemes, however, a design-build-finance-operate contract is used, where the new works are designed, financed, constructed and operated for a set period of time by private sector contractors, but where the private sector partner secures virtually no ownership rights. For schools, hospitals and the like, early schemes were let using a build-own-operate contract where the local authority partner effectively agreed to lease a fully maintained purpose-built facility, designed and procured by the private sector, for a specified period of time, of the order of say 30 years. All of these models have now been considerably developed and refined.

Problems

Early indications were that PFI was publicly widely welcomed, particularly by the construction industry (Hoare 1995). Despite the initial optimism, however, and the apparent success of several high-profile projects including the Queen Elizabeth the Second Bridge at Dartford, the Second Severn Crossing and the redevelopment of the Norwich and Norfolk District Hospital (McTaggart 1997), actual real performance in terms of new projects coming forward in the pipeline was poor. The Construction Industry Council (1998) wrote that:

> 'The impact of PFI was expected to be considerable. . . . The result was a rapid increase in the volume of PFI notices issued in the *Official Journal of the European Communities*, and a widespread expectation from the private sector that PFI would be a considerable source of work. The short history of PFI has, however, revealed that this form of procurement represents a completely new genus of work with many issues needing to be addressed before it could become a widely used procurement option.'

Part of the reason for this was plainly the fact that both the public and the private sectors were still unsure of exactly how this new method of procurement would work, and initial experience, outside the few widely publicised projects mentioned above, led some to believe that the approach had many problems. Not the least of these was the Treasury itself. Richard Saxon, Chairman of Building Design Partnership, is reported to have said (Allen 1997) that:

> 'the Treasury made it all but impossible for anyone to bring a PFI project to the starting gate. . . . PFI is not just a closed door to the smaller firms. It is a closed door, period, closed by the intransigence of the Treasury in coming to terms with commercial reality. We can continue to talk about PFI as if it existed: for us it's more hypothetical than real.'

Amongst the most pressing problems for prospective tenderers were issues of time, cost and risk, particularly where very large projects were involved.

Perhaps one of the best examples is the Channel Tunnel High Speed Rail Link, where it is reported (Allen 1997) that the original bidding costs for the four consortia invited to tender amounted to some £5 million each, following which there was a further period of 11 months of negotiation to reduce the number of bidders to two, and a further 3 months after the final contract award before the project reached financial closure. Ove Arup Partnership, a founding member of the winning consortium, report that:

> 'During the whole of that time the consortium had to fund itself. There was no outside help. And remember that all that money was at risk right up to the winning post. If we had lost then it would have been very painful. It would have hit us very hard. It wouldn't have been so bad to lose if there was a mass market of PFI projects but there isn't. The whole process has proved very, very expensive. These negotiations go on for so long, and there are so many difficulties inherent in them, that it makes them very difficult for anyone to finance.'

Similar time problems were reported (Price and Ivison 1996) for the first two PFI prisons at Bridgend and Fazakerley, where the process from the initial invitation to bid to financial closure took more than 2 years, and the Birmingham North Relief Road, where the total time from invitation to bid until completion of the public enquiry exceeded 5 years (Walker and Smith 1995).

Further problems faced by tenderers for early PFI schemes concerned the risks that were to be transferred and uncertainty over how their tenders were to be assessed. The Highways Agency, for example, in its prequalification requirements document for DBFO road schemes (Highways Agency 1995), wrote that:

> 'The Department recognises the many different views that Tenderers may have as to the scope of obligations and allocation of risk under a DBFO Contract. The Department does not have a settled view on the full extent of those obligations nor on the final allocation of risk. Nor can it prejudge what potential tenderers may be prepared to offer.'

The Agency did offer a model contract which gave some indication of the kind of 'standard' level of risk transfer against which all tenderers were required to bid, but tenderers were also at liberty to propose any 'non-standard' alternative arrangements.

There were, however, no indications in the documentation as to how bids might be assessed or what criteria were considered the most important. Whilst one must agree that it might not be possible to give a definitive list of quantifiable assessment criteria, it is difficult to escape the conclusion that government did not know what it actually wanted, and thus could not formulate any criteria against which bids would be judged. It must therefore follow that bidders could only guess at what the most important assessment criteria might be. The mental picture that emerges is of two people blundering around in a darkened room, attempting simultaneously to grasp the same slippery black ball.

The very nature of the PFI requires that government prepare an 'output specification' detailing the service that it requires to be provided. Where the project includes new construction the specification may also include details of the accommodation to be provided, and this may be supplemented in some cases by statutory requirements, for example in the case of roads or prisons. An obvious problem here is that government is not accustomed to working in this way, and the degree of prescription in the output specification is likely to have a considerable impact upon the degree of innovation that the tenderer is able to introduce. This issue apparently posed a significant problem for the Bridgend and Fazakerley prisons. Price and Ivison (1996) wrote that:

> 'The consortium's view was that, by being prescriptive, the Prison Service minimised the supplier's ability to be innovative in its approach to the provision of the prison facilities and the service. An alternative would have been for the Prison Service simply to specify the output required and to have left the supplier to develop proposals to meet the output.'

Morrison and Owen (1996) drew attention to the problems of understanding and interpreting the mass of PFI legislation. They comment that PFI is very complex within both sovereign and European Community law:

> 'The first difficulty the practitioner will find is determining which set of rules applies to a particular project. PFI is promoted on the basis that it is essentially the provision of a service by the private sector, albeit that, in most cases, such provision is preceded by a construction period. So does a typical PFI contract fall within the rules applying to the provision of (a) public services or (b) public works?'

Other critical issues of direct concern to government identified by Morledge and Owen (1998) include the following:

- The public sector must bring forward projects that are well thought through, where there is a generally perceived need, that have demonstrable viability and for which it is prepared to pay a realistic price. Projects that go through a competitive tendering process and which are subsequently cancelled because the price is too high are counter-productive to the Initiative as a whole.
- A realistic view of risk assessment, allocation and management, with government being prepared to retain those risks over which it has control and to pay a realistic market price for the risks it wishes to transfer.

- Efficient administrative systems to ease the process, control changes and settle disputes.
- A willingness by the rank and file of the public sector to accept that the private sector has a contribution to make.
- Better prequalification procedures such that no more than two to four bidders are eventually put to the expense of preparing a full bid.
- More transparent tender assessment criteria.

Many of these issues have already been identified as critical in the context of other large privately funded projects elsewhere in the world, and the only surprise is that the British Government did not foresee them when the Initiative was put in place.

A further problem arose in connection with smaller-scale projects (say between £5 and £20 million). Such projects are generally too small to attract funding on a non-recourse basis and too large for most smaller firms to take on board through 'on balance sheet' financing. In addition, bidding costs do not form a constant percentage of project value, and consequently the bidding costs for small-scale projects form a comparatively large percentage of project value. They are thus not particularly attractive unless tender lists are very small.

There have, as one would expect, also been detractors. In 1997 the Public Services Privatisation Research Unit produced a politically motivated, but nonetheless eloquent and well argued, rebuttal of PFI (PSPRU 1997), arguing that PFI did not produce good value for money and that few projects had been successfully completed. There was a clear need for an urgent review.

The first Bates review

Following its sweeping election victory in May 1997, the incoming New Labour administration carried out a wide-ranging review of PFI under the chairmanship of Sir Malcolm Bates (Bates 1997). The review made 29 recommendations, addressing many of the problems identified above. Perhaps the most important implication of the Bates review as a whole was that the Initiative should be better focused. In future, it was suggested, only the most viable of projects (i.e. those that had been assessed and approved by a new Treasury Task Force) would be brought forward. Key components in the vetting process were to be an outline business case and the development of a 'reference project' in order to compare the costs with those that would have been paid had the project been developed wholly in the public sector. The reference project (the Public Sector Comparator) has been defined (Construction Industry Council 1998) as:

> '. . . a particular solution to the output requirement. It should be a combination of capital investment, operation, maintenance and ancillary service and be in sufficient detail to provide full and adequate costing. The costs should include quantification for all the key risks inherent in the project. The reference project should also be affordable to the public sector client, with provision for all the quantified risks.'

Note that there is no requirement for the reference project to be exactly the same as the proposal submitted by the private sector developer. This becomes a significant issue in the context of buildings, where subjective issues such as the quality of the design and the contribution the building makes to the surrounding landscape become important. It is likely that, in most cases, the private sector proposal will represent the 'economy' option, and as such will almost always appear to present a better deal in purely financial terms. It has been argued (Cole 1995) that delegating design responsibility to developers in this way does not provide the best overall solution.

Bates also recognised the need for greater cost effectiveness in the bidding process. In future the number of bidders would be limited to a maximum of four, and there was also a proposal to reimburse part of the bidding cost if the project did not subsequently proceed under PFI.

The second Bates review

The second review of PFI by Sir Malcolm Bates was reported in July 1999. This time the review concentrated on institutional change and government policy. In particular Bates found weaknesses in:

- Strategic planning
- Project management
- Negotiation skills
- Financial discipline
- Management of long-term contracts

As a direct result of this review Bates recommended that the temporary Treasury Task Force be replaced by a new permanent organisation called Partnerships UK, and this body formally came into being in June 2000. Partnerships UK subsequently became a public private partnership in its own right in March 2003, when 51% of the equity was sold off to the private sector.

Current developments

The PFI has now been in place for more than a decade, is now firmly embedded at the heart of UK Government procurement practice and has been responsible for the procurement of 667 projects with a total value of approximately £42 billion between 1986 and 2004. Of these by far the greatest proportion of the total expenditure (approximately 66%) has been committed to transport, although in terms of numbers of projects, health (152) and education (102) head the list. The NHS, where a considerable proportion of new hospital provision is now delivered through this route, has carried out much work in the development of standardised contracts and service level specifications.

Further projected expenditure on PFI schemes announced in the 2005 budget is said to be approximately £3.23 billion in 2005/2006, £2.43 billion in 2006/2007 and £1.80 billion in 2007/2008.

In terms of performance, HM Treasury (2003) reports that 76% of projects surpass initial expectations, 88% of projects were delivered on time and 79% were delivered on budget. Examples of budget overruns are given in Allen (2001), where he comments that three of the PFI projects he examined, the Norwich and Norfolk Hospital, Greenwich Healthcare Trust and the Benefits Agency computer system, overran by 60%, 140% and 600% of the initial costs respectively.

Current problems and solutions

The following issues have been identified by Allen (2001) and others as being of particular interest in the further development of the PFI process:

- High bidding costs
- Refinancing
- Value for money
- Design
- The relationship between PFI client and contractor
- Concession agreements

High bidding costs

Butler and Stewart (1996) reported that bidding costs for PFI projects averaged approximately 3% of project value as opposed to approximately 1% for traditional projects. PFI projects also took much longer to procure than comparable traditional schemes, with the average time taken to achieve financial closure estimated at 26 months for local government projects and 42 months for NHS projects. In an attempt to ease these burdens, the Audit Commission (2001) published guidance for local government managers, requiring them to:

- Demonstrate a clear purpose and a strong vision of the desired outcomes from the scheme.
- Translate that vision into a simple output specification and resist the temptation to make regular changes to that specification.
- Get early commitment to the scheme from key stakeholders.
- Set up a project management structure that allows for an appropriate level of delegation to key officers and is integrated with existing decision-making processes.
- Agree a clear project plan, establish project milestones and monitor progress against the plan on a regular basis.
- Agree the key contractual terms, including payment mechanisms and risk transfers, prior to issuing the invitation to negotiate, in order to force bidders to indicate their position early on in the negotiation process.
- Be clear about how they are going to evaluate bids.

Refinancing

The financing package for a major scheme may be extremely complex, and private sector developers involved in large infrastructure schemes have long recognised that it is often possible to improve returns by refinancing projects during the development and/or operation process. Opportunities for refinancing may arise either to take advantage of international currency exchange rate fluctuations or international interest rate variations, or to reflect the changing nature of risk in a project as it moves from construction (conventionally rated a high risk activity) to operation.

The amounts involved may be substantial – as an example, consider the Shajiao B power station in southern China (estimated cost approximately US$512 million) where refinancing one year into the project to take advantage of shifts in international exchange and interest rates allegedly increased the return to the investors by approximately US$40 million. In the UK it is reported (NAO 2000) that, upon completion of the construction phase four months early and having established a track record in operation, Fazakerley Prison Services Limited (FPSL) were able to refinance the Fazakerley prison project. Refinancing here resulted in expected returns to shareholders being increased from £17.5 million to approximately £28.2 million, or approximately 61%. In this case the Prison Service stated that it believed the windfall profit was due to FPSL successfully managing the risks of procurement, and that they should be allowed to keep it (Allen 2001).

Morgan and Mathiason (2001), however, saw the issue rather differently, alleging that windfall gains were set to be taken from a wide range of projects including hospitals, roads, prisons and schools, and describing the process as 'robbery of the public'.

The UK Government's present general stance on the refinancing issue is set out in HM Treasury (2004a) as follows:

- Refinancing constitutes a material change to the agreed project structure, and the client therefore has the right to be informed of and to approve any refinancing package.
- Contractors' proposals for refinancing should be welcomed and considered positively.
- The client's long-term commitment to purchase a service at a guaranteed price for an extended period is a commercially valuable promise. Clients therefore have a natural right to share in any refinancing gains which arise as a consequence of this promise.
- Increases in returns to investors gained through increased efficiency or performance should be owned by the investors. However, the Treasury chooses not to consider improvements in loan margins and certain beneficial changes to the term and leverage of the debt finance to be due to efficiency gains, and therefore insists that such benefits should be shared on a 50/50 basis between the client and the contractor. Similar conditions apply where refinancing opportunities arise as the result of a general increase in the maturity of the market as a whole.

Value for money

Notwithstanding the fact that PFI is now firmly embedded at the heart of UK Government procurement, there is still considerable discussion about whether or not

PFI truly delivers improved value for money. In terms of the early projects, a study by Arthur Anderson and Enterprise LSE (2000) of a sample of 29 PFI projects showed an average saving of 17%, and in 2001 the National Audit Office (NAO) reported that in a survey of authorities responsible for managing 121 PFI projects worth in total some £5.73 billion, 81% ranked value for money as satisfactory or better (NAO 2001a). Allen (2001) in his review reported that 7 of the 15 projects upon which the National Audit Office had reported in detail at that time had been assessed in value-for-money terms, and appeared to show that 'if the NAO sample was representative of all PFI projects then an estimate of the savings from the £22 billion PFI projects signed up to 1st December 2001 would be in the region of £4.4 billion'. However, since two of the projects accounted for 81% of the total savings, he concluded that a more realistic estimate might be £2.2 billion or approximately 10%.

There are, however, dissenting voices; Pollock (2000) for example maintains that PFI has failed to deliver value for money in health projects, citing as examples the Kidderminster Hospital where:

> '... using PFI means that increased costs of the new hospital will be met in part by the closure of 219 inpatient beds without alternative provision. But since the catchment area of the new hospital will increase by 10,000 to 380,000, the population served will have almost one third fewer acute beds under PFI ...'

She also comments that although the first NHS Trust to sign a PFI contract stated during the outline business case that there would be no additional cost to the Trust, by the time the contract reached full business case an additional £2 million a year was required.

Allen (2001) comments that:

> 'The evidence suggests that some types of public service projects may be more suited to PFI than others. While roads and prison projects have achieved reasonable efficiency gains, projects in other sectors such as schools and hospitals have shown minimal gains.'

The Treasury is now keen to ensure that PFI projects genuinely provide real value for money, defined as 'the optimum combination of whole life cost and quality (or fitness for purpose) to meet the user's requirement', and to that end has published updated and detailed guidance for prospective PFI clients (HM Treasury 2004a, 2004b, 2004c). The guidance is seen to be required in order to reflect NAO reports on previous projects, and to take into account the Treasury's revised approach to investment appraisal. Changes include a reduction in the discount rate to be used from 6% to 3.5%, a requirement for the incorporation of an 'optimism bias' to reflect the tendency for project appraisers to be unduly optimistic about project outcomes, and consideration of the taxation implications of different procurement routes.

The intention is that PFI should only be pursued as a procurement option where it provides clear value-for-money benefits. The new procedures set out a detailed methodology for assessing all stages of a proposed PFI project, and include:

- A new test of the potential value for money of procurement options when procurement decisions are being made, to ensure that PFI is used only in those sectors where it is appropriate and has a good value-for-money case, and that departments provide sufficient budget flexibility to accommodate subsequent decisions not to use PFI.
- Reforming the public sector comparator into part of an early rigorous economic appraisal of an individual project at the stage an outline business case is produced, prior to procurement of the project, to allow projects to proceed down alternative procurement routes where they offer better value for money.
- Instituting a final test at the procurement stage of a project that would evaluate the competitive interest in a project and the capacity of the market to deliver it effectively.

Issues that should be considered in assessing value for money include:

- Good design
- Not achieving it at the expense of workers' terms and conditions of service
- Non-market factors which may affect the project, such as environmental impact and non-project-related strategic risks

HM Treasury has also developed standard spreadsheets for the assessment of PFI projects. The methodology and a worked example are set out in HM Treasury (2004c).

Design

Early PFI schemes suffered considerable criticism in respect of design issues (see for example Cole 1995), and many took the view that the PFI solution to a problem would generally represent the most 'economical' design approach. This was seen by some as a risk mitigation tactic, 'use what we know works, innovation is risky and dangerous'.

Government has, for some time, been anxious to overcome these issues and has published a significant amount of design guidance particularly in respect of health buildings, much of which is now available through the internet (see for example *How to achieve design quality in PFI projects* at www.privatefinance-1.com/resources).

This seems to have been effective. The National Audit Office (NAO 2001) reports that, in general, PFI consortia:

- Invested in good design and construction at the start of the project
- Achieved better quality buildings and a reduction in maintenance costs, while maintaining the assets to the agreed standards
- Placed more emphasis on the aesthetics of design than had previously been the case

The relationship between PFI client and contractor

Historically the relationship between the PFI client and the contractor has been seen to be somewhat confrontational. The early emphasis on a supplier/customer relationship, proven risk transfer, the client's need to ensure that levels of service were delivered in accordance with the contract, and funders' determination to protect their investment and to limit their own exposure to risk led inevitably to lengthy and complex contracts and a relationship driven by strict contractual compliance. Even as recently as 2001, when the author was involved in the development of procurement systems for the NHS, any suggestion that NHS PFI contracts should incorporate partnering principles was scornfully rejected.

Nonetheless some have consistently argued that in the development of public/private sector partnerships both parties would surely wish to maximise the chances of project success; to achieve what has been termed a 'win–win' outcome. In order that this may be achieved, however, the structure of the partnership and the working relationship between the people involved requires very careful consideration, and, as Mody (1996) points out, establishing such a partnership requires a clear understanding of the relative strengths and weaknesses of both parties in addition to an understanding of their histories. It is, however, often the case that government believes that it can only fulfil its functions by maintaining close control over all aspects of the project, but this process may be perceived by the private sector partner as an unwarranted interference with the way he or she wishes to run his/her business. Hurst (1994), for example, writes that:

> 'Even in most developed states such as the United Kingdom, there is a tendency for government to retain control of projects through a complex approvals procedure, especially where design and construction of projects is concerned. In private-sector electricity projects, especially in Asia, the state is often involved at every stage of the project, going far beyond the usual "planning" rules, treating the private-sector generators as though they were part of the public sector.
>
> This extends to requiring the right to review and approve all design drawings, the specifications for each item of equipment, the right to act as an additional *de facto* project manager during the construction phase of a project, the right to receive hundreds of "as-built" drawings for every aspect of the station, and the right to visit and inspect the station. Vast quantities of corporate information on the project company and/or the sponsors are often required . . .'

The same point is also made in connection with the first two prisons to be commissioned under the British Government's PFI programme. Price and Ivison (1996) commented that, although the focus of the contract was on the provision of a custody service rather than on the construction of a building, and thus the design and construction risk lay solely with the private sector partner, government insisted on the appointment of an independent engineer, who expected to have an input into design matters. The situation was apparently only resolved when the role of the independent engineer was redefined such that:

> '. . . government (had) the comfort of an independent adviser who was familiar with its concerns doing an initial design-check of the prison and monitoring

construction on site, but also enabled the consortium to be innovative in its design and construction methods.'

The privatised infrastructure developers' ideal market would probably be one in which they were able to provide services, or on behalf of government, to a captive market, at prices to be fixed solely by themselves, whilst at the same time being subject to an absolute minimum of regulatory or governmental control. Unfortunately, however, history shows that this kind of *laissez faire* approach does not work very well.

The major rationale usually put forward for governmental regulation and control of projects is that the government may, at some stage, become owner or operator of the utility, either because ownership of the project assets fails to be transferred to the host government at some stage during the concession or, in the event of default by the concessionaire or project operator, government may be forced to take over operation of the utility in order to preserve some essential public service. Whilst these may, in some cases, be genuine concerns, it is obvious that any increase in government 'interference' is likely to increase project risk and thus project costs and in turn therefore the cost to the user. Government is therefore faced with a considerable dilemma. The latter concern basically envisages some failure of the concessionnaire to perform his or her obligations. This concern might well be real, but as a starting position it seems somewhat negative. It is not in anyone's interest for these schemes to fail financially or otherwise once they are operational. A financial failure, or indeed any failure of the operator to deliver the required level of service, which is serious enough for government to terminate the concession contract early, is bad news for all of the parties involved. Operation of the utility by government itself is often the very eventuality that government wished to avoid, and may be the reason for privatised operation in the first place.

Smith and Walker (1994) identified those high-level factors in which success appears to be necessary in order that each of the major project participants in a privatised infrastructure project have the maximum chance of achieving their goals. They are as follows:

- A genuine desire for a 'win–win' solution with common agreement between the parties as to their mutual and individual objectives. These objectives must be explicit, realistic and achievable.
- A strong, persistent, persuasive and politically skilled project leader willing and able to fight for the scheme.
- Adequate and accurate data and risk assessment of both the procurement and operational phases, with responsibility for managing the risks given to the party best able to control them.
- An honest and accurate calculation of the project economics, including the length of the concession, which incorporates probability assessments of the influence on income and expenditure of the risks and uncertainties previously identified.
- Choice of the correct procurement methodology for the construction phase.

It is also important to remember that governments often come and go within relatively short time periods. The life of a particular administration may represent only a small fraction of the length of the concession itself, and subsequent governments (although not so far in the UK) have been known to wish to revisit agreements made by their predecessors. Two examples reviewed in Smith (1999) are the Dabhol power project in India and the Bangkok Second Stage Expressway in Thailand. In at least one of these projects the primary reason for wishing to revisit the original negotiations centred around allegations of corruption. In order that allegations of this kind may be refuted, it is plainly essential to ensure a high degree of public accountability in all government action and decision-making. The initial negotiation process must therefore be both totally transparent and accurately documented.

In the light of the above, it is suggested that a 'team' approach involving shared perceptions and shared goals between the public and private sector representatives is likely to be more effective than the very commercial customer/service-provider relationship which appears to have been used in the UK in the past.

Similar approaches have also been successfully used elsewhere. Reijniers (1994) for example analyses the working of just such a partnership in the Netherlands.

Different patterns of partnership formed with different kinds of people may be required at different stages of the process. During the initial stages, prior to the appointment of the concessionaire, particularly if a competitive tendering process is involved, governments must obviously maintain a proper distance between themselves and their tenderers, and must be seen to be impartial to all of the competitors, but once the concession is signed then it can be convincingly argued that government should perhaps take on a much more active partnership role during the construction and operational phases. It is, however, important here that government is seen to be actively contributing to the *success* of the project and not merely interfering in the affairs of its private sector partners.

The idea of PFI projects based upon collaborative principles was actively supported by the National Audit Office in their 2001 report (NAO 2001). In a major review of the management of PFI projects, the NAO concluded that some of the major issues to which client departments should pay particular attention were as follows:

- PFI projects need to be approached in a spirit of partnership. A successful outcome for both parties can only be achieved if they clearly understand each other's business and have a shared and common vision of how they can work together as partners.
- A successful partnership needs to be established at the outset. Failure to properly consider how the relationship will be managed before contracts are let is likely to lead to misunderstanding and a period of additional risk while the required service is being developed.
- The development of a successful relationship requires an appropriate contractual framework, which properly allocates risk, clearly defines the service parameters and incorporates a mechanism to deal with change.
- Having staff with the right skills is essential to good contract management. It is highly likely that additional training will be required to equip staff to work in this new way.

Small projects

Some contracting authorities have attempted to overcome the problems of making small projects viable by grouping together a number of related smaller schemes, none of which would be viable on its own, and so constructing a viable, bankable PFI package with clearly defined risks. This approach has been followed successfully by a number of local authorities for schools refurbishment, and is also at the heart of the NHS Local Improvement Finance Trust (LIFT) initiative.

Concession agreements

The drawing up of concession agreements is plainly a very skilled task. However, the following have been identified (Vintner 1996) as some of the more important questions of principle which need to be considered by governments when carrying out this task.

What provisions should government include in the agreement in order to safeguard itself against non-performance by the concessionaire? How should disputes be resolved?

It has already been established that governments would normally be wise to only consider early termination of a concession in the direst of circumstances. To do so would usually bring upon government more problems that it would solve. Vintner (1996) suggests that it may therefore be more sensible, if the government wished to be able to penalise the concessionaire in some way, for the concession agreement to incorporate a system of fines, or alternatively a system of penalty points linked to some graduated reduction in revenue.

The length of concession contracts often poses a major problem, in that the contract periods are often so long as to make it virtually impossible to predict, and thus to provide for, all of the possible changes in circumstances that may arise. This issue has been considered by a number of eminent legal writers who have questioned whether the classical approach to contract drafting, embodying fixed remedies for breach, is truly applicable. Macniel (1974), in his seminal essay on the subject, draws attention to the differences in the kind and scope of various types of contractual relationship. He makes a firm distinction between what he terms 'transactional' contracts such as buying a newspaper, which are essentially short term, where the action is affected by no past events and concerned with no future ones, and upon the basis of which the classical law of contract has developed, and 'relational' contracts where the parties anticipate a lengthy relationship and where the contract generally attempts, to some extent, to predict the future.

Campbell and Harris (1993) go further, and maintain that the foundation of classical contracting, which is based on the principle of 'presensiation' (in other words an assumption that one can predict the future with some degree of accuracy, and that a contract can be drawn up that adequately apportions the consequent risks and assigns strict liabilities for breach), is totally inappropriate for complex long-term relational agreements. Their review of the available literature leads them to believe that, where long-term contracts do include strict liabilities for future breaches, the contracting parties frequently compensate for their inadequacy by ignoring them when disputes arise in favour of '. . . a repertoire of extra-legal strategies when the liability arises'. They therefore assert that:

> 'Efficient long term contractual behaviour must be understood as consciously co-operative. We see a long term contract as an analogy to a partnership. The parties are not aiming at utility-maximisation directly through performance of specified obligations; rather, they are aiming at utility-maximisation indirectly through long-term co-operative behaviour manifested in trust and not in reliance on obligations specified in advance.'

One consequence of the increasing acceptance of the distinction between transactional contracts and relational contracts has been the increasing inclusion in relational contracts of various types of alternative dispute resolution (ADR) provisions. These mechanisms (for example mediation, negotiation, adjudication, etc.) go beyond the boundaries of classical contract law by recognising that the parties may not be in a position when they sign the initial agreement to foresee all of the possible consequences and to make provision for them. ADR therefore attempts to provide a mechanism for settlement of disputes in such a way that the parties are not necessarily rigidly bound by the terms of the original contract. Such settlements may involve modifying, or even in some cases setting aside, specific provisions in the original contract in the interests of settling the dispute quickly and, hopefully, to everyone's advantage in the light of changing circumstances. However, Campbell and Harris (1993) point out that, in order to make these provisions work, they must be supported by an attitude of trust and a willingness to co-operate, and this again reinforces all of the arguments already made above.

It would appear that in the past such options have been used reluctantly if at all. Gould (1998) reviews the various dispute resolution mechanisms commonly used in BOT contracts. The only options reported in his review are an obligation to negotiate 'in good faith' in an attempt to reach agreement on the changed circumstances, reference of the dispute to an independent third party whose decision may or may not be binding, and, in exceptional circumstances, the right of one party or the other to terminate the agreement.

How detailed should the specification be?

It has already been shown that many governments are apparently uncomfortable unless the works to be constructed and the way in which they are to be operated are closely specified in the concession agreement. The argument for doing this, particularly in the case of concession contracts, is usually two-fold: (1) government will argue that at some point it will have to take over responsibility for operation of the facility, and it therefore needs to know that it will meet whatever standards government has set; and (2) because government needs to ensure that, at the end of the concession, the facility will still be an asset and will not be either time-expired or in need of extensive, and very probably expensive, maintenance. Concessionaires complain that detailed specifications of this sort severely restrict their freedom to provide innovative solutions to problems and thus limit their competitive edge. Whilst this argument may appear to have some superficial appeal, in truth there is, usually, little really innovative engineering in PFI projects, because innovation normally involves risk, and the concessionaire will usually do all he or she can to limit the project's exposure to technological risk.

Allowing the concessionaire the freedom offered by a performance specification, especially for a complex project, may not always be wise. Performance specifications for complex projects need to be written very carefully, with no loose ends, if the host government is not to leave itself open to disappointment. In some types of facility, for example a power station, it is possible for the concessionaire to compromise the efficiency and energy saving characteristics of the plant in the interests of economy. Walker and Smith (1995) provide a good example with a case study of the Shajiao B power station in China which was provided under a BOT contract. In discussing the technical aspects of the plant they write that:

> 'The (Shajiao B) station appears to typify control of the technological risk factors through the use of design and construction methods which are both well tried and economic, although without some of the frills which might be expected in stations constructed elsewhere. . . . The general impression of the plant was that it was relatively simple and old fashioned. The turbine is designed with a relatively low basic level of efficiency and requires more frequent and extensive overhaul maintenance than up-to-date turbine technology.'

On the other hand it is, of course, much easier to specify the operation phase of the concession in performance terms, and here government should be prepared to allow the concessionaire much more freedom of action whilst still safeguarding standards of service.

The British Government, in its PFI deals, has opted for this approach, generally specifying the *service* that it expects the private sector to provide rather than the *facility* to be constructed. In other words government is quite happy to prepare a performance specification, leaving the detailed means of providing the required service to the concessionaire. This is perhaps most appropriate where there is to be no transfer of the completed facility to government at the end of the concession.

To what extent should the concessionaire's bankers be given the rights to take security over the concession agreement in the event of default on their loans?

Generally the host government will, of course, want to control the identity of the party providing services to it under the concession agreement, and it will therefore usually be reluctant to allow bankers rights of this kind. Nevertheless in many schemes the funding will not be forthcoming unless the banks have this level of security. The usual form for arrangements of this type is a direct supplemental 'step-in' agreement between the concession granter and the banks, where the bank itself agrees to take over the concessionaire's responsibilities under the main concession agreement. Vintner (1996) points out that, in setting up such an agreement, the following points merit special attention:

- The supplemental agreement should provide specifically for the banks, rather than a nominee, to step in and must require them to take over *all* of the concessionaire's obligations.
- Provision must be made for the banks to be released from their obligation at some point after their debts have been repaid, and government needs to consider very carefully how it will approach the task of maintaining the service provided by the facility when this point is reached.
- Under this type of agreement both the banks and the government may have rights to take over operation of the project. It is obvious that some provision must be made in the agreement to regulate this position and to establish who has priority rights in the event of a default.
- It may be that the banks are still required to make payments to the project after their step-in rights have been exercised. Some provision must therefore be made in the agreement to establish the legal status of such payments.

Should compensation be payable to the banks in the event of termination of the agreement by the host government?

This question is very difficult, both as to principle and as to amount. As regards the principle, it can be argued that this is one of the risks the bank assumes when it undertakes no-recourse or limited recourse financing, and that the government should not be placed in the position of having to pay a lump sum in compensation when the whole purpose of the exercise was to avoid substantial capital expenditure. On the other hand the banks argue that not to pay compensation would leave the government holding a substantial asset for which it had paid nothing whilst the banks would lose everything. Vintner (1996) sets out a number of legal observations upon this issue, but it will probably only be finally resolved through a legal test case.

Is it possible to standardise PFI contracts?

As experience with the use of PFI contracts increases it should be possible to develop 'model' forms of contract, at least for specific sectors of the market. In the UK, model contracts have already been developed for highways and for hospitals, and the UK Government now publishes a substantial manual of guidance on the preparation of all PFI contracts running to some 284 pages (HM Treasury 2004b).

The purpose of the guidance is stated to be:

- To promote a common understanding of the main risks that are encountered in a standard PFI project.
- To allow consistency of approach and pricing across a range of similar projects.
- To reduce the time and costs of negotiation by enabling all parties concerned to agree a range of areas that can follow a standard approach without extended regulations.

Topics considered are listed below. Each topic is looked at in detail, setting out the government's present position, and in many cases providing standard clauses for the use of contract draftsmen:

- Duration of the contract
- Service commencement and quality management systems
- Protection against late service commencement
- Compensation and relief events
- Warranties
- Service requirements and availability
- Maintenance
- Performance monitoring
- Price, payment for variations, payment mechanisms and set-off
- Changes in service and in law
- Subcontractors
- Assignment
- Change of ownership
- Termination
- Treatment of assets at the end of the service period
- Early termination
- Surveys at expiry or early termination
- Insurance, indemnities, guarantees, etc.
- Information and confidentiality
- Intellectual property rights
- Dispute resolution
- Step-in rights
- Land and other property interests
- Due diligence, commitment letters, etc.
- Refinancing
- Miscellaneous provisions including data protection, third party rights, etc.

Conclusion

PFI is a very complex topic, and this chapter and the previous one have simply attempted to provide an overview of some of the conceptual issues to do with the procurement of privately financed public works. Also it must be borne in mind that this whole area is constantly under development. In addition to the material covered here there are many detailed procedural matters to do with the drafting of contracts, the various negotiation procedures, etc., and those interested in the 'nuts and bolts' of PFI delivery are strongly advised to seek out the specialist journals dealing with this topic for an up-to-date review of the latest position.

References

Allen G. (2001) *The Private Finance Initiative (PFI)*. Research Paper 01/117. Economic Policy and Statistics Section, House of Commons Library.

Allen J.D. (1997) The Private Finance Initiative – the new approach to construction procurement. *The Structural Engineer*, **73** (3).

Arthur Anderson and Enterprise LSE (2000) *Value for Money Drivers in the Private Finance Initiative*. Arthur Anderson and Enterprise LSE.

Audit Commission (2001) *Building for the Future. The Management of Procurement Under the Private Finance Initiative*. Audit Commission.

Bates M. Sir (1997) *Review of PFI (Public/Private Partnerships): Summary and Conclusions*. HM Treasury.

Butler E. and Stewart A. (1996) *Seize the Initiative*. Adam Smith Institute.

Campbell D. and Harris D. (1993) Flexibility in long-term contractual relationships: the role of co-operation. *Journal of Law and Society*, **20** (2), 166–191.

Cole J. (1995) Finance over function. *Hospital Development*, May.

Construction Industry Council (1998) *Constructors' key guide to PFI*. Thomas Telford.

Gould N. (1998) Dispute resolution mechanisms to address unforeseen circumstances in long term BOT contracts. *Construction Law*, **9** (4).

Hancock D. Sir (1995) The Private Finance Initiative. In: *Proceedings of the Institute of Municipal Engineers*, vol. 109, December, Institute of Municipal Engineers.

Highways Agency (1995) *Roads DBFO Projects: Information and Prequalification Requirements*. Department of Transport Highways Agency.

HM Treasury (2000) *Public Private Partnerships – The Government's Approach*. The Stationery Office.

HM Treasury (2003) *PFI: Meeting the Investment Challenge*. HM Treasury.

HM Treasury (2004a) *Value for Money Assessment Guidance*. HM Treasury.

HM Treasury (2004b) *Standardisation of PFI Contracts Version 3*. HM Treasury.

HM Treasury (2004c) *Quantitative Assessment User Guide*. HM Treasury.

Hoare S. (1995) Let's work together. *Construction News*, 7 February.

Hurst P. (1994) BOT, BOO, BOOT – what is a true private-sector project? *Private Finance International, Asian Survey*, October, 14–16.

Macniel I.R. (1974) The many futures of contracts. *Southern California Law Review*, **47**, 691–816.

McTaggart R. (1997) PFI pioneer. *Hospital Development*, January.

Mody A. (1996) Infrastructure delivery: new ideas, big gains, no panaceas. In: *Infrastructure Delivery: Private Initiative and the Public Good* (ed. A. Mody). The World Bank.

Morgan O. and Mathiason N. (2001) PFIs bounty hunters: firms developing schools and prisons pocket millions as the 'risk' factor dwindles. *The Observer*, 8 July.

Morledge R. and Owen K. (1998) Critical success factors in PFI projects. In: *Proceedings ARCOM 98*, Association of Researchers in Construction Management.

Morrison N. and Owen N. (1996) *Private Finance Initiative*. Financial Times Tax and Law.

NAO (2000) *The Refinancing of the Fazakerley PFI*. National Audit Office.

NAO (2001) *Managing the Relationship to Secure a Successful Partnership in PFI Projects*. National Audit Office.

NEDC (1981) *Report of the National Economic Development Council Working Group on Nationalised Industries Investment*. NEDC.

Pollock A. (2000) PFI is bad for your health. *Public Finance*, 6 October, 30–31.

Price R. and Ivison A. (1996) DCMF prisons – a case of maximum security. *Project Finance International*, **90**.

PSPRU (1997) *PFI: Dangers, Realities, Alternatives*. Public Services Privatisation Research Unit.

Reijniers J.J.A.M. (1994) Organisation of public-private partnership projects: the timely prevention of pitfalls. *International Journal of Project Management*, **12** (3), 137–142.

Smith A.J. (1999) *Privatized Infrastructure: The Role of Government.* Thomas Telford.

Smith A.J. and Walker C.T. (1994) BOT: critical factors for success. In: *Proceedings of the East meets West International Conference, Investment Strategies and Management of Construction*, Brijuni, Croatia, Society of Croatian Construction Managers.

Treasury Task Force (undated) *Partnerships for Prosperity: The Private Finance Initiative.* HMG.

Vintner G. (1996) Recent developments in concession agreements. *Project Finance International Yearbook 1996.* Thomson International.

Walker C. and Smith A.J. (eds) (1995) *Privatised Infrastructure: The BOT Approach.* Thomas Telford.

Willets D. (1993) *The Opportunities for Private Funding in the NHS.* Occasional Paper, Social Market Foundation.

15 Construction procurement: Europe and China

Introduction

It is perhaps no surprise that the early development of construction procurement in the UK was heavily influenced by the prevailing legal system, the social and business environment and the evolving needs of the public sector; hence the gradual evolution from the trade-centred construction industry of the Middle Ages, through the development of the independent architect as artistic project designer, the development of those civil and structural engineering skills, many derived from military engineering practice, required to drive the industrial revolution, and the rise of the general contractor in response to the building booms of the nineteenth century. Furthermore it is equally evident that those same procurement approaches would have been exported to the developing world as a by-product of nineteenth and early twentieth century colonial development. We might therefore expect to see clones of the British approach being used in many Commonwealth countries, ranging from parts of Africa (for example South Africa, Zimbabwe and Ghana), through the Far East (India, Sri Lanka, Hong Kong, Malaysia and Singapore) to Australasia and the western areas around the Caribbean such as Trinidad, Tobago, Jamaica and Bermuda.

It is also evident that similar processes would occur in other 'old world' countries, and that the procurement processes that they would evolve might differ in many important respects from those developed in Britain.

In this section of the book we therefore explore some of these alternative approaches, and the ways in which they vary from the British approach, through a review of developments in the most significant of our European neighbours. We also include an overview of the position in China, a communist country attempting to come to terms with a capitalist world and one of the world's largest consumers of construction resources. Chapters 16 and 17 examine the position in the USA.

One of the problems we face in attempting to compare procurement systems across the world is that whilst there have been a number of attempts to develop taxonomies of one sort or another, there is no single generally accepted method of comparison. We have therefore resisted the temptation to develop yet another, and have instead attempted to simply highlight the major variations in regional practice in order to provide a flavour of the procurement that has developed.

Europe

Although it is common to talk of a European construction industry, in fact there are several distinct groupings of countries whose construction industries share common characteristics. Observers have identified four basic groupings as follows (Oliver-Taylor 1993):

(1) The traditional British system (basically the UK and Ireland)
(2) Countries using the French system
(3) The northern European approach
(4) The Mediterranean approach

The British system has already been considered in some detail, and the other systems are considered below.

The French system

France, Belgium and Luxembourg all use variants of the system historically developed in France. The French system has some similarities with the UK approach, but there are important differences.

Statutory and legal requirements are achieved through a more complex and more highly regulated series of relationships than apply in the UK. The architect and contractor are jointly responsible for the condition of the building (excluding fair wear and tear) for 10 years following completion, and decennial (10-year) defects insurance is almost universal. The client's insurer therefore exercises a substantial degree of control over both design and construction. The process can be seen as analogous to the 10-year guarantee offered by registered housebuilders in the UK.

Although under the Roman law based French legal system the term 'contracting system' is not strictly appropriate, it is used here to describe the system that regulates the relationships between the various parties involved in the construction industry. Under the French system the obligations and responsibilities inherent in the relationships between the parties are clearly stated and standard contractual procedures are in place.

In the public sector the designers' contracts often include target costs for buildings which, if under- or overestimated, trigger burdensome penalty clauses. Building contracts are usually fixed price, based on drawings and specifications. The apportionment of risk between the designer and contractor is dictated by the scope of the designer's contract, which may be limited to scheme design or may include full working drawings. It has been claimed that this situation creates defensive attitudes and generates a large number of disputes.

The architect

The role of the architect under this system is usually limited to preparation of general outline drawings and a specification. For public sector projects the preparation of the working drawings will be carried out by a group of specialist designers and engineers working in a *bureaux d'études*. These specialists translate the architect's concept designs into working drawings, and engineers are seen as key players in this process. For the private sector a version of the develop and construct approach is common, whereby the contractor prepares the working drawings from the architect's preliminary sketches to the architect's approval. Concrete and joinery designs are commonly prepared by the contractor.

Contractors

Contractors tend to fall into two main classifications – (1) *gros oeuvre*, specialising in reinforced concrete shell construction, and (2) *second oeuvre*, specialising in fitting and finishing trades – and this division is reflected in French construction education. There is no separate quantity surveyor in the French system. Measurement is conventionally carried out by a 'measurer/checker' (*métreur/vérificateur*) employed by the contractor, whilst cost monitoring is the architect's responsibility.

The use of approval systems for contractors is widespread, and particularly for public works contracts where only properly qualified and registered contractors will be permitted to tender.

The client

The role of the 'client's principal representative' is assigned to the *maître d'oeuvre*, and his or her role may also incorporate aspects of what, under the UK system, would be considered to be the role of either the architect, the construction project manager, the quantity surveyor or the main contractor. On simple projects he/she will usually be an architect or an engineer, but on more complex projects is more likely to be a body composed of several consultants each of whom is responsible for part of the overall role. On major projects there may also be a separate consultant (known as the *pilot*), reporting to the *maître d'oeuvre*, who is responsible for overall construction planning, site organisation and motivation. The fragmentation produced by separate trades contractors each producing their own drawings often ensures that much of the *pilot*'s work is concerned with making sure that all of the pieces fit together and with resolving clashes and dimensional anomalies.

Contractual approaches

Contracts may be let either (1) to general contractors (*enterprise générale*) – in which case it is common for the main contractor to sublet most of the work except the main structure – or (2) as separate trade packages (*lots séparés*) – in which case co-ordination is carried out by the architect.

The latter is considered the traditional French approach. A variation of this system is a grouping of contractors (the *groupement des entreprises*), under which a group of separate trade contractors come together in a joint venture for a particular project.

Advantages and disadvantages of each system are described below.

Lots séparés

This is considered to be the traditional French construction method, and is the most widely used. It resembles the British construction management system in its organisational structure; however, the contractual relationships between the parties differ. Generally the project is divided into separate work packages or lots, which are let to individual trade contractors and are each supervised by the client's representative, the *maître d'oeuvre*, who on smaller projects is usually also the architect. Each trade contractor has a direct contract with the employer, and is selected on the basis of a submitted tender from a list chosen by the client or architect, who will also accept recommendations from the rest of the design team.

The normal method of selecting a contractor is by competitive tendering, the criteria used being much the same as in the UK. Despite the European Union's intention to create a more open market across Europe, there is little evidence of foreign construction firms operating in France in their own right, although it is well recognised that the European construction industries are very closely interconnected, with virtually all of the major construction groups owning substantial shareholdings in each other.

The popularity and continued use of the *lots séparés* system is said to be due to the advantages it can offer to the employer:

- It allows the employer or his/her representatives the opportunity to limit those tendering for a particular package, thus ensuring that it goes to a known and experienced contractor. The concept of nomination is not known in France; there is no contractual mechanism obliging a contractor to work with a particular subcontractor as in the UK. This is important where specialist work is required or there is an element of contractor design.
- It allows employers potential savings, in that by dealing directly with the trade contractors they do not have to pay the additional percentage profit that a main contractor would demand were they the employer's subcontractors. Employers thus benefit directly from subcontractor pricing.
- It allows the potential for time saving, due to the ability inherent in this system to stagger the preparation and letting of the works packages.

The main disadvantage of this system to the employer is the workload. As the employer has separate contracts with each trade contractor, the employer (or his/her *maître d'oeuvre*) must produce contract documentation, drawings and specifications for each trade. This in reality can mean over 100 tender packages, which must be analysed and reported on as they are returned. Once the contractor has been selected, each must be treated separately, with guarantees, retention and payments individually administered. The system is criticised for its labour-intensive administration, and consequently this means that the fee component of the project costs are high, and employers are required to have an organisational structure capable of dealing with this task.

From the contractor's point of view, this system is advantageous because each contractor has a direct contract with the employer. If there are any problems or discrepancies the contractor goes directly to the employer for a response. The contractor is also considered to be financially more secure as his or her payment comes directly from the employer, rather than via a main contractor as in the UK. The main disadvantage for the individual contractors is the contractor's responsibility. If one of the trade contractors causes a delay in a project, it is standard procedure that he/she should bear the full costs of liquidated damages for late completion. This usually offers a powerful incentive to make up time once delays have occurred.

Enterprise générale

This system is very similar to the UK traditional method of procurement. The *enterprise*, the main contractor, agrees to undertake the project for a fixed price. The main contractor is chosen on the basis of producing the most attractive tender. The employer has a variety of options available with this procurement route:

- Full drawings and specifications are prepared before the tender stage, and the contractor is responsible for ensuring that the project is completed in accordance with these. The advantages of this approach are similar to those offered by the traditional system in the UK.
- The tender may be let on the basis of a brief and outline drawings and possibly specifications, and the contract is awarded to the contractor who offers the best proposal. This method requires the contractor to come up with a full design solution, taking the role of the architect and the *bureaux d'études*. This is essentially the same as a variant of the UK develop and construct approach.

Between these two forms, there is a whole range of contractor–design options.

Specific advantages offered by the *enterprise générale* approach over the *lots séparés* approach are as follows:

- The employer and his or her representatives have a single point of responsibility and point of contact.
- There is only a single contract to administer, thus reducing the amount of tender documentation.
- The *enterprise générale* takes responsibility for the subcontractors, and also if required can assist with the co-ordination of detailed design and construction.
- The *enterprise générale* can assist with the design of a project and offer the advantages of buildability when a develop and construct approach is taken.

Despite these advantages, the *enterprise générale* system remains an unpopular option, although its popularity has increased during recent years.

The perceived disadvantages of the system are:

- It is considered to be an expensive way of acquiring a building. Generally, the contract costs approximately 10% more than that for *lots séparés* projects.
- The system requires greater pre-contract preparation time. Usually, the design and documentation must be completed before the tender is sent out. The contractor is therefore not usually involved in pre-contract design.
- Due to the production of detailed pre-contractual documentation, there is a comparatively long initial tender period.

- The process tends to be very sequential, with little opportunity for fast-tracking.
- The contractor is considered to have too much power in this form of procurement.

Groupement des entreprises

The *groupement* is a group of contractors who come together to undertake a project (essentially a kind of joint venture). There are two types of *groupement*:

(1) *Groupement des entreprises solidaires* – a joint venture in which each of the contractors takes an equal responsibility for the whole works, and jointly bears the risk should any of the members default.

(2) *Groupement des entreprises conjointes* – each contractor is separately responsible for a section of the works, but combines with others to submit a single tender for the project as a whole. The works are formally divided up, with no contractor bearing any responsibility for the default of any of the others. This is a more common approach and ranks with *lots séparés* as one of the most popular methods.

The advantages of *groupement* for the client include:

- Reduced documentation
- Shorter tender period
- Savings of cost and time in post-contract administration
- Only requires the analysis of one tender document

The principal disadvantage is the length of time required to produce pre-contract documentation.

The growth in popularity of this system has led to a situation where a client may ask a group of *lots séparés* contractors to come together to form a *groupement* for a project. This is attractive to contractors because for major projects it allows the pooling of resources, and because the *groupement* can take on work that no single contractor could take on individually. The pooling of resources also allows them to achieve savings in such items as site overheads, facilities and insurance premiums. In addition the spirit of greater co-operation often leads to smoother progress on site.

Tender documentation

The client (*maître d'ouvrage*) will normally have decided in advance of issuing tender invitations which contracting system he or she wishes to use. As in the UK, clients may have a list of possible contractors (selective tendering), or they may choose open competition, in which case the tender will be advertised and any contractor may bid. Tender documentation is usually prepared by the architect or the engineer. Because tenders are often let on some form of develop and construct basis, it is not always practical to accept the lowest tender price. In the public sector it is usually the tender that is 'economically the most advantageous' that is accepted.

Contracts

French public sector contracts are based on the public sector procurement code, the *Code des Marchés Publics* (CMP). For private sector contracts the *Association Française de Normalisation* (AFNOR) provides a standard form of contract under the Civil Code. It is important to note that under the CMP and the law relating to the public sector client (*maîtrise d'ouvrage publique*) the client has not only an economic relationship with the contractor but also a set of fundamental obligations which he or she must fulfil. For the public sector client, these include financing the investment, putting in place those responsible for design, technical control and execution of the works, and the letting of contracts for architectural and engineering design.

Contracts incorporate general administrative clauses – *Cahier des Clauses Administratives Générales* (CCAG) – and general technical clauses – *Cahier des Clauses Techniques Générales* (CCTG). To this may be added specific administrative and technical matters relating to the project – *Cahier des Clauses Administratives Particulières* (CCAP) and *Cahier des Clauses Techniques Particulières* (CCTP). These documents together effectively represent all of the information included in the preliminaries and specification sections of the British bill of quantities without any quantities, and are collectively known as the *Cahier des Charges*.

The contract documents, which are not necessarily part of the tender documentation, typically include:

- Articles of agreement
- Particular conditions of the contract relating to the project as above
- Specifications
- Drawings
- A programme of works
- Insurance cover notes
- The contractor's qualification certificate

Contracts are frequently lump sum, with virtually all responsibility for construction placed upon the contractor. Variations are possible if the architect or the client changes his/her mind or specified materials or products are not available. Where variations are ordered contractors are normally required to produce an estimate before the work is put in hand, and it is common for this to be used as a basis for negotiation of the actual price.

Contract sums may be subject to fluctuations according to price changes, especially for contracts over 12 months' duration. Adjustments may be made with the use of indices. Contracts include a retention clause. Retentions are by law limited to a maximum of 5% and are held for one year after handover.

The most popular forms of procurement approach used in France for large construction projects (i.e. those exceeding about £5 million) are as follows:

Public sector:

- Public works contract (*marché de travaux publics*) – the standard CMP contract with employer's design.

- Agreement for the preparation and supervision of public works (*contrat de maîtrise d'ouvre*) – a contract for the preparation and supervision of works generally on the basis of a lump sum price.
- Contracts for the design and realisation of public buildings and works (*marché de conception réalisation*) – a design and build contract. The use of this method is strictly controlled. Basically, design and build may only be used where necessary for technical reasons, for example where the work poses particular technical difficulties and where the skills required are only possessed by the contractor.

Private sector:

- Constructor's agreement (*contrat de louage d'ouvrage*) – uses the standard AFNOR form of contract.
- Property development contract (*contrat de promotion immobilière*) – generally uses bespoke agreements based on a lump sum turnkey contract.
- Project manager agreement (*contrat de maîtrise*) – a bespoke agreement where the project manager takes full responsibility for final delivery of the facility, and enters into construction contracts as the employer's agent.

The situation in Belgium is somewhat similar in that the most popular forms of procurement approach for large construction projects (i.e. those exceeding about £5 million) are:

Public sector:

- Public works contract (*marchés publics/overheidsopdrachten*) – the standard public works contract with employer's design.
- Public/private partnership contracts (*association publique privée/publieke en private samenwerking*) – generally a bespoke joint venture contract used for long-term construction/facilities management.

Private sector:

- Traditional contract (*contrat d'enterprise/aannemingsovereenkomst*) – an independent architect and engineers with contractor are responsible for construction. Ten-year defects liability.
- Design and build (*contrat de promotion/ontwikkelingsovereenkomst*) – a turnkey contract with the building site administered by an independent architect.
- Construction management arrangement (*contrat de maîtrise d'ouvrage délégué/ overeenkomst tot delegatie van bepaalde taken van bouwheen*) – a version of the *lots séparés* approach where the works are co-ordinated by a construction manager employed by the client under a construction management agreement.

Generally in France there seem to be far fewer disputes over contractual matters between the contractor and the client and there seems to be a greater degree of mutual trust than exists in the UK. It is considered that this may well have its basis in the way the construction process is organised and in the insurance system. The predominance of a develop and construct approach means that the contractor cannot blame the client when things go wrong. Additionally, the contract documentation is weighted more in favour of the client and has a higher weighting than is usual with standard contracts in the UK.

Contemporary developments

In France, over the past decade, there has been substantial evidence of concentration in the sector as larger provincial companies have bought out smaller rivals specifically to target more effectively the public sector market. Initially larger firms sought to develop clearly differentiated marketing strategies largely aimed again at securing work in the public sector. Examples include the development of an 'all-in service' including engineering design, financial appraisal and improved mechanisms for project co-ordination and control. Subsequent developments involved further mergers and acquisitions between some of the larger players, and a renewed concentration on the private sector, with construction firms diversifying into property development, property and facilities management and urban design.

At the same time, a variety of new procurement methods have been introduced, including the revival of an ancient form of contract, the *Marché d'Enterprises Travaux Publics* (METP). The METP is in fact a form of build-operate-transfer contract under which local authorities can let contracts requiring the contractor to both construct a specified facility and manage the service that the facility provides, with payments spread over a number of years, after which the facility is returned to the local authority. An example is the use of this approach in schools refurbishment with a 10-year contract period.

The northern European approach

This approach is used in northern Europe, primarily by Sweden, Denmark, Germany, Austria and the Netherlands. Germany differs from other countries in the grouping in that it is a federal republic comprising 16 separate states (*Länder*), each of which has some degree of political and legal autonomy. Law-making and the exercise of regulatory power is therefore split between the federal government and the local administrations.

The northern European group is characterised by very comprehensive sets of national standards which tend to act as 'best practice' guides for the industry. Germany probably has the most highly structured approach, with Austria adopting the least structured. There is a saying that 'in France everything is permitted; in England everything is permitted unless it is expressly forbidden; in Germany everything is forbidden unless it is expressly permitted'.

German regulations governing construction projects (the *Verdingungsordnung für Bauleistungen* (VOB)) are published by the federal government, and include a method of tendering for public works, model conditions of contract, standard specification clauses and performance standards for all major building elements, together with guidelines for measurement. These form the basis for virtually all building contracts in both the public and private sectors.

Control is achieved through strict adherence to statutory requirements, rigid checking by the responsible authorities and by clear definition of roles and responsibilities of the parties in the industry. Standard contractual procedures are in place (in effect there are only two major forms of contract) and the balance of risk appears to act rather more in the contractor's favour than is usual in the UK system. Contractual relations are enshrined in the German Civil Code (BGB), and under German law penalties are enforceable for late completion.

Many projects are let on the basis of plans and specifications, and the architect traditionally carries out most of the pre-contract processes including design, preparation of an approximate cost estimate, preparation of the building performance specification (*Leistungsverzeichnis*) and control of the tendering process. The architect also prepares the contract programme and supervises the work on site on behalf of the client. For smaller projects bills of quantities are sometimes prepared by the architect or the engineer. Quantities may be either firm or remeasured on completion. In Germany contractors are often asked to give details of labour, material, plant and overhead costs for each item. Larger contracts will almost always be let on a lump sum basis based on a performance specification. On the rare occasions when bills of quantities are used for larger projects the quantities will be provided on a 'no liability' basis and contractors use them at their own risk.

In Germany separate trade contracting (*Einzelvergabe*) (similar in many ways to the French *lots séparés* system) is still widely used, with co-ordination being carried out by the architect, and although in recent years the general contractor (*Generalunternehmer*) has grown in importance it has been reported that a high proportion of commercial clients stated that they would prefer to use a 'tender by lots' approach. General contracting is, however, characterised by a high level of subcontracting. Contractors who specialise in working in this way are termed *Generalübernehmer* or management contractors. Also, as in France, there is a distinction between the trades associated with construction of the structure (*Bauhauptgewerbe*) and the finishings (*Ausbaugewerbe*). There are echoes here of the French *gros oeuvre* and *second oeuvre* approach. For major German projects where the contractor is acting as main contractor, the 'responsible site manager' (*verantwortlicher Bauleiter*) has ultimate contractual, managerial, financial and legal responsibility for the work on site. It is claimed by some that this emphasis on the importance of site management has led to an extremely professional approach to construction management, which has in turn given rise to high quality work and very competent staff.

Sweden also has historically used a form of the bill of quantities for some projects. If quantities are used they are usually measured by a 'quantity reckoner' (*massberäknar*). It is common practice for the contractor to be given the right to check the quantities for a project on a stipulated number of days, after which only authorised variations will be accepted. On large schemes the 'building controller' (*byggnadskontrollant*) has replaced the architect as project co-ordinator and is also responsible for site co-ordination and control, progress payments and variations. The building controller is often a building engineer with special training in cost accounting.

In countries using this approach there appears to be a less confrontational approach to contracting, and this may perhaps be due to the fact that the high degree of regulation leaves less scope for claims and disputes to arise. As a consequence it appears that disputes are more often settled by informal negotiation based upon the working relationships developed by the parties, rather than by reference to contract documents. It would appear that there is a greater expectation of team working, and that contractors wish to develop longer-term relationships with clients.

Contemporary developments

Recent years have seen significant developments in the construction industries in this group. In Denmark and Sweden, construction management and design and build contracts have made significant inroads into the market, and whilst partnering and other collaborative forms of procurement are effectively non-existent in Germany, these arrangements have increasingly been used successfully in Scandinavia, in particular in Denmark, Finland and Sweden.

In Germany during the period since reunification, contractors have attempted to seize more control of the development and construction process from the clients by providing more integrated services. This process has seen the development of some design and build contractors (*Totalunternehmer*) and design and manage contractors (*Totalübernehmer*). The limit of this model is the evolution of the *Projectentwickler*, literally a project developer, who takes on the full range of tasks involved in integrating the whole development and construction process.

Public procurement legislation has also been substantially revised in order to bring the legislation into line with EU and GATT requirements. The new government procurement code (*Vergaberechtsänderungsgesetz*) was introduced in 1999.

The Mediterranean approach

Here we group together the construction industries of Greece, Italy, Portugal and Spain. Although there are similarities between them, particularly in philosophy, there is also more diversity in this group than in any of the others. The similarities between them lie more in the ways in which they differ from the UK system rather than in their similarity to each other.

Public administration, except perhaps to a lesser degree in Spain, tends to be very bureaucratic and slow to respond. Procedures for public tenders tend to be lengthy, inefficient and expensive. The private sector is characterised by networks of personal and company relationships which influence how the work is shared out.

Contractual relationships are governed to a large extent by national civil codes. The concept of standard conditions of contract outside of the public works area is often seen as a limit to the flexibility of the parties involved and has not been widely taken up, although standard forms of contract do exist in some areas.

Business relationships are carefully nurtured and maintained, and disputes are therefore subject to very lengthy negotiation. The leisurely pace of the administration of justice in these countries combined with significant inflation rates makes it generally uneconomic to resort to litigation for the settlement of private disputes.

In all of these countries, networks of relationships are fundamental to the way in which business is carried out. Politics and business are very close and those in the network exercise significant control.

Greece

Contractors for public works must be licensed, and the rules governing tender procedures and public works contracts are enshrined in law. Tender documents usually comprise drawings, general specifications, technical specifications and a list of work items that the contractor is required to price. Bills of quantities are not used.

For private sector contracts there is no licensing requirement. Many contracts are negotiated with a single builder, and where competitive tenders are used they are usually taken as a starting point for negotiation. The lowest tender may not therefore be accepted.

All construction work must be supervised by a licensed engineer, who is also responsible for financial matters such as interim payments and settlement of the final account. Quantities of work 'as built' are measured by the contractor and used by him or her to compile a final account which is then subject to approval by the engineer. Prices for variations are generally calculated in accordance with a set of rules for the analytical cost of building works published by the Ministry of Public Buildings and Works. Many contracts incorporate a performance bonus. Retention is typically 10% held until the end of the maintenance period (normally 12 months).

Italy

The majority of the industry is trade based but with some general contractors, one or two of whom are very large worldwide companies. Each site is usually supervised by a director of works (*direttore dei lavori*) who is often an architect.

All contractors are required to be licensed with a central licensing authority, and when tendering must show that the project is within the types that they are licensed to carry out. Before submitting tenders, contractors must usually pay a deposit to the employer, which is refunded to unsuccessful tenderers less a nominal tax. Tenders are let on plans and specifications.

Contracts may be let to a general contractor (who will almost always sublet virtually everything) or on a trade-by-trade basis similar to the French *lots séparés* approach. In this case the client's consultants are responsible for co-ordination. Any disputes are referred to a panel of three arbitrators, one appointed by the client, one by the contractor and one by the court. The arbitrators' decision is binding.

For public work, contractors may be asked to tender either on the basis of their own prices or by quoting a percentage on or off a figure quoted by the client, often derived from a bill of quantities priced in accordance with an official published price list. The successful tenderer is not always the lowest. Procurement law provides a formula for the calculation of a lower threshold, and if the discount offered is lower than the threshold then the tender may be rejected on the basis of a qualitative assessment. If the discounted price is higher than the threshold then it must be accepted provided that all of the tender requirements are met.

No standard forms of building contract exist, although there are 'regional' standard forms published by engineering organisations which are often used as a basis for specific projects.

The Italian construction industry was rocked by corruption scandals involving the procurement of public sector works in the 1990s, as a result of which the industry has undergone substantial change. In addition the government has enacted new legislation designed to control public procurement and to improve the quality of completed projects. It is, however, difficult to judge the extent to which matters have really improved.

Spain and Portugal

Most contracts are let to general contractors who then subcontract extensively. Public works contracts are governed by law, and tender documents must include conditions and specifications, a general description of the works, priced bills of quantity, a contract programme, drawings and the contractor's qualification details. The law also covers procedures to be followed in awarding contracts, setting out and approval, penalties for non-completion, variations, determination, handover, final account preparation and a variety of other more minor issues.

Traditional procurement is by far the most popular approach, but in recent years design and build and management contracting have gained ground, as has the use of PFI techniques for infrastructural development.

Public sector contracts in Spain are governed by a section of the Civil Code (*Ley de Contracto del Estado*), and the public works contracts regulations are published as the *Boletin Oficial del Estado*. Although private sector contracts are also governed by the Civil Code, there is considerable scope for negotiation in the contract terms. The architectural colleges (*Colegios Oficiales de Arquitectura*) publish a standard contract document, the *Pliego de Condiciones Técnicas y Económicas del Colegio de Arquitectos*, for use in the private sector, but evidence suggests that this is mainly used for small projects. Larger projects almost always use bespoke purpose written contracts.

In Portugal, bills of quantities are often provided as part of the tender documentation, and standard conditions of contract and specifications (*caderno de encargo*) have historically been commonly used.

In Spain, contractors are frequently provided with a copy of the official estimate with quantities. Each tenderer then measures his or her own quantities and prices the project in order to arrive at a tender figure that is expected to be lower than the official estimate. Bills of quantities are occasionally used, but there is no standard method of measurement.

Contracts may be awarded either to the lowest tenderer (*subasta*) or, particularly if design is involved, to the most suitable contractor bearing in mind experience, etc. in addition to the tender figure (*concurso*). In some cases the client will estimate the contract value and then approach a suitable contractor to ask if he or she will be prepared to do the work for that price (*contracción directa*).

Private sector contracts in Portugal and Spain are not subject to the same degree of regulation as contracts elsewhere, and most contractual provisions are open to some degree of negotiation. Spanish contracts may be of three types:

(1) A single figure bid stating the length of the contract period.
(2) A tender including the contractor's priced bill of quantities.
(3) A tender incorporating both of the above together with proposals for financial savings.

Tenders may be either 'open' – in which case the work is remeasured and valued on completion – or 'closed' – in which case the bill is taken to be accurate. Contractors will often expect lengthy conversations with the client and his/her technical consultants before submitting a bid.

Architects in Spain have an extremely high status in society, enjoy a considerable degree of legal protection, are highly regulated and on construction projects have almost unlimited power. Both architects and engineers are organised into a series of provincial professional organisations (*Colegios Oficiales de Arquitectura*), and must be members in order to be able to practise. The professional colleges take a substantial role in the procurement process, in that:

- Clients must advise the college when an architect has been selected.
- Before a project may proceed beyond basic sketch design, the client must pay all design fees to the college, which retains a percentage and pays the architect.
- The college, on behalf of the authorities, checks the project for compliance with the appropriate regulations.
- Clients may not apply for a building licence without the college's approval (*visado*).

The colleges are grouped together under a national council (*Consejo Superior de los Colegios de Arquitectura de España*), which has established professional standards, codes of conduct and fee scales, although the latter are being phased out. Separate project managers are rare, with the architect usually undertaking the project management role.

There is no separate quantity surveying profession in Spain; perhaps the closest title and role is the *aparejador*, a kind of technical architect who is conventionally employed by the project architect to produce drawings and bills of quantities for use in tendering. Officially an *aparejador* must also be employed to provide technical on-site direction of the works, but this is often not done. They are, however, often appointed by the client to provide post-contract quantity surveying services such as interim valuations, cost and quality monitoring. As with architects, *aparejador* are organised into provincial colleges (*Colegios Oficiales de Aparejador*).

By contrast, in Portugal, the engineer is the dominant professional, with architects having virtually no role in project supervision.

The People's Republic of China and the Hong Kong Special Autonomous Region (HKSAR)

Historically there was a considerable amount of Western investment in Chinese infrastructure during the period following the Opium Wars in the middle of the nineteenth century, but investment reached a peak following the Sino-Japanese war in the 1890s when the so-called 'Great Powers' (Britain, France, Germany, Belgium, Japan and the USA) all forced their attentions on China. The first flush of investment in Chinese construction ended in 1911 with the establishment of the Republic of China, but foreign investment of one kind or another continued until the Japanese invasion of China in 1937. After the formation of the People's Republic in 1949, when China allied herself with Soviet Russia, virtually the only foreign investment came in the form of Russian aid. Germany, however, still maintained a quiet presence, and the present Chinese legal system was originally largely based upon the German legal code.

Foreign investment really began to take off again in the late 1970s and early 1980s when China decided she needed to modernise her industry and infrastructure and introduced the open door policy. Investment really boomed following the collapse of communism in Russia.

China is a paradox. Officially she is still a communist country, but unofficial capitalism is rampant, and gained something approaching official blessing when the then premier Deng Xiao Ping confirmed that 'to be rich is glorious'.

The Chinese construction industry

Modern mainland China has two separate and distinct construction industries: (1) the internal industry, which is still officially state controlled through the Ministry of Construction, and (2) the external industry, which is largely not.

China is also a major exporter of construction services in the world market, competing for large projects and earning valuable foreign exchange through companies such as the China State Construction Engineering Corporation, which has been involved in more than 400 projects in more than 50 countries.

The construction industry in the HKSAR still continues to generally follow the British approach, although European, Japanese and American construction companies working there have all brought elements of their own approach.

The external construction industry

The external construction industry consists of foreign-funded projects where investors prefer to see their projects built by firms they know and trust, and controlled by techniques they understand. Many projects, including those by external Chinese investors, are handled from Hong Kong, and therefore tend to follow the conventional British system using modifications of the Hong Kong Standard Method of Measurement of Building Works (HKSMM) and standard forms of contract. Construction may, however be carried out by Chinese firms, and project documents will most often be prepared in Chinese. Other nationalities (e.g. American, Japanese and German) sometimes use their own preferred methods, although many use Hong Kong-based consultants and thus follow the traditional British approach. A number of Hong Kong consultants, including quantity surveyors, have now established branch offices in China. One major difference between China and other countries is that China insists that, although foreign architects may prepare designs, they must work in partnership with a state-controlled local design institute. This is seen as essential in order to ensure that buildings comply with Chinese regulations and it also aids technology transfer.

Many foreign construction companies now operate in China, although mostly in joint ventures with Chinese firms. In 1995 the American company Bechtel became the first overseas construction company to obtain a construction licence allowing it to tender for projects without a local partner.

The indigenous construction industry

The indigenous Chinese construction industry works rather differently. Traditionally, under the communist system, private ownership of land and buildings was not permitted. Projects were therefore identified by either central or local government administration or by a 'development unit'. An initial budget price for the work was calculated by one of the state-run design institutes using a standard published schedule of rates, and funds were allocated in the form of a loan from the People's Construction Bank. Once funds were allocated, the project was designed by the design institute, which also prepared schedules of material and estimates of cost by reference to a standard method of measurement and a schedule of rates sanctioned by the Ministry of Construction. Once the pre-construction estimate had been approved, the work was then given to a state-organised construction work unit, materials were supplied according to the schedule and construction went ahead.

This centrally planned approach has been criticised in recent years as being cumbersome, bureaucratic and uncompetitive, and a number of attempts have been made to introduce some form of competitive tendering.

Recent years have also seen the development of the Chinese economy away from communism and towards a system of 'a market economy with Chinese characteristics'. This system appears to be designed to allow China to benefit from capitalist competition while still allowing the state to retain an iron grip on political and economic power. Chinese land law has also been changed, such that some limited private land and asset ownership is now permitted, particularly in the Special Economic Zones, and the beginnings of a development industry are emerging.

The construction professions

China is also beginning to liberalise its policies towards the construction professions. In the past all construction professionals involved in the indigenous industry were employed by the state, all Chinese university courses were geared towards use of the Ministry of Construction standard documentation, and professional associations were banned as subversive. Graduates therefore tended to be technologists and technicians, following rules laid down by someone else, rather than professionals as the term might be understood in capitalist countries, taking decisions that involve professional judgement.

China is very aware that development of its construction industry towards some kind of competitive model will require a truly professional class. Annual state examinations are now being held for land administrators and construction estimators, and anyone with more than 5 years' experience can apply. The intention is to turn existing state employees into 'professional' general practice and quantity surveyors/cost engineers. As far as quantity surveyors are concerned, China claims to have 800 000 estimators whom it would like to turn into professional quantity surveyors (as a comparison the Royal Institution of Chartered Surveyors (RICS) presently has somewhere around 70 000 members worldwide across all specialisms). China is also keen to see its professional examinations and degree courses recognised by professional institutions outside China, and has been encouraging input from organisations such as the RICS, the Institution of Civil Engineers (ICE) and the Chartered Institute of Building (CIOB) for some time.

Opportunities

China appears to offer many opportunities, although a number of studies have been carried out that indicate that Western construction firms are not likely to get very far on their own. Some Western firms have worked in China, but only generally in construction management or where they have skills or expertise not readily available locally (e.g. building services engineering). Many go through Hong Kong, and most overseas joint ventures try to get their affairs governed by Hong Kong law.

The potential market in China is huge, but they also have a very large domestic construction industry which they want to keep fully employed and which they jealously protect from outside intervention. China has also historically had a reputation for never buying more than one of anything. Many businessmen over the past 150 years have dreamed that if only they could crack the Chinese market they would be rich – few have succeeded.

Reference

Oliver-Taylor E.S. (1993) *The Construction Industry of the European Community*. Stem Systems Ltd.

16 US delivery processes critically reviewed

Introduction

The construction industry's performance across the world has been under discussion for some time. Although many improvements have been proposed, there appears to have been very little documented change in results, or success (Post 1998; Hindle and Mbuthia 2002; Jaggar et al. 2002). For the past 10 years, the International Council for Building Research Studies and Documentation (CIB) has been unsuccessful in finding a mechanism to optimise construction performance in Europe, the Middle East, Asia, Australia and Africa (CIB-Programme Committee 2003). The CIB has moved from one potential solution to another, among them: continuous improvement, total quality management, partnering, business process re-engineering and lean construction. The CIB and other construction organisations have not been able to fix the inherent problems of construction non-performance.

Major issues include:

- The inability to finish on time, on budget and to meet the expectations of the building client/user
- A shortage of competent entry-level personnel
- The diminishing value and profit of high-quality construction services
- The declining quality of construction (Post 1998; Green 2001; Rosenbaum 2001)

These issues led to the CIB's latest subject of attention: identifying the construction industry's value.

The construction industry in the USA has fared no better. According to the *Engineering News Record*, 'Although owners/clients were satisfied with their construction quality, many would not hire the contractor again' (Post 1998). No contractor sector – general, mechanical or electrical – received a 'hire back' rating based upon performance of 70% or more. Of these contractors, 42% did not finish on time, 33% were over budget and 13% had litigation pending against them. The gravity of the situation can be quickly recognised when the construction non-performance rate is compared with the rate of non-performance in the manufacturing sector, which is less than 1%. Cost overruns, quality issues and late completions have made the construction industry in the USA very unpredictable (Post 1998, 2001).

Many modifications in US construction delivery processes have attempted to increase construction performance. This chapter discusses the differences between the various construction delivery systems. In the USA these delivery systems include:

- Design-bid-build
- Construction management
- Design and build
- Construction management at risk
- Indefinite delivery, indefinite quantity (IDIQ)
- Job order contracting (JOC)
- Time and materials
- Performance contracting
- Best value procurement (the performance information procurement system (PIPS) is a type of best value procurement that forces a full information environment; see Chapter 17)

Indefinite delivery, indefinite quantity, job order contracting and time and material contracts have been around for the last 10 years (Badger and Kashiwagi 1991). Design and build and construction management at risk are two processes that have had better results than design-bid-build (Konchar and Sanvido 1998). Performance contracting and best value contracts are the latest processes to be tested in the construction industry.

Low-bid award

The design-bid-build (traditional) delivery process awards to the lowest priced contractor. The rationale involves the following processes:

- The designer identifies the commodity by determining products, means and methods, and design.
- All contractors are treated as a commodity, they are all the same, and therefore the best value is the lowest price.
- The client's professional has to minimise the risk caused by the specification and impact of the low price award through management and inspection.

The problems with the low-bid process are that:

- Minimum standards motivate the contractor to lower performance levels. No two contractors offer the same performance. The higher performer, who is on time, on budget and who meets quality expectations, usually has more experience and employs better craftspeople. However, the high performer receives no credit for good performance. Instead, this contractor must reduce quality in order to compete with the lowest price options.
- The process places the client and the contractors on opposite sides, each with different objectives (to be covered in detail in the following section).
- Material and workmanship and methods specifications in combination with owner management and inspection allow inexperienced contractors to bid, where, if they were not directed in detail in what to do, they could not bid or compete on the project. The document that is meant to minimise risk ironically attracts the risk, motivates contractors who bring risk to the client

and puts experienced contractors and craftspeople in a disadvantageous position.

- The professional who tells the contractors what to do, then manages and inspects in order to minimise risk, has actually become the 'magnet' for non-performing contractors. If the non-performing contractors did not have the means and method specifications from the professional, they could not even bid, due to their lack of experience with the project.
- Management and inspection force the client's representatives to make subjective decisions on acceptability. When decisions are made by the client, liability shifts from the contractor to the client.

Best interest of the client vs best interest of the contractor

The low-bid process puts the client and contractor in an adversarial position. Figure 16.1a shows that the client sets a minimum standard, but expects higher performance. The client feels that the minimum standard is required in order to ensure that he or she is not being cheated by low bidding contractors.

Because of the low-bid award, the contractors and manufacturers transform the minimum into a maximum, so profit can be made in three ways:

(1) Lowering the level of performance from what they have offered in the past.
(2) Offering something that does not meet the intent of the specifications but, because of a lack of information and inconsistent standards, is accepted as meeting minimum specifications.
(3) Identifying errors in the requirement and bidding the project low, with the knowledge that in the low-bid system the client will have to issue an instruction or a change order to get the required work done. Change order/instructions are advantageous to the contractor because at the time of the change order/ instruction, there is no competition and the risk is extremely high to the client.

The contractor is in business to make a profit. A contractor who does not make a profit brings risk to the client, because the lack of profit may force the contractor

(a)

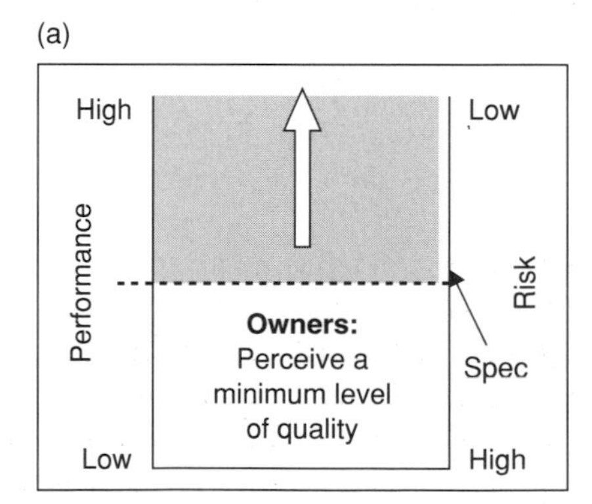

(b)

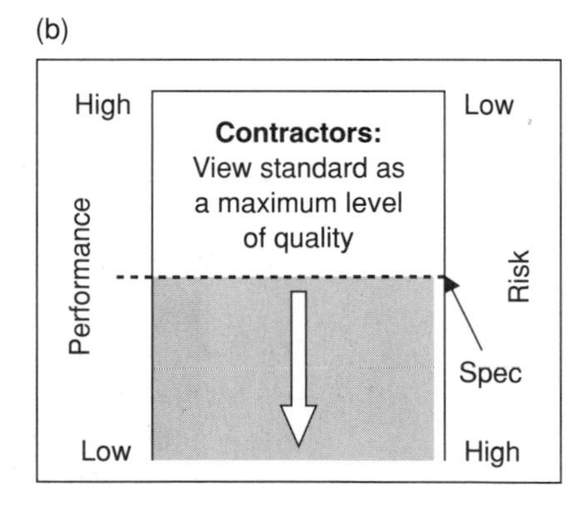

Figure 16.1 Client vs contractors: differences in objectives.

to go out of business. A contractor who goes out of business brings the biggest risk to the client in the long run in terms of delays and unrecoverable costs.

The contractor can increase profits by dropping the performance as low as is acceptable. Contractors will not use their profit to pay for additional quality and performance. If the client/user decides to use a professional to manage, control and minimise the risk of non-performance, he/she is attempting to do something that has been proven to be a very unsuccessful business practice. It is illogical, it is inefficient and it is argued that it has never worked.

The difference in objectives creates a gap. There is no overlap.

Building clients have been dissatisfied with the traditional design-bid-build process. One way to overcome the finger-pointing between the contractor and the designer is to adopt the design and build concept. One party (a general contractor) can be chosen at the beginning of the process and handle both the design and the construction, minimising the amount of change orders. However, to select the design and build contractor so early in the process forces the client to chose a contractor based on a specific product and price.

Once the contract is awarded, any changes to the design product become change orders. As most clients do not know what they want, this becomes a problem. The client loses flexibility, the contractor is being forced to change the design and the parties distrust each other because everyone has different expectations. A common result is that the client appoints another designer to watch the design-builder.

Clients have realised that design and build does not solve the problem of meeting their high expectations. They want more control over the design, but are still convinced that the contractor can help the designer. A process is created whereby the designer is hired by the client to design. The client then hires the contractor before the design is complete in order to assist the designer in making the design constructable. The contractor is then called a construction manager at risk with a guaranteed maximum price (GMP). The process does not motivate the designer or the contractor to act in the best interests of the client.

This chapter discusses the various alternative US delivery processes and compares them with acknowledged successful business practices. They can all be successful, however, if the client understands that the problem is not in the grouping of functions, but in proper selection (using performance information) and outsourcing (minimising management, control and inspection).

The design and build process

Figure 16.2 shows the construction industry in terms of competition and performance. In quadrant I, competition is high and performance is low, representing the results of the traditional design-bid-build (DBB) process. A designer is chosen to design the project. The designer designs the construction requirement as a commodity and awards to the lowest bidding contractor. Under the assumption that all of the options are the same (commodity), the lowest bidder is the best value.

The design usually includes:

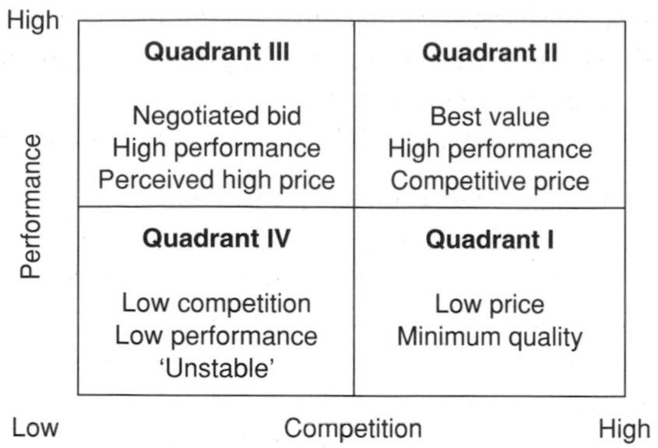

High

Performance

Quadrant III	Quadrant II
Negotiated bid High performance Perceived high price	Best value High performance Competitive price
Quadrant IV Low competition Low performance 'Unstable'	**Quadrant I** Low price Minimum quality

Low Competition High

Figure 16.2 Construction industry structure.

- General conditions, which are contractual terms that identify punitive actions if the contract terms are not met
- Construction requirements that meet the client's intent
- Means, methods and materials
- Drawings
- Specifications

The DBB process in quadrant I also requires construction management and inspection, which is usually provided by the design/engineering firm. The design and specifications are a regulatory document which, when combined with the contractors forced to bid the lowest possible price, encourages the contractors to submit change orders and deliver the lowest possible performance. The construction manager and inspector are required to be experts in construction in the DBB process.

Contractors identify the reason for construction non-performance as the low-bid environment (Erdmann 2002). Due to the poor construction performance of the DBB process, the designer's value has been questioned, leading to reduced fees and design functions (Post 2001). This results in fewer complete designs, which may lead to more risk owing to change orders and lower performance. Oblivious to this cycle, clients continue to look for the lowest design costs.

This trend was highlighted by the *Arizona Business Journal* (Burrough 1995), which concluded that more than 40% of all design projects were won through bidding or price competition. Up 5% from the previous year, 80% of all design firms surveyed said that they participated in some type of bidding or price competition. In addition, more than 90% of architect engineer (AandE) firms surveyed believed fee competition was on the rise. In a market environment of increasing fee competition, AandE firm compensation is likely to decrease. The DBB process awarding to the lowest bidder (quadrant I) results in an adversarial environment for designers, contractors and clients, and has had a very poor performance record.

Price pressure of the worldwide competitive marketplace demands competition in all services. Quadrant II (Figure 16.2) is a performance and value based quadrant in which competition is determined by price and performance. In this quadrant, competition among high performers drives continuous improvement. Quadrant II requires minimal management by the client. Partnering is a perception instead of a function. Quadrant II is a true outsourcing quadrant. The requirement for performance also compels the contractors to be performers, to quality control their own work, to be efficient and to train their personnel.

Movement into quadrant II requires the following process attributes:

- Construction performance information
- Competition based on performance and price
- Client gets best value and contractors maximise profit
- Designer liability and risk are minimised
- Contractors are motivated to continuously improve
- Clients accept liability for their actions
- The risk of construction non-performance is transferred to the contractor

The construction industry's designers and contractors have proposed several processes in order to move construction into quadrant II, including:

- For capital investment projects:
  — construction management
  — design and build
  — construction management at risk
- For maintenance and repair projects:
  — indefinite delivery, indefinite quantity (IDIQ)
  — job order contracting (JOC)
  — time and materials

Figure 16.2 identifies the difference between quadrants I and II as a level of performance and not the functions of the contractor. In quadrant II, the high level of performance requires identification of the difference between competitors. Performance information (on time, on budget, without increases in cost, change orders, meeting quality expectations) must be used. If there is no performance information or the performance information is not used, the process is not in quadrant II.

The performance information must be used at the beginning of construction or during the selection. If it is not, the process is not quadrant II. The use of performance information in the selection process does not differentiate any of the above listed alternative delivery processes. Therefore, all the above processes can be moved to quadrant II or remain in quadrant I. Interestingly, the design-bid-build process can also be moved to quadrant II. It is not the design-bid-build process that causes problems; it is the concept of treating construction as a commodity and awarding to the low price option.

Construction management

Construction management (CM) is when a professional designer, engineer or consultant group represents the client and manages the contractor who is chosen by the client. CM of a low-bid contractor has not greatly improved performance of construction in quadrant I. The additional management of a non-performing contractor does not increase performance, neither does it minimise the client's risk, as the construction manager always passes the risk to the client. Added management does not create a 'win–win' environment.

Under a CM approach construction managers have no real responsibility. If construction goes bad, they report to the client that the contractor is not performing. If there is something wrong with the design, then they can blame the designer. The construction manager carries no real risk. When groups have no risk but are making decisions and directing others, they form a bureaucratic relationship between organisations.

In order to justify their value to the client, construction managers are motivated to identify what the contractor is doing wrong and require that the contractor carries out non-productive functions such as additional paperwork and passing on information that is not needed. This is verified by the poor performance of the low-bid environment and by the movement towards alternative delivery processes (Edrich et al. 2002).

Clients who use CM perceive the need for CM as being due to the low performance of contractors (Figure 16.3a). However, if the client initially hires a performing contractor, the need for a CM team decreases (Figure 16.3b). The client who uses CM is bureaucratic, has a poor performing contractor base and has a very inefficient delivery system. CM is used in quadrant I but never in quadrant II.

It should be used only:

- When the political risk is high.
- When the contractor is not believed to be capable of satisfactory performance without help.
- When the construction risk is very high (not enough funding, not enough time or the contractor cannot meet the employer's requirements).
- When not using CM will provide an opportunity for those with no construction knowledge to criticise the client's project management team.

Construction managers have had minimal impact on construction performance.

Design and build

In the design and build (DB) process, the client hires a single entity called a design and build firm, which is typically a partnership between a contractor and a designer.

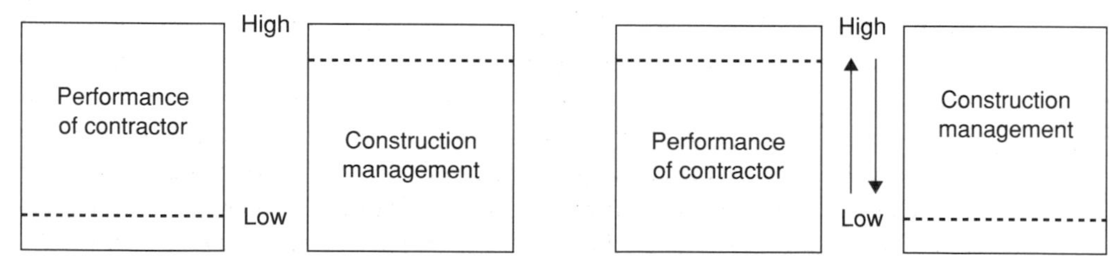

Figure 16.3 (a) The perceived need for construction management. (b) Effect on the need for construction management by hiring a performing contractor.

The contractor is the lead entity due to his or her bonding and insurance capability, which is needed to legally meet the requirements of the construction contracts. The advantage of the DB process is the reduction in delivery time due to the capability of simultaneous design and construction work. Also, procurement of services is done in just one time. Other advantages include one point of contact for the client to deal with and the fact that the selection can be based upon factors other than price in most government projects (Krizan 1998; Crane 1999; Eickmann 1999; Grammer 1999; Gray 1999; Stortstrom 1999; Edrich et al. 2002; Carpenter 2003).

Disadvantages of the design and build process include (Krizan 1998; Grammer 1999; Crane 1999):

- *High bid preparation costs*: many clients do not know how to differentiate between designers, and therefore ask for partial designs, design and construction detailed cost estimates and extensive presentations. Not only are the designers involved, but also the contractors are now involved.
- *Biased selection*: the selection may be more biased due to the lack of performance information.
- *Inflexibility*: clients must select using performance and price and do not know how to differentiate performance. They usually do not use performance information. Therefore they lock in the delivered construction product requirement. The process becomes very inflexible for the client. Any change leads to change orders.
- *No motivation to act in the best interests of the client*: the DB process gives no motivation for the DB contractor to act in the best interests of the client.

DB teams are usually required to prepare costly preliminary designs. Due to the heavy investment required by each DB team, users often prequalify DB teams to a maximum of five bidders. The prequalified DB teams then propose a preliminary design and price. After a team is selected, the process becomes fairly inflexible, unless the client decides to increase the budget.

Owing to the lack of information, inflexibility in the DB contractor's option once awarded a contract and the perceived lack of checks and balances in the process, clients sometimes hire additional consultants to regulate the DB team. The DB process motivates partnering between the contractor and designer but not always between the DB team and the client. Design liability is minimised by the DB team relationship. The process transfers the risk of non-performance to the contractor only if the user can clearly indicate that the requirement was present at the beginning of the project and has not been changed. The client's propensity to change his or her mind in the course of design is the biggest problem in the design phase. Change of the client's requirement now raises the cost due to accelerated and simultaneous construction. A risk of the DB process is that the contractor makes a preliminary cost estimate before a completed design is done, and then attempts to meet the client's expectations within the cost constraint. This will lead to the same problems experienced in the design-bid-build, low-bid award process (Krizan 1998; Ricketts 1999; Angelo 2002; Samad 2002).

The DB process should only be used when:

- The client has a good idea of what he or she wants
- Construction is required in a very short time period
- The client has performance information on alternatives and major components and can use the performance information
- There is no other way to select based on performance and price and move the construction into quadrant II

Construction management at risk

The construction management at risk project delivery method is similar to that of design-bid-build. However, under construction management at risk, the client hires the contractor early in the design phase to assist the designer. Although it is a very subjective decision, the contractor is chosen on a performance basis. The contractor is responsible for providing value engineering and constructability reviews during the design phase of the project. The contractor then takes control of the project once the design is complete, and a guaranteed maximum price is set.

Major advantages of this process are as follows:

- The designer is chosen by the client, giving the client time to make design changes, which do not raise the construction price
- The contractor can be selected on performance and price
- The designer is involved very early in the design stage to assist the contractor
- The process is more like the traditional process than design and build, making traditional clients more comfortable
- The two points of contact offer some checks and balances for traditional building clients

The contractor is chosen later in the process, allowing more competition between contractors. The later the contractor selection, the more information is available, optimising the selection of the contractor.

The disadvantages of this process are that:

- There are two points of contact (designer and contractor)
- There is an absence of competition after the initial selection is made
- There is minimal incentive added by the process to make the contractor perform in the client's best interest

Construction management at risk may not be as widely used as design and build because there are still two points of contact and the requirements to minimise design liability and increase contractor profit are incongruent. Construction management at risk is a contradiction in terms because unless it is done in quadrant II, construction management at risk is not at risk. The guaranteed not-to-exceed price (GNP) or guaranteed maximum price (GMP) has the same connotation as a low-bid contractor who is awarded a DBB contract. If contractors bid too low, they will use change orders to recoup their money or reduce quality. The GMP is a misnomer and gives the client a sense of false security. GMP is a quadrant I term. If construction management at risk is done in quadrant II, it has very few advantages over quadrant II DBB. If the client is in quadrant II, the choices are DB or DBB.

Construction management at risk should be used when:

- The time period is sufficient, otherwise design and build is the method of choice
- When quadrant II design-bid-build cannot be achieved
- When the client does not know exactly what he or she wants to be built
- In quadrant I, when performance information is not used in selection, but selection can be made on factors other than price

Therefore, although construction management at risk will be more successful in quadrant II, rarely will it deliver better performance than DB or DBB in this quadrant.

Indefinite delivery, indefinite quantity (IDIQ)

IDIQ is a contracting mechanism for awarding a contract to a contractor for a well-defined, repeated repair and maintenance function, usually at a fixed price. Contractors bid for the work on a unit price. Once a contractor wins a contract by having the lowest bid, the contractor can be issued repeated work at the unit price if he or she performs well. IDIQ has the following advantages (Anderson et al. 2003):

- It minimises procurement functions: after a contract is awarded based on unit price, the construction can be procured through a work order instead of being bid out.
- It bypasses the procurement functions of design, advertising, bidding and award and thus aids faster delivery.
- It motivates the contractor to act in the best interests of the client. If the contractor does not perform, the client does not have to give the contractor any more work.

Disadvantages of IDIQ include:

- Lack of competition after first award
- It may not bring best value to the client
- It does not encourage contractors to raise their level of performance; furthermore, it encourages contractors to not respond

IDIQ has evolved, and currently clients may prequalify a minimum number of contractors and then bid each work order. This evolution is attributable to the client's mistrust of the IDIQ contractor in quadrant I. It minimises whatever advantages the system has over the traditional DBB or DB processes. This distrust originates from a lack of performance information and the lack of information on best value.

Strengths of the IDIQ process involve the ability to minimise procurement actions. Instead of multiple procurements, there is just one procurement. IDIQ is strengthened by moving into quadrant II and using performance information. Procurement actions are simplified even more due to the minimisation of change orders, negotiations, construction problems and the legal paperwork to protect the client.

However, if the client moves into quadrant II with DBB, the client can get better results because the minimised procurement requirements allow more competition. Whenever there is more competition it brings better value to the client. The advantage of the reduced procurement functions of IDIQ is nullified by the minimised procurement requirements in quadrant II. However, if the client cannot minimise his or her procurement functions, IDIQ is the clear choice of delivery of maintenance and repair or renovation construction in quadrant II. If the client is in quadrant I, IDIQ becomes a 'lose–lose' proposition, as motivation is to minimise price, profit, value and quality. However, when compared with DBB in quadrant I it delivers construction faster with similar quality.

Job order contracting (JOC)

JOC is an IDIQ contract with different types of repair and maintenance work. The contractors bid on a unit price book by estimating the type of work (combinations of functions with unit prices) that will be required by the user over the length of the contract, usually one to three years, with an option to extend (Anderson et al. 2003; Schreyer and Mellon 2003).

The contractors bid by estimating the type and amount of work, using prices from the unit price book, and then adding overhead and profit as a coefficient. The client usually accepts the lowest coefficient.

Once the contract is awarded, the client pinpoints a requirement and has the JOC contractor scope and price the work. The client reviews and approves the estimate and directs the contractor to do the work.

JOC is an IDIQ contract; because it requires management at site, there is a minimum amount of guaranteed work. The advantage of JOC is similar to IDIQ. The JOC contractor is required to maintain an on-site staff that manages the contract. A guaranteed minimum amount of work motivates the contractor to perform, and first-rate performance can earn the contractor additional job orders.

The weakness of JOC lies in the client's misunderstanding of the process, which results from a lack of information about construction and may lead the client to mistrust the contractor. The majority of JOCs are awarded to the lowest bidder. Many contracts fail due to the contractor's low profit margins. Also, clients may not issue enough work orders to cover the contractor's overhead and profit expectation. If they do not receive enough work, contractors may inflate work units and charge higher prices on unlisted items. As a result, clients hire consultants to manage the JOC or award multiple JOCs, making the contractors bid against each other on work orders. These actions minimise the effectiveness of JOC.

Time and materials contracting

The simplest type of contract is the time and materials contract. The client directs the contractor to do work and pays for overhead and profit, time and materials. The process limits procurement actions, encourages the contractor to do good work in order to obtain more work, and allows the contractor to maintain a higher profit margin. Two disadvantages of this process are that the client's representative must manage the contract, and it is very difficult to determine a contractor's performance. This process is optimised if it is moved to quadrant II. If it is a quadrant II process, it becomes like a design and build process (Carpenter 2003).

Performance contracting

Performance contracting is defined as the process whereby contractors bid on a client's performance requirement such as a building, roofing system or maintenance and repair items. There are no detailed directives or means and methods specifying how work has to be accomplished. One objective of performance contracting is to procure performance. This means completing the job on time and on budget while meeting quality expectations. Another objective is the minimisation of management of the contractor and the contract, and the transfer of performance risk to the contractor: '. . . [Performance contracting] is designed to ensure that contractors are given freedom to determine how to meet the Government's performance objectives, that appropriate performance quality levels are achieved, and that payment is made only for services that meet these levels' (OFPP/OMB 1998). Performance contracting is the outsourcing of functions.

One of the major obstacles to using performance contracting is the lack of performance information and the lack of a consistent methodology and structure to compare performance and price. Thus, performance contracting has stayed in quadrant I, and has been relatively ineffective.

Best value selection

The US federal government, one of the largest clients of facilities procuring construction, has had difficulty procuring performance construction through the low-bid design-bid specification procedure. The federal government, through the Federal Acquisition Regulation (FAR), allows best value awards in order to move to a performance-based environment (considering performance and price).

However, the inability of procurement agents to compare value has restricted the use and diminished the effectiveness of best value processes. Best value procurement allows the client to consider factors other than price, such as past performance and a contractor's value, when selecting a contractor. Many procurement agents have difficulty using the best value process because determining a contractor's value can be highly subjective. Procurement agents also have difficulty using performance information, since the information gathered on a unique project may not apply to another project. Many construction and procurement personnel still think that the best value is the lowest initial price (Winston 1999).

The transition from the price-based bid to a performance-based bid has been difficult due to the following (Mather and Costello 2001; personal conversation with FPPO deputy):

- Inability to change the current bureaucracy, where a lack of performance information leads to a low price award. A low price award always requires a minimum specification or means and methods directive.
- Difficulty in identifying performance due to a lack of performance information.
- Inability to minimise subjectivity in the selection.
- Inability to minimise the risk of non-performance (on time, on budget, meeting quality expectations).
- Inability to influence the contractor's performance based on the potential of future work.

Although the US federal government recommends using best value and performance contracting for construction, the process of moving from quadrant I to quadrant II has been unsuccessful due to a lack of understanding about performance and lack of knowledge on how to use performance information without bias.

If best value can be implemented, performance contracting can ensure that best value is optimised. All alternative delivery processes can be used with best value selection (using performance information) and performance contracting (identifying the requirement and allowing the contractor to perform).

Performance Information Procurement System (PIPS)

PIPS is a newly developed information-based procurement system which uses best value selection and a performance contracting approach. It is designed to: minimise management and inspection; minimise the liability and risk of the designer and client; maximise the profit of the contractor; use partnering as a perception but not a function; and transfer the risk to the contractor. It is developed in a way that motivates contractors to improve on every project and do their own quality control. PIPS has a methodology to find the best available value and motivates the contractor to assign the best performers to the project. PIPS uses performance information to select the best value contractor, and then uses the rating on the best value on the project to alter the contractor's future competitiveness and performance. PIPS is fully discussed in Chapter 17.

Comparison of procurement processes

Table 16.1 compares the different procurement processes as they are currently practised, using the seven requirements necessary to move into a performance environment. Process ratings include '2' if the process always involves the criteria, '1' if the process possibly includes the criteria due to the client's representative understanding of quadrant II concepts and '0' if the process cannot incorporate the criteria.

It is interesting to note that when PIPS is used in this analysis the process appears to score very highly.

Table 16.1 Comparison of procurement processes.

No.	Criteria	DBB	DB	CM@R	CM	IDIQ	T&M	JOC	PC	BV
1	True competition for construction services after design	0	1	1	0	0	0	0	1	1
2	Profit is maximised for the contractors	0	1	1	0	0	0	0	1	1
3	Design liability is minimised	0	1	1	0	0	0	0	0	0
4	Contractors are motivated to continuously improve	0	0	0	0	0	0	0	0	0
5	Forces client to be liable for their actions	1	0	0	1	0	1	0	1	1
6	Forces partnering	0	1	1	0	0	0	0	0	0
7	Transfers the risk of non-performance to the contractor	0	1	1	0	0	0	2	1	1
Total		1	5	5	1	0	1	2	4	4

Another comparison of procurement processes was made by using successful business principles developed by eight noted business experts (see Table 16.2):

(1) Buckingham and Coffman, authors of *Break All the Rules* (Buckingham and Coffman 1999)
(2) Phillip Crosby, author of *Quality Is Free* (Crosby 1980)
(3) W. Edwards Deming, originator of continuous improvement and author of *Out of the Crisis* (Edwards Deming 1982)
(4) Peter Drucker, management guru and author of *Post Capitalist Society* (Drucker 1994)
(5) Paul Friga and Ethan Rasiel, authors of *The McKinsey Mind* (Friga and Rasiel 2002)
(6) Jack Trout, co-author of *The New Positioning* (Trout and Rivkin 1986)
(7) Jack Welch, former CEO of GE and author of *Straight from the Gut* (Welch 2001)
(8) James Womack, co-author of *The Machine that Changed the World* (Jones et al. 1991)

Table 16.3 lists the procurement processes that contain the successful business practices. The rating scheme is '2' for the practice embedded in the process structure, '1' if the practice can be used in the process and '0' if the practice cannot be used in the process.

Again if PIPs is entered it scores a very high 40 as it rates 2 for each identified business principle.

Table 16.2 Successful business practices (identified by the experts cited in the text).

No.	Successful business practices	Buckingham and Coffman	Crosby	Edwards Deming	Drucker	Friga and Rasiel	Trout	Welch	Womack
1	Use only the best past performers	X						X	
2	Structure of interactions based on logical goals/expectations	X				X	X		
3	Minimise control, direction, inspection (management)	X	X	X					X
4	Quality is achievable, measurable, and profitable		X						
5	Assist others to be successful		X						
6	Faith in system that works/minimise experts		X						X
7	Quality means customer satisfaction		X	X					X
8	Transactions should be rated relatively (benchmarking)		X						
9	Quality environment is set by the user		X						
10	Zero defects		X						X
11	Quality is measured by the price of non-performance		X						
12	Continuous improvement			X					
13	Do not award on price			X					
14	Break down barriers (team orientation)			X				X	X
15	Minimise minimum standards			X					
16	Knowledge workers have responsibility and liability				X				
17	Identify key factors					X			
18	Look for direction instead of precision					X			
19	Worker or contractor minimise risk					X			
20	Structure should be set up to listen and not dictate					X			
21	Minimise information, maximise responsibility					X	X		X
22	Understand goals in clear terms (performance)						X		

Table 16.2 (*cont'd*)

No.	Successful business practices	Buckingham and Coffman	Crosby	Edwards Deming	Drucker	Friga and Rasiel	Trout	Welch	Womack
23	Simplicity is sophistication					X	X		
24	Minimise functions of insecure who cloud the issues						X	X	X
25	Information reduces uncertainty						X		
26	Competition is required							X	
27	Pay the best more, because they minimise total cost			X				X	
28	Change requirements to change the workers							X	
29	Differentiate people in A, B, C classes							X	
30	Proactive and preventative approach								X
31	Fix problems by finding the cause								X
32	Increase performance instead of dropping price								X
33	Feedback loop for incremental improvement								X

Table 16.3 Characteristics of procurement processes using selected criteria from Table 16.2.

	Successful business practices																				
Process	**1**	**2**	**3**	**4**	**7**	**8**	**12**	**13**	**14**	**15**	**17**	**18**	**19**	**20**	**21**	**22**	**25**	**26**	**32**	**33**	**Totals**
DBB	0	0	0	0	0	0	0	0	0	0	0	0	0	0	0	0	0	0	0	0	**0**
DB	1	1	1	1	1	1	1	1	1	1	1	1	1	1	1	1	1	1	1	1	**20**
CM@R	1	1	1	1	1	1	0	1	1	1	1	1	1	1	1	1	1	1	0	1	**18**
CM	1	0	0	0	0	0	0	1	0	0	0	0	0	0	0	0	0	0	0	0	**2**
IDIQ	1	1	1	1	1	0	1	1	1	0	1	1	1	1	1	1	1	1	1	1	**18**
T&M	1	1	1	1	1	1	1	1	1	0	1	1	1	1	1	1	1	1	1	1	**19**
JOC	1	1	1	1	1	1	1	1	1	0	1	1	1	1	2	1	1	1	1	2	**21**
PC	1	1	1	1	1	1	1	1	1	1	1	1	1	1	1	1	1	1	1	1	**20**
BV	1	1	1	1	1	1	1	1	1	1	1	1	1	1	1	1	1	1	1	1	**20**

Table 16.4 Relative success of the processes.

Process	Total
Design-bid-build	0
Construction management	2
Construction management at risk	18
Indefinite delivery, indefinite quantity	18
Time and materials	19
Design-build	20
Performance contracting	20
Best value	20
Job order contracting	21
Performance information systems	40

Table 16.4 lists the relative success of the processes based on their characteristics of successful business practices.

Outsourcing

The objective of outsourcing is to transfer functions that are not among the company's core competencies. In other words, the functions that are not core competencies of one company can be performed more efficiently by another company (which already performs those functions as its core competencies).

Successful outsourcing is the ability to find companies that are efficient and performance oriented to a specific task, particularly, completing work on time and on budget while satisfying quality expectations. Successful outsourcing is accomplished by providing the experts with the requirements and allowing them to perform, transferring the risk to the expert.

Unsuccessful outsourcing occurs when companies outsource functions to non-performing companies, causing them to manage and control their functions. All the components of a company that manages an outsourced function should be eliminated since these components offer no value.

Instead of hiring another expert to manage and control the outsourced function, the user needs a process to determine the requirement, identify a performing vendor (expert) and communicate the requirement to the vendor.

Industry consensus contends that poor construction is the fault of non-performing contractors (Butler 2002). However, research confirms that the reason for poor construction is the client's inability to competently outsource (Elliott 2001). Clients outsource construction to contractors and then hire managers to oversee the outsourced construction, and since these 'experts' representing the client manage them, contractors are not compelled to accept risk. These 'experts' shift all of the risk back to the client. When the construction is late, exceeds budget or does not fulfil quality expectations, the managing 'expert' is blamed, although the expert rarely accepts any liability.

The low-bid system forces the client to hire managers to extract performance out of low-bid, non-expert contractors. This is the most inefficient way to outsource work because it forces the contractor to use the most inexperienced, inexpensive personnel. The system increases the risk of non-performance and makes it useful to justify the need for a managing 'expert'. Instead of paying a managing 'expert' 10% to manage the contractor, the client can pay a performing contractor (performing expert) 5% more upfront and receive high-performance results at savings of 5% of the delivery cost. The odd man out in optimising efficiency is the managing 'expert'. Clients need to save on cost without increasing risk. They need to dispose of experts who manage and hire experts who perform.

Conclusion

Tables 16.1–16.4 coupled with the documented performance of the delivery processes lead to the following conclusions:

- The design-bid-build process, which uses specifications and low-bid, is not congruent to successful business practices. This process creates an adversarial environment and results in non-performing construction.
- Construction management, the first reaction of facility clients to combat the construction non-performance, is an ineffective method of increasing construction performance.
- Facility clients use alternative delivery mechanisms (design and build and construction management at risk) to increase construction performance. These processes work much better, as shown by the higher use of correct business practices. However, in the long run, these processes do not motivate all parties to continuously improve.
- Facility clients are using IDIQ, JOC and time and materials to minimise bureaucratic actions – that is, repetitive procurement time and control. This in itself improves performance. However, these three processes still do not fully utilise all of the recognised successful business practices.
- Performance contracting and best value contracting are procurement methods that compare performance and price, but still lack some of the characteristics required to force a performance environment.
- An information-based or performance contracting process has the best probability of increasing construction performance. This is discussed in Chapter 17 and reviewed against the results of over 350 tests.

The cause of construction non-performance is not considered to be a construction issue but a business issue. Every other industry has improved performance by using performance information and correct business practices (Tulacz et al. 1997).

References

Anderson N.R., Asmar C., Des Sureau E.P., Kane J.W. and Skolnick S. (2003) Job order contracting. *Washington Building Congress Bulletin*, http://www.jocinfo.com/default.htm.

Angelo W.J. (2002) Team should be contractor led. *Engineering News Record*, **249** (7), 47.

Badger W.W. and Kashiwagi D.T. (1991) Job order contracting: a new contracting technique for maintenance and repair construction projects. *Cost Engineering*, **33**, 21–24.

Buckingham M. and Coffman C. (1999) *Break All the Rules*. Simon and Schuster.

Burrough D.J. (1995) Builders dispute ASU study. *The Business Journal*, **15** (42), 21, 30.

Butler J. (2002) Construction quality stinks. *Engineering News Record*, **248** (10), 99.

Carpenter D. (2003) *General Contracting*. David L Carpenter, Inc: Fine Carpentry, http://www.davidlcarpenter.com/general_contracting.htm.

CIB-Programme Committee (2003) Re-valuing construction. In: *Proceedings of CIB 2003, International Council for Research and Innovation in Building and Construction*, Manchester, http://www.revaluing-construction.com/.

Crane J. (1999) Who says you can't use design-build? *Military Engineer*, **598**, 46–48.

Crosby P. (1980) *Quality Is Free*. McGraw-Hill.

Drucker P. (1994) *Post Capitalist Society*. Harper Collins.

Edrich J., Richard B. and Bruce S. (2002) Weighing the options. *Civil Engineering*, **72** (8), 48–51.

Edwards Deming W. (1982) *Out of the Crisis*. Massachusetts Institute of Technology.

Eickmann K.E. (1999) Secrets of design-build? *Military Engineer*, **598**, 53–54.

Elliott V. (2001) Performance-based outsourcing. *Facility Management Journal*, September/October, 15–16, 18, 20.

Erdmann R. (2002) *The Relationship Between the Design-Bid-Build (DBB) System and Construction Non-Performance*. Unpublished Masters Thesis, Arizona State University.

Friga P.N. and Rasiel E.M. (2002) *The McKinsey Mind*. McGraw-Hill.

Grammer M.E. (1999) Design-build in the Corps of Engineers? *Military Engineer*, **598**, 59–61.

Gray J.A. (1999) Design-build in the public sector? *Military Engineer*, **598**, 67–68.

Green S.D. (2001) Towards a critical research agenda in construction management. In: *Proceedings of CIB World Building Congress, Performance in Product and Practice*, Wellington, http://www.personal.rdg.ac.uk/~kcsgrest/critical-research-agenda.htm.

Hindle B. and Mbuthia G. (2002) From procurement system to delivery system, an important step in the process of construction business development. In: *Proceedings of CIB W92 Procurement Systems Symposium, Procurement Systems and Technology Transfer*, Nottingham Trent University, pp. 169–178.

Jaggar D.M., Ross A., Love P.E.D. and Smith J. (2002) Towards achieving more effective construction procurement through information. In: *Proceedings of CIBW92 Procurement Systems Symposium, Procurement Systems and Technology Transfer*, pp. 179–193.

Jones D.T., Roos D. and Womack J.P. (1991) *The Machine that Changed the World*. Harper Perennial.

Konchar M. and Sanvido V. (1998) Comparison of U.S. project delivery systems. *Journal of Construction Engineering and Management*, **124** (6), 435–444.

Krizan W.G. (1998) Big tests ahead for design-build. *Engineering News Record*, **240** (24), 47, 49–50.

Mather C. and Costello A. (2001) An innovative approach to performance-based acquisition: using a SOO. In: *Proceedings of Conference, Acquisition Directions Advisory, Acquisition Solutions*, Chantilly, Virginia, May, pp. 1–10.

OFPP/OMB (1998) *Statutory and FAR requirements.* OFPP policy letter 92–5, http://www.arentfox.com/quickguide/businesslines/govcont/govcontrelatedarticles/statfarreq/statfarreq.html.

Post N.M. (1998) Building teams get high marks. *Engineering News Record*, **240** (19), 32.

Post N.M. (2001) Bumpier road to finish line. *Engineering News Record*, **246** (19), 56–63.

Ricketts M.L. (1999) Design-build not draw-build. *Military Engineer*, **598**, 65–66.

Rosenbaum D. (2001) No fix for craft labor shortage. *Engineering News Record*, **247** (5), 14.

Samad S. (2002) Managing change orders. *Cost Engineering*, **44** (10), 13.

Schreyer P.R. and Mellon H.H. (2003) *Partner Postner and Rubin, the Gordian Group*. Job Order Contracting. Consulting specifying engineer. http://www.jocinfo.com/default.htm.

Stortstrom G. (1999) Advantages of design-build? *Military Engineer*, **598**, 51–52.

Trout J. and Rivkin S. (1986) *The New Positioning*. McGraw-Hill.

Tulacz G., Krizan W. and Tanner V. (1997) The top clients. *Engineering News Record*, **239** (21), 30–32.

Welch J. (2001) *Straight from the Gut*. Warner Books.

Winston S. (1999) Pentagon pumps up performance. *Engineering News Record*, **243** (14), 10.

17 Performance Information Procurement System (PIPS)

Introduction

The Performance Information Procurement System (PIPS) developed by Dean T. Kashiwagi, Director of the Performance Based Studies Research Group (PBSRG) at Arizona State University (ASU), Tempe, Arizona, USA (Kashiwagi 2004, www.pbsrg.com), is an information-based procurement system that uses best value selection and a performance contracting approach. It was developed specifically for US construction markets but may have the capacity to be adapted for UK and other markets. In broad principle there is some similarity between PIPS and the selection methodologies developed in the UK by the Construction Industry Board and the Construction Industry Council (Construction Industry Board 1996, 1997a, 1997b). PIPS is, however a much more sophisticated and ambitious approach with potentially wider uses, particularly in the maintenance and repair sector where so much UK work is carried out. It uses an information technology that can:

- Minimise management
- Create an environment where the amount of information being shared is minimised, and accountability is maximised
- Transfer risk and force accountability to the lowest levels
- Create a performance environment that encourages craftspeople to work as entrepreneurs and not as hourly workers

What is claimed to make PIPS unique is that it has been tested and documented over a 10-year period of time (1994 to the present) on over 400 projects with a construction value of US$230 million. The documented results include (www.pbsrg.com) the following:

- Ninety eight percent customer satisfaction has been achieved.
- There are no contractor-generated cost increase change orders.
- Projects are completed on or before contract time.
- It can reduce the client's construction management and inspection role by as much as 80%.

Material used in this chapter has been adapted from Kashiwagi (2004). For more information on PIPS, see www.pbsrg.com

- Construction first costs or transaction costs are not related to performance; higher performance did not cost more.
- The process worked in both the private and public sectors.

PIPS is suitable for both maintenance works and capital projects. It is designed to minimise management and inspection, the liability and risk of the designer and client, and to maximise the profit of the contractor, using partnering as a perception but not a function and transferring the risk to the contractor.

It is designed to motivate contractors to improve on every project and do their own quality control. PIPS has a methodology to find the best available value and motivates the contractor to assign the best performers to the project. It uses performance information to select the best value contractor, and then uses the rating on the best value on the project to alter the contractor's future competitiveness and performance.

PIPS is a non-technical, logical process. However, when clients start to use PIPS after years of using a very inefficient process, simple, logical processes are counterintuitive, commonly raising questions such as:

- Why shouldn't we prequalify the contractors and then seek bids?
- What use is it to ask contractors for only their best references?
- Won't the contractors charge more for performance?
- Aren't the project managers the minimisers of risk?
- Why do we have to rate the performance of subcontractors?
- Why do we need so many references?

If a client follows the PIPS process, the following should be logically expected:

- Projects that are delivered on time and meet the client's expectations of quality.
- Higher contractor profit and performance.
- First costs the same or less.

New processes mean change, and the majority of clients' representatives are averse to change. Most managers will not risk forcing change if it does not bring a huge payback.

PIPS has been successfully run in the following areas and by the following clients in the USA (for an updated list and points of contact visit the website www.pbsrg.com):

- Private entities around the Phoenix metropolitan area (Intel, Motorola, IBM, Honeywell, Boeing, etc.)
- The states of Wyoming, Hawaii, Utah and Georgia
- United Airlines
- The Federal Aviation Administration (FAA), US Coast Guard and US Army Medical Command
- Dallas Independent School District (DISD)
- Denver Hospital

The PIPS process comprises a number of phases:

- Set-up of the process and training
- Deciding on the performance criteria to be evaluated
- Collection of past performance information (PPI), bidding and analysis of bids
- Selection of the best value contractor
- The pre-award phase
- Construction and rating of construction

The major reason for the success of PIPS is that performing contractors are motivated to compete with their best craftspeople. The best way to cultivate performance is to hire contractors who can prove previous performance and then let them do their job. PIPS therefore minimises client functions during construction. The contractor's performance is rated at the end of the construction phase and this rating will have an effect upon the contractor's future competitiveness. PIPS creates a documented, closed-loop information system that becomes self-regulating over time The contractors will maintain the process, regulate it and ensure that the client is running the process correctly.

PIPS:

- Is a method of performance information collection and use
- Achieves competition between contractors based on their ability to perform over a period of time on numerous projects and their ability to minimise risk on a specific client's project
- Turns over all risk to the contractor during the pre-award phase, ensuring the minimisation of risk before construction
- Allows contractors to monitor their performance based on the minimisation of risk
- Measures contractors' performance in the critical areas and modifies their future performance

Set-up and training phase

The set-up phase is one of the most crucial phases. If it is done correctly, it will minimise problems during the whole process. The design of PIPS should be carried out by personnel who know how to make it work. When the managers do not understand how to implement PIPS they create opportunities in the process for decision-making. This brings risk to the success of the process. Potential users should:

- Seek the assistance of Arizona State University or a certified PIPS training group. The PIPS/IMT (Information Measurement Theory) technology is licensed and trademarked
- Train a potential core team in the client's organisation
- Define the constraints of the user in policies that may require modifications to the PIPS process
- Identify the performance criteria

- Identify the relationship of the criteria in terms of weights
- Convey to the contractors who wish to bid that the process is an information-based one that will be awarded to the best performer within the budget and that the client is intending to use the process and the performance information in the future

Once the above functions are completed, the client is ready to move on to the bidding and collection of past performance information.

Core group

The client should identify a core team. Success of the implementation depends on having a core group. The implementation has to be done on a test basis, and the core group will be the test implementers.

Client or client policies

PIPS is a selection methodology. It is housed within the policies of the client. For example, the theoretical PIPS encourages hiring the best available performer identified by the selection model.

Policies can be implemented to minimise the risk of the client. Past history of PIPS identifies that there is no construction risk resulting from PIPS. In other words, performing contractors are not a source of risk. Risk (not on time, not on budget and/or not meeting the client's expectations) is caused by three things:

(1) Contractors being surprised, then surprising the client
(2) The client having expectations that do not match reality and then being surprised
(3) Unforeseen conditions

One issue that has been raised is that PIPS will award to the same contractor over and over. However, the environment should prevent this from occurring. If one contractor's performance is very high relative to all the others, one of the following should occur:

- The weights of the performance vs the price should be changed to make the environment more competitive.
- If a project has low risk, the weights on past performance should be minimised, depending more on current ability to do the project.
- The areas of differential should be weighted less, forcing all contractors to continually improve.
- The performing contractor will have an advantage if a project is of sufficient risk and the risk can be defined in terms of past non-performance and the possible cost of non-performance, the performing contractor will have an advantage.

Selection of performance criteria

PIPS selects contractors based on performance and price, and these then constitute the main choice criteria. Performance is made up of the following criteria:

- Past performance of the general contractor and critical subcontractors: site manager, electrical subcontractor, mechanical subcontractor, waterproofing contractor including roofing and waterproofing; other critical systems (high cost, high risk items).
- Current capability to minimise risk on the unique project (which may include assessment through interviews and risk assessment plans).

PIPS is a structure for delivering construction. It covers the design, the competitive bidding, the selection of the contractor, the minimisation of risk before award, the construction and the rating of the contractor, which affects the contractor's future competitive opportunities. Past performance criteria affect the entire cycle, and not only the selection of the contractor.

The objectives of past performance criteria include: (1) identifying whether a contractor has performed in the past; and (2) motivating a contractor to pay attention and perform on the project at hand. The ratings on a project will count towards a minimum of 25% against the future performance rating.

These two objectives must be kept in mind when selecting criteria. Project managers must think ahead to when the project is awarded and constructed, and the contractor is rated. By having the performance criteria on items such as 'close out documentation' and 'documenting risk information during the project', the contractors are motivated to perform and eliminate the risk of the client.

Ten years of testing has led to the generic past performance criteria for all contractors. The criteria below were developed from test cases, contractor expert opinions and analysis of past performance data:

(1) Ability to manage the project cost (minimise change orders)
(2) Ability to maintain project schedule (completed on time or early)
(3) Quality of workmanship
(4) Professionalism and ability to manage: minimisation of the client's representative's time, including responses and prompt payment to suppliers and subcontractors
(5) Close-out process, as-builts, operating manuals and so forth submitted promptly
(6) Communication, explanation of risk and documentation (construction interface reporting on time at least 90% of the time, and accurately at least 90% of the time)
(7) Ability to follow the user's rules, regulations and requirements such as those involving housekeeping, safety and more
(8) Overall customer satisfaction and comfort with rehiring the contractor on the basis of performance
(9) Total number of different jobs surveyed
(10) Total number of different customer responses

Contractors do not have to be resurveyed every time a different type of project is undertaken and the detail of the criteria may be adapted to different circumstances, but will need further testing. The first eight areas identify whether the contractor can perform (performance determined by non-technical terms that the client understands). Criteria 4–7 place the contractor who gets the project at risk. Past tests have shown that most past references give a contractor consistent ratings for all the criteria. However, the contractor who gets the project, and who will have his or her future competitive performance rating affected by the performance on the project, will have to do the tasks in criteria 4–7 well. If the contractor does not do it well, he or she will be downgraded, thus affecting his/her future competitive rating. The last two criteria are related to the amount of experience of the contractor. They minimise the risk to the client by giving credit to contractors who have better past performance.

PIPS considers the number of jobs and the number of different references against the subjective performance ratings on a relative basis. The bottom line is that a contractor with forty references on forty projects and a high rating will prevail against a contractor with only two references on two projects with the same rating. The system will minimise the client's risk. Therefore on risky projects the client is telling the contractors, 'The contractor, who has done this type of work successfully, over and over, will minimise the risk'. This approach forces contractors to show both the ability to perform and repeated performance.

When purchasing specialist systems (roofing, painting, waterproofing, mechanical, etc.), objective past performance information may be considered in addition to the subjective criteria. For example, when purchasing roofing systems, the following performance criteria have been used to minimise risk:

- Documented service period of the proposed roofing system
- Documented performance of the contractor
- Percentage of roofs in database that do not leak
- Percentage of roofs that never leaked
- Percentage of roofs that still leak
- Amount of traffic on the roof system
- Size of the roofs installed

Performance criteria should not be technical. They should be in terms of what clients can express and understand. Any effort to force clients and client representatives to understand technical terms or capabilities is inefficient if the client is attempting to outsource. If something is technical, it forces decision-making (which does not abide with the goals of PIPS).

Bidding phase and past performance information phase

Once the project is selected, the next two phases are run simultaneously. The past performance information (PPI) and the bidding process begin at the same time. The PIPS process is similar to the usual procurement cycle. The only difference is the training meeting where the contractors are instructed on how to submit the PPI.

The next steps in the process are:

(1) Prepare the request for proposal (RFP) (see below)
(2) Include the client's definition of performance, which is the weights on the various criteria, in the RFP
(3) Send out the invitation to bid to the contractors who will attend the initial training meeting
(4) Hold the training meeting
(5) Give the contractors directions on how to turn in the past performance numbers

Then:

(6) The contractors submit the performance information to the client
(7) The prebid meeting is held
(8) The contractors turn in their bid
(9) Interviews are conducted of key personnel
(10) The procurement agent makes the decision on which contractor to select
(11) The best value contractor minimises the risk during the pre-award phase, ending with a pre-award meeting
(12) The contractor is awarded the project and enters the construction phase

Request for proposal (RFP)

The RFP includes the following:

- Request for contract documents (bid price sheet and contractor team composition)
- Design and general conditions for a fully designed project, requirements only for a design-build project
- A short description of PIPS
- Weights defining performance
- Request for a risk assessment plan

Weights for the criteria

The assumption behind PIPS and the prioritisation model is that the client and his or her project are unique. Performance is always a trade-off among the different criteria. Value is a trade-off between the client's definition of performance and price (Figure 17.1). Weighting the criteria allows users to describe to the contractors who they are and their expectations. The weights are unique for each client and each project. They should be weighted as such, and not on anticipated project risks.

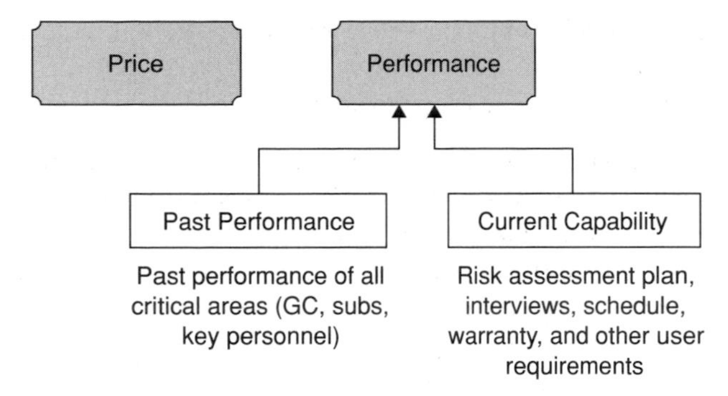

Figure 17.1 An illustration of the weighting scheme.

A general construction project example can explain how the weights are set. The first decision a client or the client's representatives must make involves weighing price and performance. Clients predominantly want to weight price at at least 50% and performance at 50%. The PIPS model analyses relative distances in performance. If there are large differences in performance, the model results will override small differences (less than 10%) in price. If the differences are not major differences, the client's contract person can override the model's prioritisation and award to the second highest performer. Therefore, regardless of the weighting, the following will happen:

- If the top-performing contractor is minimising identified risk through high performance, he/she will be prioritised first even if the weighting favours price.
- If the top prioritised contractor is too expensive when compared with the next highest performing contractor, and the contracting officer cannot justify the differential, the next performing contractor will be awarded the project.

Regardless of the weighting, the contractor who has *best perceived value* will be awarded the contract. The process allows the client to be totally comfortable with the best value contractor. If the client is uncomfortable, and is more comfortable with another contractor, he or she should choose the second rated best value contractor. Value is always a trade-off of performance and price.

The options listed below for weighting performance and price are recommended for projects displaying the following characteristics:

- *Performance based*: performance/cost – 70 and 30%.
- *No bias*: performance/cost – 50 and 50%.
- *Cost conscious*: performance/cost – 40 and 60%. Cost is more important, although performance has a significant impact.
- *Very cost conscious*: performance/price – 30 and 70%. The contractor does not have to be the low bidder to acquire the project but must be very competitive in price. If the contractor is incapable of identifying and minimising risk and has poor past performance, that contractor is non-competitive unless the weight on price is dramatically higher.

- *Price-based*: performance/price – 0 and 100%. The contractor must have a past performance rating but not necessarily as a high performer. Performance on the job for which the contractor is currently bidding impacts his or her future performance rating if the contractor is awarded the project.

The client's second major decision concerns the relative importance of the different performance criteria, which includes: (1) past performance of critical components (explained later in this chapter; see 'Identification of past performance references') and (2) current construction capability to minimise risk on the unique project, including a risk assessment plan (see later in this chapter).

It is proposed that current capability to minimise risk on the specific project that will be completed is much more important than past performance on projects that are different. Current capability involves:

- The unique requirements of the construction project being considered.
- Matching the critical personnel to the project.
- First person information: the client sees the potential matches, whereas past performance is determined by other people who may have different perceptions.
- Comparing the best available values, while past performance involves comparing a contractor against whom the reference has had experience with.

The next decision is to weight the past performance information (PPI) in areas such as (Figure 17.2):

- Site superintendent
- Project manager
- General contractor
- Electrical subcontractor
- Heating, ventilation and air conditioning (HVAC)
- Plumbing
- Roofing contractor
- Waterproofing contractor

The last decision necessitates weighting the current capability areas (proposed weights are in parenthesis):

- Schedule (10%)
- Risk assessment plan document (40%)
- Interviews with the site manager (50%)

The schedule can be weighted more heavily, if schedule is an issue. However, contractors will maximise their profit by finishing as soon as possible. If the required construction time is too short, they will take all the time and try to minimise risk by putting more resources into the project. If the required time is too long, they will shorten it, and maximise quality. However, the chance that one contractor will finish far ahead of his or her competitors is remote.

The risk assessment plan ratings will knock out anyone who cannot identify risk to the client, minimise risk and act in the best interests of the client. The plans should have an impact on the selection, but are not the most influential factor. To be able

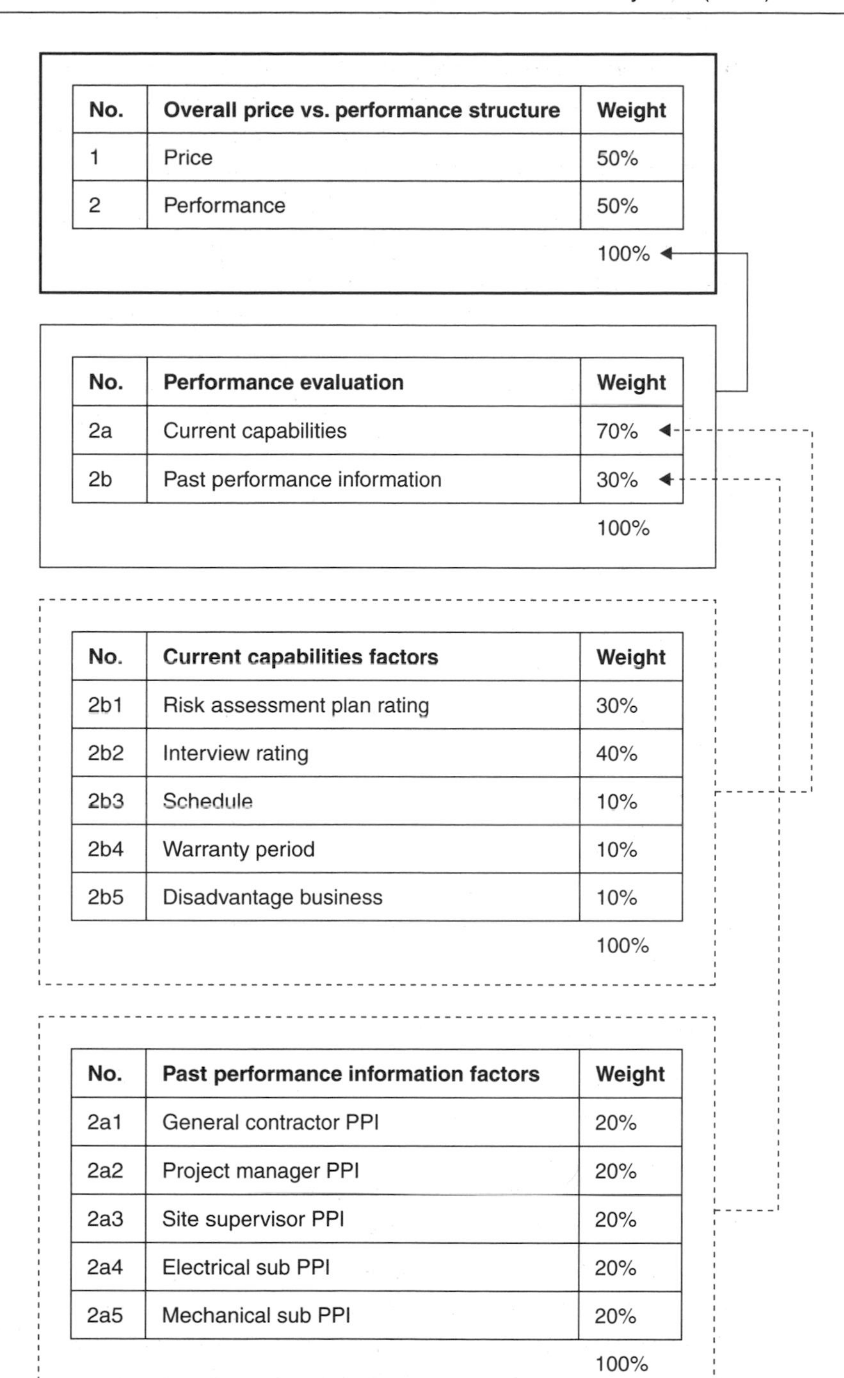

No.	Overall price vs. performance structure	Weight
1	Price	50%
2	Performance	50%
		100%

No.	Performance evaluation	Weight
2a	Current capabilities	70%
2b	Past performance information	30%
		100%

No.	Current capabilities factors	Weight
2b1	Risk assessment plan rating	30%
2b2	Interview rating	40%
2b3	Schedule	10%
2b4	Warranty period	10%
2b5	Disadvantage business	10%
		100%

No.	Past performance information factors	Weight
2a1	General contractor PPI	20%
2a2	Project manager PPI	20%
2a3	Site supervisor PPI	20%
2a4	Electrical sub PPI	20%
2a5	Mechanical sub PPI	20%
		100%

Figure 17.2 Example of a weighting scheme.

to identify and manage risk is a requirement for a performing contractor. However, to have a person on the project that understands and can minimise the risk is even more important. If the key person does not understand risk, the person will have to be managed. If management of the project team is required by either the contractor or the client, the system will be inefficient, value is minimised and risk is increased, because it is not truly a 'win–win' situation.

The interview should be rated the heaviest because it provides the most information. The interview, described later in this chapter, is not a traditional 'marketing and sales' interview where senior officers of the contracting company and marketing personnel give a marketing pitch. It is an in-depth look at the site superintendent or project manager's understanding of risk, minimisation of risk, experience with like projects, and level of perception and ability to predict the future outcome.

Training meeting

All critical contractors (contractors who are required to give references) should be invited to the training meeting including the critical subcontractors. It is recommended that the meeting is mandatory for all contractors desiring to participate. Results of tests show that the subcontractors impact on the performance of construction more than the general contractors. If clients do not invite the subcontractors, they will be increasing the risk of non-performance. The general contractors may often suggest to the client that it is unnecessary to have the subcontractors trained. This is in keeping with the status quo. They do not see it as in their best interest to go to a PIPS system and lose their control over the project.

The objectives of the training meeting are to convey to the contractors the following concepts about PIPS:

- It is not a price-based process. It does not serve the contractor to minimise price at the expense of risk.
- The competitive advantage will go to the performers.
- The performers must minimise the risk of not being on time, not being on budget with change orders and not meeting the quality expectations of the client.
- The client is going to use this information over and over.
- If the contractor cannot minimise risk, he/she will be eliminated in the risk assessment and interview phase or the pre-award phase.
- If the contractor can compete and win based on performance, he/she will have to be able to manage the project by documenting and minimising risk.
- If the contractors do not perform once, they will probably not have another chance.

Past performance information (PPI)

Past performance information identifies whether a contractor has high performing personnel who have performed in past construction projects. Because all construction projects have unique characteristics and requirements, it is obvious that information other than past performance is required to ensure performance on a unique future construction project. Therefore, the past performance information does not have to directly relate to projects very similar to the current project being bid on.

The PPI phase is the second phase in the PIPS process. Although some consider it a qualification phase, it is not a qualification process. No contractors are disqualified or unable to bid unless they do not meet the legal requirements to bid (insufficient bonding or lack of proper licences). PIPS does not unfairly penalise contractors for lack of performance. It simply gives credit to contractors who have more experience.

Identification of past performance references

There are various ways to identify the past performance of a contractor or individual. Some of these options may include:

- Surveying all projects done for a particular client
- Selecting a random number of past projects (good and bad)
- Surveying the last ten projects or the projects completed within the last 2 years
- Surveying only projects that are similar to the type of project being procured
- Allowing contractors to select their best references

It is suggested that contractors should be allowed to identify only their best performing projects.

The difference between PIPS and the design-bid-build or any other process is that PIPS awards to the best value performer. All other processes require minimum qualifications. Minimum qualifications are always subjective, require regulation, are price based, motivate contractors to perform to the lowest possible level and are time intensive.

By allowing contractors to choose their references, the client minimises the following problems:

- Deciding whether the contractor or client is at fault if a project did not result in performance
- Being accused of picking the wrong projects
- Being accused of not being fair with all the contractors
- Being questioned that the number of references required is too many or too little
- Clarifying any 'bad' ratings
- Deciding whether the reference is valid or not

The client instructs the contractors to:

- Pick only their best references and submit no references where there is a chance of disagreement over the performance of the contractor.
- Call the referees before submitting them to verify their performance.

The contractor should turn in at least one reference. However, because the objective of the process is to minimise risk, the process will give the advantage to those with more references and better performance ratings. Once references are turned in, they will not be discarded. They will be used repeatedly in the future to select projects. Past performance is not, however, the only contributing factor in contractor selection.

Contractors have three types of projects: successful projects, unsuccessful projects where the contractor is at fault and unsuccessful projects where the client is at fault. If the contractor is forced to submit jobs that may have been unsuccessful, the client must clarify which party was at fault. Whatever the case may be, this forces the user to do more work and to make subjective decisions.

When selecting the best references, a contractor chooses from a population of projects where:

- The contractor performed – high performance (HP)
- Something went wrong and the contractor was responsible for the non-performance – contractor fault (CF)
- Something went wrong and the client was responsible due to the client's direction, management or inspection – client fault (OF)

The objective of the references is to have contractors show their optimal performance. If contractors send in references with suboptimal performance, they will lose the competitive advantage in past performance. There is no minimum number of references. However, if a contractor has no references, he or she will be severely penalised. A contractor should attempt to turn in at least one reference in performance data, up to a maximum of 40 references (many clients allow contractors to turn in more references at a future point in time).

Contractors with a limited number of performing personnel and high performance projects have a decision to make:

- Do they turn in more references knowing that their average scores may go down? However, staying competitive in past performance will give them a chance in the current capability factors.
- Or do they stay with a low number of references, knowing that the model will identify them as not consistently doing high performance work?

How poorly a contractor does in any criteria depends on the competition. Contractors are placed in a position of making their own decisions, so they must put the best possible combination together.

Even though contractors may not be eliminated by poor past performance, they will rarely win the bid.

In most general construction projects, the subcontractors are ultimately responsible for the quality of the work. A high performing general contractor with poor performing subcontractors may be as risky as a lower performing general contractor with high performing subcontractors. The relationship between the general contractor and the subcontractor is no different than the relationship between the client and the contractor. This is another 'win–lose' relationship, which may cause construction non-performance and the instability of the industry. Any requirement to manage another party's performance leads to inefficiency, and it would have been better to pay more for performing subcontractors than trying to manage performance of cheaper subcontractors.

For large or complex projects, past performance information should be collected on all of the critical components, for example:

- General contractor
- Site manager
- Electrical subcontractor
- Mechanical subcontractor
- Roofing/waterproofing subcontractor

These components may not be critical on every project since each project is unique. Some of the benefits of collecting performance information on all critical components are in order to:

- Ensure that all critical elements consist of performing groups and individuals, minimising the risk of non-performance.
- Allow clients to give all contractors the same rating at the end of the project when all elements have a performance rating. This makes everyone responsible for everyone else's work. It assists the project to have many 'eyes' instead of just those of the general contractor.
- Motivate subcontractors to do their work in a timely fashion.
- Bring together a group of talented people who have pride in their work.
- Develop a synergistic relationship between the general contractor, subcontractors and key personnel.

The PIPS process

The PIPS process:

- Forces contractors to accept responsibility for their references
- Compels contractors to know and find out about their performance from their references
- Transfers the liability of data collection from the project manager to the contractor
- Prevents client bias entering into the past performance numbers
- Defuses contractor protests and arguments about the inequity of the process

The only way that contractors can improve without outside regulation is by knowing their own performance, and improving their performance in order to increase their profit margins.

The client's representatives should only verify the references that put the client at risk. These include:

- The winning contractor's team references
- Any team that is close to the winner
- Any reference score that has a large impact on the spread of values

Rating the past performance of contractors

The subjective rating of a contractor should range from a high of ten to a low of one. It is proposed that the rating system should be very straightforward. The following rules apply:

- If the previous client has no problem about hiring the contractor again based on the contractor's performance, the rating should be a 10.
- If the previous client is not sure about hiring the contractor again, the rating should be a 5 or somewhere in between 5 and 10.
- If the previous client does not want to hire the contractor again based on the contractor's performance, the rating should be a 1.

These instructions are given to all past referees. The contractor could also explain this verbally to the referees. Many referees will not give 10s. The contractor is responsible for choosing clients who will give 10s and other very high numbers, which show that the clients would hire the contractor back without any hesitation.

If 1–10 ratings cannot be used, whatever process is used (colours or descriptive words) is transformed into relative numbers and fed into the prioritisation and information generation tool.

Prebid meeting

The client's representative can pass out the RFP at or after the training meeting. The prebid meeting should be mandatory for general contractors and highly recommended for critical subcontractors. Over time, the high performance subcontractors will start attending the prebid meetings. If it is a very complicated project, which requires substantial work by a critical subcontractor or where the risk in a certain area is high, it should be mandatory for the critical subcontractor to attend. The prebid meeting should include the following:

- A brief project description by the designer
- A review of logistics, operational conditions, schedule and clients in the area
- Q&A with the contractors regarding any information that the contractors may request
- A site walk if possible
- A quick review of the risk assessment plan requirements, the key personnel interviews and the pre-award phase for the selected contractors

It is imperative that the contractors understand that the risk is being passed to the contractors in order to be minimised. It is also important for the bidder to realise that this is not a price-based selection, even though price is important, will be considered and may be the deciding factor if all contractors 'look the same'.

Capability to perform the required construction

Past performance information reduces the field of performers without using minimum requirements. If contractors' past performance information shows non-performance and they wish to compete, they must first prove that they can perform by performing on projects and submitting new references. However, past performance does not match up performing contractors with their performing personnel and the unique project requirements. To accomplish this using past performance information requires too much work, too many data and too much management and maintenance of the information.

Tests have shown that if past performance information is used without subjective translation, and past performance will be used in the final selection and affect their future competitiveness, contractors will bid on a project only if they can provide the expertise and perform to bring about a 'win–win' outcome (best value for the client and profit for the contractor).

The combination of past performance, risk assessment, interview and the responsibility to accept and minimise all the risk is an impossible task for non-performers. Once the field is reduced to those contractors who can perform on the unique project, the contractors compete under PIPS based on their current ability to identify, prioritise and minimise risk (based on time, expectation and budget).

In specialised or one system retrofit areas such as roofing, painting, waterproofing and mechanical divisions, the proposed value of the system, which includes the perceived value of what is being offered, may also be factored into the current performance.

It is important to impress upon contractors that if they do not have good past performance, a risk assessment plan and/or key personnel, the only way to win the project is by bidding low.

The bidding phase

There are very few differences in the bid phase compared to the traditional process. The bid includes:

- Legal requirements.
- A risk assessment plan.
- Price.
- A list of critical components including name of site superintendent, project manager and critical subcontractors. This cannot be altered unless there are unforeseen conditions as determined by the client.
- A general schedule including total number of construction days (total days).

The biggest difference with PIPS is the time between the bid and the award. This process takes a couple of weeks to identify the top prioritised contractor and another 1–3 weeks for the pre-award phase before the contract is actually awarded.

Value added features by critical subcontractors

There is frequently a problem in the price-based environment where subcontractors may identify value added concepts and pass them along to the general contractor with whom they are bidding. The general contractor then passes the ideas to a price-based contract, eliminating any competitive advantage of the performance-based subcontractor.

In the PIPS process, all contractors can submit value added concepts directly to the client's representative before bid day. The concepts are reviewed, and the concepts that are not approved are published to everyone. As a part of their bid to general contractors, critical subcontractors can now submit the added value of their bid (in monetary terms) plus their price. After being selected, the general contractors who hire the value added subcontractors can incorporate the concepts into their risk assessment plan.

The risk assessment plan

The risk assessment plan (RAP) includes five major components:

(1) Identification of the chance of the contractor finishing on time, on budget and meeting quality expectations
(2) Method of risk minimisation
(3) Value engineering (VE)
(4) Cost breakout
(5) Construction schedule

The entire document should be no longer than two pages for a design-bid, systems project such as roofing, waterproofing, painting or modification. It should be no longer than ten pages for a £100 million project. This is a high level business plan and should be concise, well organised and showing an understanding of the magnitude of risk.

The RAP should identify how the contractor knows that the risk is a risk, how the risk can be quantified in terms of cost, time and quality, and how the contractor will minimise the risk. For example:

- Commissioning is a problem. The contractor identifies that four out of five similar projects have commissioning problems. The site superintendent proposed has a record of 95% success with commissioning, is the contractor's expert in commissioning and performed commissioning on five other recent projects.

- The relative cost of a similar facility is £85 per square foot. The cost can be reduced by an accelerated schedule. These factors must be applied. The site superintendent has finished five similar projects, with cost savings of £10 per square foot.

The RAP identifies current capability. Current capability is considered to be more important than past performance for the following reasons:

- The risk assessment plan identifies if the contractor has the right people, the right solution and can do the unique project at a specific time.
- Each construction project is unique. As similar as projects may seem, there are always major differentials in terms of critical staff, the client's staff and expectations, delivery schedule, extenuating circumstances, complexity and subcontractor performance.
- Contractors are asked to submit only their best references.
- The contractor project manager and site superintendent may be different on the subject project.
- The type of procurement – low-bid, negotiated or performance based – impacts on performance tremendously.
- Weighting the current capability more heavily than past performance compels contractors to submit their best personnel on the project.

An RAP review committee is selected at the beginning of the procurement process. It should consist of two or more individuals, depending on the client's rules. The user of the building and the building client's project manager should be on the committee. The committee should be the same committee that rates the interviews. Committee members should rate the RAPs individually and should be briefed on how to do this. The definition and purpose of an RAP should be explained to them.

An RAP should:

- Be a concise, well organised business plan
- Include identification of risks, minimisation of risks, prioritisation of risk and added value
- Prioritise the risk in terms of cost, time and expectations
- Be in non-technical terms
- Identify differential

The committee reviews and rates the RAPs in terms of relativity.

Interviews

Interviews are not mandatory in the PIPS process. In some cases, the amount of risk minimised is not worth the user and contractor's time and effort, since risk is already minimised by the other steps in the PIPS process. However, for major, complex projects, the interview is a key component. The factors to consider when deciding whether to interview the contractors are:

- The number of bidders
- The amount of time in the bidding schedule
- The number of committee members and the time each member has to participate
- The amount of risk involved
- The complexity of the project
- The number of subcontractors involved
- The scope of the project in terms of price and value to the client

Questions that should be asked in the interview may include:

- What is your background with the current company?
- Why were you selected for the project?
- What are the most critical components of the project?
- What are the largest risks on the project?
- How do you prioritise the risk?
- How do you minimise the risk?
- How do you know that they are risks?
- How did you choose the subcontractors for this job?
- How much information will you pass on to the user's representative?
- What do you expect from the user's representative?
- How is this project different from previous projects on which you have worked?
- Are you uncomfortable managing the construction project and having no construction manager from the client's side?
- What are your company and individual goals on this project?
- What will you do differently on this project from previous projects and why?
- What are your strong and weak points, and what are you doing to minimise your weak points?

The questions should be the same for all interviewed personnel. However, the client's representatives should be allowed the freedom to ask further questions that force the interviewers to elaborate on their answers if the interviewers feel the site superintendents and project managers have not answered the question to their satisfaction.

The interview serves as an opportunity to assess how well the site and project manager understand the risk assessment plan. It is also an opportunity to see how close their perception mirrors the risk assessment plan. Moreover, the interview will show their experience in doing the unique type of construction required.

The rating should be done in a similar manner to the risk assessment plan. An average site manager should be given a 5; a superior site manager should be given a 10; if a site manager is not desired, he or she should be given a 1. Ratings can be given between high, average and low performers.

The interview is the most critical information step. It should be weighted heavily. No contractor should be awarded a project without key personnel with good interview ratings. The best value should also have great risk assessment scores and past performance. If the best value does not have the best interview score, the best risk assessment plan and good past performance, this should be identified as a risk. If the price differential is great and a lower value is selected based on price, this decision also brings risk. If the price of the contractor is far below the average price of the other top performers, this can be identified as risk. Risk occurs when contractors, in giving the client best value with too low a price, cannot make a decent profit. If this is the case, the price-based contractor who is selected must have his or her plan down to the *n*th degree, because there is little leeway for surprises or inefficiencies.

The pre-award stage

The pre-award period is designed to shift all the construction performance risk to the contractor. If he or she is a performing contractor, he/she will know how to minimise the risk. The pre-award phase is very important for contractors. It is their project to lose or forfeit. The objective of the pre-award period is to allow contractors to verify that they can perform on the project and meet the user's expectations. In traditional construction, too many times the designer, project manager or contractor will want to dive into the construction, only to discover that better planning would have saved money and time. Diving into a project without having the construction experts review and coordinate with the designers results in confusion, non-performing work and a lack of responsibility.

The following activities should be carried out by the contractor:

- Coordinate the design and drawings with all critical participants.
- Clarify all conflicting drawings and any construction that cannot be constructed. Force the designers to finish their design work.
- Check the critical lead time items.
- Identify what is required from the user.
- Be able to minimise all risks identified in the risk assessment plans.
- Ensure understanding of the contractor interface.
- Introduction to the key user, client and project manager personnel. Confirm and understand their expectations. Communicate the risks and expectations concerning the project.

The pre-award phase could last 2–3 weeks. It should culminate with a non-technical presentation by the project team to the user, client and the client's project manager. All questions by the user should be given to the contractor in advance. The pre-award presentation should be attended by:

- The client's project manager
- The contractor team and key critical components
- The designer
- The facility user
- The client

The project manager should control the meeting. The following should be reviewed:

- Members of the construction team and reason for their selection
- Risk
- Minimisation of risk
- Prioritisation of risk in terms of time, money and expectations
- Schedule
- Key roles of critical components
- Major systems
- Value added items
- Requirements and assistance of the user
- Construction interface
- Contractor, site superintendent and project manager goals

Contractors must understand that they regulate the performance of the project. By identifying risks in risk assessment plans, a contractor forces all other contractors to minimise that risk. The contractor who receives the project is required to minimise *all* of the risk identified by the competing contractors. If a contractor perceives that another contractor may lower the price by not performing the 'same' job, the contractor can force the low bidding contractor to perform the function by listing it in his or her risk assessment plan. This process quickly forces non-performing contractors to either perform or withdraw.

The contract

The construction contract should include the following:

- The client/designer's requirement and intent (design)
- Design drawings and specifications with modifications
- An approved list of submittals
- Information interface format and frequency
- The contractor's final risk assessment plan
- Pre-award meeting minutes
- All clarification documentation
- Schedule
- Cost
- User's legal contract documents, such as bonding and insurance

In order to minimise changes in the status quo prescriptive specifications, the contract may include the client's general conditions with the statement that the client can modify the general conditions to assist the contractor to perform in the best interest of the client as determined by the client's representative.

During construction

During construction, the client, project manager and designer's objective is to assist the contractor to be successful. If the contractor is a high performer, the work of the client's representatives is minimised. If the contractor is less performing, the work of the client's representative is increased. There are two options for the client's representative: if the contractor does not perform, (1) penalise the contractor financially or (2) assist the contractor to become better performing.

Option (2) leads to the client's representatives understanding how to bring change. The PIPS process environment will minimise the amount of work for the client's representative. By working with the contractors, the client's representatives will begin to understand that there is a difference among contractors. The objective of PIPS is to deliver performance. Inspectors can still inspect; however, they should be more like facilitators. The contractor should make most of the decisions and should ask the opinion of the client's representatives only when it involves information that the project manager has, and which has not been given to the contractor.

Updating performance information

The most important duty of the PIPS core team is to update and maintain the contractors' and manufacturers' performance information in the PIPS system and in the price-based environment. The following data should be compiled:

- Change order rates with and without PIPS
- Customer satisfaction ratings with and without PIPS
- Design costs with and without PIPS
- Project costs with and without PIPS
- Project performance with and without PIPS

Conclusion

In summary, the objectives of the PIPS approach are to:

- Select the best value contractor who can minimise the construction risk of the client
- Assist the best value contractor to minimise his or her risk of non-performance
- Ensure that the contractor uses the best practices principle of reviewing the project in detail
- Minimise any misconceptions and potential problems before construction

PIPS uses performance information to select best value contractors and should result in projects delivered on time with higher performance and within budget. Examples of the use of the PIPS system are illustrated in the Appendix.

References

Construction Industry Board (1996) *Selecting Consultants for the Team: Balancing Quality and Price*. Thomas Telford.

Construction Industry Board (1997a) *Code of Practice for the Selection of Main Contractors.* Thomas Telford.

Construction Industry Board (1997b) *Code of Practice for the Selection of Subcontractors.* Thomas Telford.

Kashiwagi D.T. (2004) *Best Value Procurement*, 2nd edn. Performance Based Studies Research Group.

18 Summary

This book has highlighted the importance of construction to quality of life and to the economy of a country. The review of the UK construction industry has shown a diversity of type of demand and the complexity of supply.

Most customers (usually referred to as clients) are inexperienced in procuring construction. Only a minority of clients are regular purchasers of building work and are able to use their relatively consistent and sizeable buying pattern to develop relationships with the supply side and to lever real advantage through collaboration and strategic alliancing.

The supply side is dominated by small firms, often specialist in character, who focus only upon their part of the project and usually work as subcontractors to larger firms. Larger construction firms act as contractors and tend to take on the role of coordinating these subcontractors. This diversity results in an industry where market conditions can be very varied and where the majority of customers are disadvantaged due to their relative inexperience and ignorance of the processes involved in procurement.

In most cases investment in a building project is potentially a very high risk, particularly to this inexperienced majority of clients who seek to procure construction relatively rarely and on whom the extent of expenditure has a significant impact in terms of their financial structure. Nonetheless real estate has proved to be better than most other forms of investment over time and the value of an efficient and functional building can be significant to most clients. Saxon (2003) illustrates the enormous multiplying factor where good design delivers high value.

Performance of the industry in terms of benchmarked project delivery suggests a need for improvement. Government, the industry and its representative bodies have been seeking performance improvement over the last 12 years and enormous debate has taken place, with a number of reports and proposals being published. All these have focused upon process and there is now a wealth of useful advice available to clients and their project teams, emphasising amongst other aspects the benefits of collaborative procurement processes in achieving value for money. An excellent example of the sort of advice available is the ClientZone section of the Department of Trade and Industry-funded website of the Constructing Excellence programme, where the industry's key performance indicators can also be viewed (www.constructingexcellence.org).

Construction professionals appear, however, to be somewhat resistant to change and there has been a tendency to pursue traditional practices, which are known and well documented, even if they are less relevant in the market or to the economic conditions that exist at the time. Equally, investment in research and innovation is much less than that in the manufacturing sector, perhaps as little as one tenth of that expended elsewhere, and mostly emanating from the work of product suppliers or specialist firms rather than design or construction teams.

In this scenario the difficulty associated with achieving the right building delivered at the right time for the right cost is significant. The key to a successful outcome rests with the client who must:

- Identify the key performance indicators against which project performance can be judged, consistent with the business case for the project.
- Attempt to establish realistic time and cost parameters.
- Adopt an appropriate procurement strategy that reflects the above two points as well as the client's attitude to risk.

Procurement strategies that are most commonly adopted have been summarised and reviewed, but as Hughes (2005) points out, there are so many potential variables in each case that there are a wide range of amendments made to the standard model.

There has been a tendency for inexperienced clients to focus upon initial capital cost and to ignore both long-term output value and use costs. Perhaps this is understandable when most clients purchase products rather than projects. Additionally they have been less aware of the importance of establishing a bespoke procurement strategy for their project. This has resulted in the dominance of price-led procurement approaches to both design and the selection of consultants and constructors, which has not encouraged collaboration or much focus upon a best value output.

Europe has an increasing influence on UK law and to some extent upon how things are done. There is no doubt that the UK construction industry can learn from some of the commonly adopted practices in those European countries with large and sophisticated construction industries. This book has summarised these practices and particular attention is drawn to the way in which the interests of clients are protected by mandatory responsibility and insurance.

Dean Kashiwagi has provided a section to the book that illustrates how the USA commonly approaches procurement but also some radical new approaches which have been successfully applied. Perhaps the strongest message from Kashiwagi and his team is that clients should focus upon the importance of selecting a project team with a proven track record and paying a fair price for the service provided. If they do, the project team may also increase its focus on the client as customer.

There are subtle differences in European practices and in terms of the way Kashiwagi has expressed the new thinking he recommends, but both emphasise less the cost of the project and more the importance of the value to the client of the output, through facilitating increased collaboration throughout the supply team. These are important messages which may improve client satisfaction, but therein is the problem. Clients do not consistently belong to organisations where advice for occasional clients (the majority) is readily available. Many clients wrongly assume that they can obtain reliability in terms of early estimates, programme and functionality in the purchase of a new building even though design is incomplete. They rely upon construction professionals to advise them and confirm that, as the Construction Clients' Forum (1998) expressed, 'clients do not like unpleasant surprises'.

Building procurement is a project-based activity using a temporary project team. The client provides the rationale for the project, initiates the process and usually will be the end user. Wherever possible the client, as the consistent influence throughout the life of the project, should be empowered to achieve a successful output through appropriate and timely advice from carefully selected advisers.

It is likely that future construction supply will increasingly be provided by 'fix it' specialist builders able to employ and train their employees, by smarter builders focusing on niche markets and by an increase in the development of design and build joint ventures brought together for larger projects.

Where clients are able to develop an increased awareness of the importance of their role in the process and a consequential empowerment in terms of setting project strategy they will be increasingly able to achieve project success in terms of their own key performance indicators.

References

Construction Clients' Forum (1998) *Constructing Improvement*. Construction Clients' Forum.

Hughes W. (2005) Keynote address. In: *Proceedings of 5th Annual Construction Marketing Conference*, London. Marketing Works Training and Consultancy Ltd.

Saxon R. (2003) *Povey Lecture*. Joint Contracts Tribunal.

Appendix
PIPS case studies

Case study no. 1: United Airlines

Introduction

The United Airlines (UAL) San Francisco Maintenance Center is responsible for performing many high-risk functions 365 days a year. The facility department is faced with the difficulty of maintaining approximately 5 million square feet of office space, 135 acres of land, seven hangars and various other buildings. Traditionally, projects were awarded using the low-bid process, resulting in costly change orders and low quality work.

Ron Campbell (Project Manager of UAL) became aware of PIPS and the Performance Based Studies Research Group (PBSRG) in 1996. After hearing about the process and results from the Federal Aviation Administration (FAA) storm damage repair and roofing results from the Phoenix metropolitan area, he decided to test the system at his facility. He became the visionary at UAL, procuring the highest quality construction work that was ever seen on the site. Campbell realised that a properly maintained facility, rather than repeatedly replacing poorly installed work, could save UAL thousands of dollars. The process was run on 32 projects with outstanding results.

Construction results

The Dock-7 renovation project convinced Ron Campbell (and other personnel at UAL) that the PIPS process was far superior to the traditional low-bid process that they were accustomed to using. Traditionally, UAL used specifications, consultants and inspectors in an attempt to obtain high-performing products and services.

The Dock-7 building had the following problems:

- The current roof was riddled with leaks
- The exterior metal was decaying due to the poorly painted surfaces
- The floor was unfinished

The dock was constantly being used to maintain aircraft and the building could only be shut down for very short periods of time (30 days). The building needed three different types of work done (roofing, painting and flooring), which would all have to be accomplished in the *same* time frame.

Ron Campbell used the PIPS process to procure three high quality contractors. The immediate obstacle was the absence of a general contractor coordinating work between the three trades (UAL partnered with three independent subcontractors). The challenges that the project manager and the contractors faced were that:

- The building could only be shut down one time. All the work had to be accomplished during the shutdown. None of the contractors knew the work would be simultaneous work.
- The roofing work caused debris to fall onto the floors.
- The power-wash performed by the painter caused the floors to get wet.
- The floorer could not work on the floor with debris falling down from the roof or water on the floor.
- The new roofing material was placed near the hangar doors, which prevented access to an entire wall for the painter.

The contractors were not initially aware that other work would be carried out at the same time. They quickly realised that they had a problem. Without any intervention from the project manager, the contractors proposed the following:

- The flooring contractor proposed to work during the evening
- The roofing contractor agreed to work during the day (during normal hours)
- The painting contractor adjusted his schedule in order to work at the weekends

The project manager was surprised to see that the contractors quickly identified the problems and also provided solutions to the owner without forcing the owner to make a decision for them. This was unlike anything that the project manager had experienced. All three contractors worked together to minimise the destruction of each other's work. The shift changes that were proposed by the contractors occurred without any intervention from the owner. All three trades were able to complete their work on time (actually 1 day ahead of schedule), at very high quality and, most impressively, *without any change orders.*

This project illustrates how performing contractors can work together to minimise the risk for themselves as well as for the owner. The owner was surprised at the high quality of work. He was also impressed when the contractors did not ask for any change orders in order to accomplish the work during non-business hours. The contractors understood that doing the job once (and properly) would result in greater profit than trying to delay their work and issue change orders.

High quality painting

The most important aspect of any painting project is preparing the surface, or 'prep work'. In low-bid work, it is fairly difficult to specify how to do high quality prep work. Ron Campbell was no stranger to this concept. He stated that previous painters simply sprayed over the original paint, having done very little prep work. The painters sprayed over rust and unclean surfaces, which caused the paint to peel off soon after.

The project manager stated that he was forced to paint and repair previous paint work every other year. He decided to use PIPS to hire a high quality contractor to paint approximately 115 000 square feet of the hangars. The project manager awarded to the highest prioritised contractor that was within budget.

The awarded contractor proposed to do the work over a longer time period than it took most of the other bidders, but this was due to the intensive amount of prep work on the surfaces. The contractor power-washed all the surfaces and then prepped all of the walls. In certain areas on the parapet walls, large cracks had formed, which caused water to infiltrate the building. The painter stated that painting over the crack would not prevent water infiltration. The painter cleaned the cracks and then injected the cracks with high-strength epoxy to seal them from water infiltration. The painter then painted over the epoxy. Forming a smooth surface with the existing wall. It was the amount of prep work that made the difference between what UAL had previously seen and what they received under PIPS. The contractor completed the work on time, with no change orders and to a very high quality standard.

In 2004, the site was reinspected by PBSRG. Although the hangar walls were dirty (since they had not been washed), there were no signs of deterioration after 5 years of infield service. This was impressive since the site sits a couple of hundred yards from the San Francisco Bay, which creates very salty and damp conditions. Current employees at the facility stated that this was the best work that the facility had ever procured. Ironically, Ron Campbell stated that it took the PIPS painter less time and less money to do the high quality work than if UAL had awarded to the low-bid contractor.

High quality roofing and flooring

Campbell reroofed almost every roof on the UAL maintenance site. He procured built-up roofs, modified bitumen roofs and sprayed polyurethane roofs. None of the roofs currently leaks. He also coated five hangar floors. Four years later, an inspection shows the floors having very little deterioration. Between the roofs, floors, painted hangar doors and waterproofing of all the tanks, Ron Campbell has left the mark of performance on the San Francisco International Airport (SFO) UAL site.

Conclusion

The United Airlines maintenance facility implemented PIPS for over 2 years. The project manager (Ron Campbell) was the key to making the system work. During this time period, he ran PIPS on every type of project that he was involved in, including:

- General construction
- Roofing
- Painting
- Flooring
- High-speed roll-up doors
- Elevator installation

Table A1 Results of PIPS projects at United Airlines.

Criteria	Results
Total number of projects	32
Total award amount	$13 million
Users comparison – low bid vs. performance based	3:09
Overall satisfaction of PIPS	100%
Overall quality of construction procured using PIPS (10 max)	9%
Percent of PIPS contractors that would be hired again	95%
Percent of project finished on time	98%
Percent of projects finished within budget	100%
Percent of projects with no change orders	100%
Number of companies that were evaluated (PPI)	75

- Underground storage tanks
- Submersible pumps
- Environmental projects

The results of the projects were far superior than anything that had been procured at the facility. Lessons learned included the following:

- Quality construction can be procured with less management and inspection.
- The quality construction lasted twice as long as the low-bid awarded construction.
- Performers gravitated to UAL.
- The performing construction minimised maintenance and repair.
- By using performance information at the right time, UAL minimised the amount of technical specifications it used.

The results of the PIPS projects are shown in Table A1.

The research carried out at UAL shows that by minimising management, inspection and control by the user, quality actually increases. The Dock 7 case study also identifies that high quality contractors are focused on creating 'win–win' relationships and minimising the risk for themselves as well as the owner. The research illustrates the importance of contractors working with a type A project manager. Most managers fear that PIPS will eliminate their control and functions. Without the visionary beliefs of Ron Campbell, the system would never have been tested at the site.

Case study no. 2: the State of Utah

Introduction

Until 1999, PIPS had not been run on any large projects. Late in 1998, another visionary facility/construction manager, Richard Byfield, Director of the Division of Facilities Construction and Management (DFCM) for the State of Utah, attended the autumn PIPS conference at the Arizona State University (ASU). He had previously heard of the PIPS process (called the Performance Based Procurement System (PBPS) at that time) through presentations at the National Association of State Facilities Administrators (NASFA). Byfield saw PIPS as the only possible methodology to increase performance.

He implemented the process on five multimillion dollar construction projects (totalling approximately US$86 million in construction). The projects were unique since:

- They were the largest projects ever tested using PIPS
- This was the first test on a large general contractor job
- The projects were complex, and required many performing subcontractors

This section will review the state of the industry in Utah, concentrate on the University of Utah housing project case study and include the details and results of all the Utah tests and lessons learned.

The State of Utah construction environment

The State of Utah construction environment exhibited the characteristics of the low-bid environment (Kashiwagi and Byfield 2002a, 2002b):

- Change orders other than scope and unforeseen site conditions were at 5.7% (1.2% over the contingency construction budget measured over a 10-year period).
- Construction was not being completed on time.
- There were construction problems that had led to litigation or the State of Utah paying the contractor for delays and change orders.
- Use of DFCM construction managers' time was ineffective due to problems caused by low-bidding contractors.
- Low-bid contractors were extremely successful at getting contracts.
- Subcontractors were selected on the basis of price only. Some general contractors claimed that price was not the only selection criteria, but the selected subcontractors almost always had the lowest price.
- There was a lack of liability of designers due to confusion over construction problems. Therefore, all costs flowed to the owner.

The University of Utah housing project (2000 Winter Olympic village housing)

The DFCM was facing one of its most critical projects in the autumn of 1998 – construction of the University of Utah Housing Phase II, which would provide the housing and beds for the participants in the 2002 Winter Olympics. The US$131 million project was divided into two phases: Phase I, which included a portion of the housing and utility installations for Phase II; and Phase II, the balance of the housing.

The University had selected the master planner and designer for the project, despite DFCM's suggestion of hiring a different designer. When problems occurred due to the designer, the University of Utah staff blamed DFCM. It was one of the most political situations the author had come across.

From Phase I, it was evident that the master planner was having difficulty with the overall design and coordination of the multiple contractors on site. The following problems were documented:

- Phase I was behind schedule (by 3 months). The only contractor on schedule was a design and build contractor who had more control over the design and was not selected using low-bid.
- Design services for Phase II were behind schedule (by 3 months) and would not be completed in time for the contractors to bid.
- It would be very difficult to meet the deadline for completion of Phase II construction of May 2000 using the low-bid procurement delivery system.

The process was implemented in an environment of political unrest between the University of Utah staff and DFCM, between DFCM and some of the contractors in Phase I, and between the master planner and designer, and DFCM and the contractors.

PIPS implementation

There were a number of obstacles to overcome in the Phase II project. The following problems were occurring in Phase I:

- The architect was having difficulty meeting the design dates, and the construction in Phase I was not being completed as initially scheduled.
- The architect/construction coordinator could not stop the slipping schedule and delays.
- The architect/construction manager, general contractors, subcontractors and the State of Utah and University of Utah construction managers required a new 'way of doing business'.
- The construction environment was low-bid, litigious and non-trusting.
- The project would have to be awarded in 3 months, with the specifications and drawings not yet completed.
- The design and coordination phases had already used almost the entire contingency fund, resulting in very little room for error in Phase II.
- The project was due to be completed before June 2000, making the installation of the landscaping very difficult.

There were unique pressures on the PIPS test. The University of Utah housing project was a substantial project (US$53 million). Due to the lack of a similar 'performance-based' system, the largeness of the project and the radical change that PIPS brought to the State of Utah construction procurement philosophy, the following occurred:

- Highly successful contractors who did not usually bid on government work were attracted to participate.
- 'Successful' contractors were attempting to define 'performance' for the State of Utah. This resulted in these contractors having a difficult time understanding the

simple PIPS and Information Measurement Theory (IMT) concepts taught in the educational seminars. Some of the contractors took a 'personal' approach that if they did not win, the system was flawed.

- The openness of PIPS allowed the participating contractors to set the performance criteria and show their performance capability. The opportunity seemed to stun the contractors. Contractors were used to 'marketing' themselves and not differentiating based on performance. A few of the contractors made 'obvious' errors in their proposals, which supports the above idea of 'not being ready for a performance-based approach'.

Performance data collection

Past performance information was collected on the general contractor, critical subcontractors and the project manager and site superintendent. Performance lines were collected on 86 contractors.

The contractors were instructed to give only their best references, and were told to contact their references to inform them of the importance of the survey. Despite these instructions, some contractor references resulted in poor performance numbers. This illustrated that some contractors do not understand whether they perform or not. The following describes the unusual level of success of the data collection process:

- Number of surveys: 1931
- Total number of contractors surveyed: 86
- Percentage of surveys returned: 69%
- Average number of criteria per survey: 42

The weights for the performance criteria were set by the user and are shown in Table A2. The price and performance of the general contractor were weighted the heaviest. The management plan and interview were also weighted heavily.

Table A2 Criteria weights.

No.	Criteria	Weights
1	Price	20
2	Management Plan	10
3	Contractor Interview	10
4	Site Superintendent Performance	5
5	Site Superintendent Interview	5
6	General	20
7	Electrical	5
8	Mechanical	6
9	Framing	4
10	Plumbing	4
11	Masonry	3
12	Drywall	3
13	Roads	3
14	Landscaping	2

Table A3 Contractor raw scores.

		Firm 1	Firm 2	Firm 3	Firm 4	Firm 5
1	Price	$55.6M	$50.3M	$48.8M	$52.6M	$49.9M
2	Management plan	8.43	6.21	4.24	5.31	5.73
3	Contractor interview	8.94	9.06	9.13	7.25	6.52
4	Site superintendent PPI	9.71	9.46	9.09	9.24	9.36
5	SS interview	8.97	8.75	9.10	5.57	6.75
6	General PPI	9.09	9.19	8.40	8.96	8.52
7	Electrical PPI	8.93	8.41	8.41	9.02	8.41
8	Mechanical PPI	8.44	8.45	8.45	9.10	8.45
9	Framing PPI	7.80	8.41	8.52	8.30	8.33
10	Plumbing PPI	9.30	8.65	8.65	8.66	8.65
11	Masonry PPI	8.15	8.41	8.50	9.01	9.01
12	Drywall PPI	8.72	8.36	8.00	8.00	8.72
13	Roads PPI	9.23	8.23	8.23	8.80	8.80
14	Landscaping PPI	8.20	8.20	8.20	9.41	8.20

PIPS results

One major 'successful' contractor did not bid on the project due to political considerations connected to the 2000 Olympics. Five general contractors did bid. Table A3 shows the performance information and price for each firm.

The contractors were told that there were two major requirements that had to be met: (1) the project had to be complete by 1 May 2000, including the landscaping; and (2) the project budget was US$53 million. Any bid over $53 million would be non-responsive.

The highest performing contractor's bid was $55 million. This contractor was a joint venture between the design build contractor who had worked on Phase I of the housing project and another general contractor. The contractor gave the following reasons for its high priced bid submittal:

- There were flaws in the design, which would raise the project cost over the budget of $53 million. The major flaw was the omitting of the Federal Housing Administration (FHA) requirements of the housing units, which eventually ended up costing the State of Utah approximately $3 million for Phase I. The contractor claimed that its price included the FHA requirements.
- The contractor knew the expectations of the owner because of its personal experience in the Phase I work.
- At the time of the bid, the design for the village centre (the cafeteria facility for the new housing units) was not completed and bid as an allowance (this later ended up costing $11 million instead of the estimated $7 million).

The contractor requested to change the bid a day after the bid submittal. It did not 'fully realise' that the over-budget proposal would eliminate its bid. The contractor should have done the following, which is allowable under the PIPS rules: (1) identified the design flaws or unstated requirements and priced them as additives;

(2) identified the risk of working with the State of Utah and set rules and requirements, which would reduce the cost.

However, because its bid was above the stated budget requirement, the contractor was eliminated as non-responsive.

Contractor #2 was the second highest performer with a bid of $49 million. However, its projected completion date for landscaping was 2 months beyond the 1 May 2000 deadline set by the University of Utah. Contractor #2 gave the following arguments: (1) substantial completion of the project would be by 1 May; (2) it was impossible to finish the landscaping in the spring by 1 May as the spring season was not long enough to do the landscaping effectively.

The contract documents called for substantial completion, including landscaping, by 1 May 2000. The University wanted no landscaping to occur once students occupied the facility. Additionally, it wanted no landscape material tracked into the units. Contractor #2 was eliminated as being non-responsive. It protested the bid award and the bid protest was denied due to the following facts:

- Contractor #2 specifically asked in the preproposal meeting if the deadline for landscaping could be beyond 1 May. The addenda listed the question with the answer 'no'.
- The preproposal meeting presentation, which was also part of the addenda, covered the requirements that would eliminate a bid as non-responsive. The requirements were to be within budget and finishing all landscaping by 1 May.
- The University requested that the landscaping be done by 1 May due to the projected immediate use of the housing units.
- The successful bidding contractor (contractor #3) had proposed a way to minimise risk and finish on time. It proposed to install much of the landscaping the previous year (1999), and repair any 'dead' or damaged items at the end of the construction period (2000). Its submitted landscaping budget included the cost of repairs.

Contractor #3 was awarded the contract. It should be noted that the landscaping and finish of several buildings actually were completed beyond 1 May; however, the contractor was delayed through no fault of its own, resulting in the landscaping being finished in July 2000. Contractor #2 did not understand that the legal requirement was 'to have a construction plan finish by May 1, 2000, if the contractor is allowed to perform on their proposal'.

Contractor #1 and #2 were both disqualified due to a lack of understanding of the difference between PIPS and the policy requirements of the State of Utah. Requirements cannot be altered by the PIPS. PIPS has to work within the constraints of the user's requirements. The objective of PIPS is to measure differences in order to prioritise alternatives to meet a unique requirement.

Contractor #3 had the following differentiating characteristics:

- It had not bid on State work in the past.
- It assigned the 'best' personnel to the project. Both the project manager and the site superintendent were extremely knowledgeable about the project. They were quick thinking, logical and passed on information very quickly.

- It had a workable plan that finished within the contract period.
- It was creative in handling the critical areas of framing, sheetrock and landscaping.
- Additionally, it was the low-bid on the project.

Construction of the University of Utah housing

The following occurred during the construction period:

- The Phase I contractors (particularly the infrastructure contractor) were not finished on time, delaying the start of the PIPS contractor.
- The design was not complete, specifically related to utilities and site engineering requirements. Critical design components such as finished grade were not available until 4 months into Phase II. One of the issues was the fact that the designer thought that the project was a design and build project once the contractor was awarded the contract. A major point in PIPS is that where there is a full design, the designer remains responsible for design. However, the performing contractor is responsible for asking for clarifications for any issue causing confusion or designs that are not constructible.
- The performing contractor changed his 'proposed construction plan' twice to attempt to get construction under way immediately. In both cases, the designer had information that was not passed on to the contractor. A pre-award meeting would have solved this issue before the contract was awarded. However, the issues dragged on for 3 months, preventing the contractor from starting.
- One of the critical construction requirements, the $7 million village centre design, was completed 5 months late. The contractor was asked to submit a $7 million allowance in its bid. However, the final cost of the village centre was $11 million.
- The designer had difficulty staffing the construction management/engineering support functions.
- DFCM and the University released the lead design firm, which was responsible for overall construction management. The University of Utah personnel took over the coordination and management of daily construction. It is interesting to note that the designer was not selected under the PIPS process.

The village centre project was substantially complete on 11 May 2000. The contractor was paid a $350 000 acceleration fee to make up the lost time (3–4 months). The contractor identified to the author that the entire fee went to the specialty contractors. All but one of the remaining ten buildings, the landscaping and punch list items were completed by the end of August. The State of Utah made the determination that the contractor was delayed due to incomplete project drawings and a failure of the designer to respond to the contractor in a timely manner throughout the project.

A major lesson learned on this project is a procedural step in the PIPS. The PIPS has a 'pre-award and partnering' meeting before the award. Before the meeting the following is accomplished:

- The contractor reviews the drawings in detail with its critical subcontractors.
- The contractor identifies all items that cannot be constructed or have incomplete information.
- The contractor submits a list of requests for information to the designer.
- They discuss the responses at the pre-award meeting.
- A construction information interface should be agreed to and implemented.
- The pre-award meeting minutes and agreements become a part of the contract.

The pre-award meeting has the following advantages:

- The contractor can carefully coordinate with its critical subcontractors.
- Errors or issues in the design are identified before construction.
- The designer is forced to respond to the contractor's requests for information (RFIs) in a timely manner.
- The information at the pre-award meeting identifies responsibilities, ensures that information will be passed on in a timely fashion and creates a partnering environment.

This meeting was not held before the award for the University of Utah housing project. The partnering meeting was held after the award. The contractor did not receive needed information from the designer. It is significant to note that the designer did not alert the contractor during the partnering meeting that its construction plan could not be accomplished based on the late finish of Phase I contractors, and other such information was not passed to the contractor. As previously stated, the project took 3–4 months to begin construction. The lack of an information system led to confusion, difficulty in solving the problems and finally the transfer of the construction coordination from the lead designer to the University of Utah.

Analysis of the PIPS implementation

The project was awarded within the scheduled two-and-a-half month procurement time. This included the two free education sessions, data collection, management plan and interview reviews, and the prioritisation based on all the information. It also included the review of two protests on issues that did not address the PIPS.

The process selected an innovative contractor with a clear plan of success. The successful contractor was well within budget, had excellent personnel and had a plan to meet the user's needs by creative contracting. The contractor addressed the landscaping and framing by using innovative scheduling and prefabrication.

The construction finished on time and on budget. All change orders were owner-directed scope changes and time acceleration. The contractor did all that could be done to finish on time, within the bid price. The contractor, with the aid of one of the subcontractors, finished the village centre (at $11 million) in 11 months and opened the centre on schedule.

An analysis was performed to compare the results of the PIPS contractor with the Phase I low-bid and design and build contractors (see Tables A4 and A5).

The PIPS contractor when compared with the other two:

Table A4 Delivery system performance results.

	Low bid	Design build	Performance based
Days added to schedule	234	332	105
Actual days added until substantial completion	279	184	105
Percent change in scope	12.9%	19.4%	8.9%
Percent change in unknown events	1.3%	0.8%	0.9%
Percent change in error	2.4%	4.8%	1.3%
Percent change in omissions	1.5%	0.8%	1.6%

Table A5 Contractor comparison.

	Low bid	Design build	Performance based
Change in scope	12.9%	19.4%	8.9%
Change in unknown events	1.3%	0.8%	0.9%
Change in errors	2.4%	4.8%	1.3%
Change in omissions	1.5%	0.8%	1.6%
SUB TOTAL	18.2%	25.8%	12.6%
CONTRACT	81.8%	74.2%	87.4%
ADJUSTED TOTAL	100.0%	100.0%	100.0%

Table A6 Project manager evaluation of performance.

No.	Criteria	Unit	Score
1	Is the project currently running on schedule?	Y/N	Y
2	Is the project currently running on budget?	Y/N	Y
3	How many contractor change orders have been issued?	(#)	0
4	Please rate your overall satisfaction with the contractor	(1–10)	9

- Had the smallest change in schedule
- Finished the earliest in respect of the schedule
- Had the smallest change in scope (additional cost)
- Had the smallest charge for errors and omissions

The facility manager rated the contractor (see Table A6). The user and facility management program director were very satisfied with the contractor's services.

Overview of the State of Utah PIPS tests

Four other major projects were procured at the State of Utah:

(1) Southern Utah University Project: (budget: $17.3 million) construct a new three-story building composed of classrooms, a gymnasium with an elevated track, a central hall and a competition size indoor swimming pool.

Table A7 State of Utah results.

No.	Criteria	Utah results
1	Total number of projects procured	5
2	Awarded cost	$80 506 376
3	Budget	$85 770 000
4	Percent under budget	7%
5	Users comparison of low-bid vs. performance based low-bid: Performance based (10 is the max)	4:9
6	Percent satisfied with PBPS	90%
7	Percent of users that met higher performing contractors using PBPS	100%
8	Percent of users that procured a higher performing contractor using PBPS than previous methods of procurement	88%
9	Overall quality of construction that was procured using the performance-based systems.	9.2
10	Performance rating of the contractor/system your company procured	9.9
11	Percent of users that would hire the contractor again	100%
12	Percent of users that would use the performance-based systems to procure another project	90%
13	Percent of projects that finished on time	80%
14	Percent of projects that finished within budget	80%
15	Number of contractor-caused change orders	0
16	Number of companies that were evaluated (PPI)	357

(2) Gunnison Correctional Facility Project: (budget: $9 million) construct a new 288 bed dormitory style correctional housing unit, a guard tower, a hazardous storage unit, a kitchen and support building.
(3) Richfield Youth Correctional Facility Project: (budget: $3.5 million) construct a facility that includes administrative offices, a service wing with a kitchen, a detention wing, a gymnasium, a secured housing wing, a security yard, parking and landscaping.
(4) Bridgerland Applied Technology Center (ATC) Project: (budget: $3 million) renovate an existing technology building at Bridgerland ATC.

Table A7 shows the overall results of the State of Utah implementations.

Lessons learned

The following lessons were learned from the above projects:

- Despite the modifications, the PIPS system procured outstanding results on large projects.
- The site superintendent and the project manager are key components in a large construction project.
- The pre-award period is critical before the award of a contract.
- A high performance contractor will fix a design that is less than optimal.

- High performing contractors minimise the risk and think in terms of the best interests of the owner.
- Contractors need to be educated on the differences between legal requirements and PIPS requirements.
- PIPS is robust.
- It is important to create and educate a 'core team'.
- Partnering contractors worked very well together (the system allowed for high performing subcontractors to partner with high performing generals).
- The management plan ratings can be skewed if the raters know the origin of the proposal. However, a trained PIPS administrator can catch any possible mistakes.
- The raters of the management plan should sign a document that states they do not know the origin of the proposals and do not have a conflict of interest.
- Users need to be educated more on the system and on the theory. Users should minimise decision-making and control.
- Designers are not minimising the risk of non-performance. In the Gunnison Project, the user stated that the biggest problem was the architect. The user stated that the contractor could have added $200 000–300 000 to cover unforeseen events. Ultimately, this project did not finish on time or within budget.
- An information interface, which is posted on the internet, is required, which quickly identifies participants who are not passing on information. Information flow stops the bureaucratic process of pointing fingers when something goes wrong.

Value-based selection process

Richard Byfield was a visionary who brought innovation to a very conservative State of Utah. The author commends him on his integrity, vision and accomplishments in the delivery of construction. Without his trust in the process, PIPS may not have been implemented in the State. After the five projects were run, contractors and DFCM modified PIPS and called it the value-based selection (VBS) process. The major changes were as follows:

- Deleting the critical subcontractors' past performance.
- Minimising the number of references from 40 to 10, and only requiring verification of 3–5 references.
- Replacing the non-biased artificial intelligent prioritisation tool with the subjective decision of a review panel who saw the past performance information, the risk assessment management plan, interview scores and the bid price.
- Not using the risk minimisation factor of the number of references.
- Not using the rating on the latest performed project to modify the future rating by 25%.
- Not using the pre-award phase (technically this was not a change, since they did not use this in their initial tests).

The modifications made the process more subjective, more difficult for high performance contractors to differentiate and harder to justify. Over time, the author predicted that they would return to the design-bid-build low-bid award. This was based on the hypothesis that if decisions are made with subjective bias, performance information becomes less important and price becomes more important. This prediction is being fulfilled as 3 years later, in the December 2003 DFCM building board meeting, the policy is being forwarded to not use VBS in design-bid-build projects (Utah State Building Board Minutes 2003). In the other delivery systems (design and build and construction management at risk), where subjectivity can be used in the awards, VBS will be continued.

Comments by the State of Utah participants

The following are comments made by various people that were involved with the PBPS projects:

> 'Ten to twelve months have been cut off of our completion schedule. Our students will have access to the facility one full year sooner. The bids did not go over estimation, leaving equipment money for this project intact. . . . Bridgerland has enjoyed the greatest and most rapid successes that we have ever experienced . . .' (Richard Maughan, Superintendent, Bridgerland ATC)

> 'The contractor took full responsibility for the project and made it run very smoothly . . . there were certain instances when the contractor went beyond their "responsibility" by proposing solutions to problems that arose during the process of the project . . .' (Lyle Knudsen, Program Director, State of Utah DFCM)

> 'In my twenty years of project management experience I have never had a better experience or outcome to a project [Southern Utah University Project] . . .' (Frank McMenimen, Program Director, State of Utah DFCM)

> 'When an owner chooses a contractor on the sole basis of cost, the General Contractors in turn have no option but to use any and every low subcontractor bid that comes across the fax machine. Therefore the contractor's main focus on the project will be the cost. When this method is used, the quality and schedule will become secondary issues at best . . .' (Dennis Forbush, Project Manager, Hogan & Associates Construction)

> 'The results of PBPS have given us the best construction results in ten years . . .' (Richard Byfield, Director, State of Utah DFCM)

Conclusion

Richard Byfield was personally responsible for bringing PIPS to deliver large construction projects. Even though the State of Utah may not have been able to sustain the innovation and see the vision of 'outsourcing' construction, his efforts led to successful projects in the State of Hawaii, the Dallas Independent School District, the

Denver Hospital and current projects with the US Coast Guard, the Federal Aviation Administration and Harvard University. Like his peers Gordon Matsuoka, Steve Miwa, Chris Kinimaka, Miguel Ramos, Ron Campbell, Marcos Costilla and Charlie Serikawa, his visionary efforts led to the following conclusions:

- PIPS can work on large and complex general construction projects.
- Four out of five projects were completed on time and within budget. The overall cost of the projects was approximately $5 million below the estimated budget.
- The State of Utah received a much higher level of quality (PIPS scored a 9 and low-bid scored a 4). The users also rated the contractors that they procured a 9.9 out of 10.
- The State of Utah had difficulty grasping the concepts of IMT. It began making decisions and trying to alter the system without understanding the theoretical impacts they would have.

The PIPS process worked at the State of Utah using a less developed, constrained process. Since these tests PIPS has been improved as follows:

- The data collection process has been refined so that it is now done by contractors.
- The number of criteria has been minimised to eight for all contractors.
- The pre-award phase has become the most critical phase. Clients are instructed that if this phase is not run, there is risk.
- Explanation of the designer and contractor functions has become very clear. The designer is always responsible for design.
- General contractors are given the average ratings of the specialty contractors.

References

Kashiwagi D. and Byfield R. (2002a) State of Utah performance information procurement system tests. *ASCE: Journal of Construction Engineering and Management*, **128** (4), 338–347.

Kashiwagi D. and Byfield R. (2002b) Testing of minimization of subjectivity in 'best value' in procurement by using artificial intelligence systems in State of Utah procurement. *ASCE: Journal of Construction Engineering and Management*, **128** (6), 496–502.

Utah State Building Board Minutes (2003) *Utah State Building Board Minutes*. Committee Room 129, Salt Lake City.

Case study no. 3: University of Hawaii

Introduction

In 1998, the State of Hawaii's Department of Accounting and General Services (DAGS) implemented the Performance Information Procurement System (PIPS). DAGS immediately began noticing a difference in attitude from the contractors who were bidding the projects. As the projects were completed, the State observed significant improvements in performance, which were unlike anything that DAGS had seen before. The results of the 4-year implementation of PIPS were as follows:

- PIPS resulted in 3% savings on overall project costs
- PIPS resulted in greater contractor accountability
- PIPS resulted in a lower number of change orders
- PIPS provided the State with higher quality construction

The University of Hawaii (UH), like most other agencies, awarded projects based on the lowest cost. This had been the accepted method of ensuring that UH was receiving quality work at an affordably low cost. However, the marginal performance was requiring a high level of owner management. Projects were not completed on time or budget and did not meet quality expectations.

In 2000, Allan AhSan, Director of Facilities at the UH, implemented PIPS after recognising the increase in construction performance that the State of Hawaii had received using the PIPS process. UH also wanted to know if PIPS could do the following:

- Increase the quality of work.
- Complete the projects on time, within budget and with no contractor-cost change orders.
- Shift risk from the UH to the contractors (that were performing the work).
- Provide justified documentation on using best-value procurement to select a contractor or system.
- Minimise management, regulations, qualifications, user specifications and inspections.
- Optimise UH's project management personnel. The UH wanted to see if it could become more efficient by doing more construction work with fewer project managers (or by increasing the number of projects each project manager could successfully manage).

What made the implementation at the UH successful was the identification of Charlie Serikawa as the PIPS project manager. Serikawa was the visionary at the UH. He was searching for a process to ensure work was completed in the best interest of the UH. He had many years of experience working as a project manager for a performance-oriented general contractor, and was looking for a solution that would minimise the tremendous amount of management, control and decision-making of the UH engineering and construction management staff. He quickly fell in step with the PIPS process, due to his logical understanding of the construction process. He was a leader, coordinator and facilitator of successful construction projects. When the UH decided to discontinue PIPS, and all efforts failed to implement a process like PIPS, Serikawa retired and is now working as a consultant.

Implementation of PIPS by the UH

The UH ran PIPS in three different areas: painting, roofing and gymnasium sports flooring. The UH PIPS project manager (Serikawa) noticed a positive change in contractor attitude soon after PIPS was implemented. He strictly followed the three basic rules of the PIPS process, which are:

(1) Minimise decision-making
(2) Minimise the amount of work performed by the UH project managers
(3) Minimise risk

Painting projects

No technical specifications were issued on eleven painting projects. The project manager's only requirement was the desired colour coordination of the buildings. The award would go to the contractor who proposed the best value, longest warranty with a proven past performance, and could offer the work within the UH's budget and time constraints. This was a complete shift in the manner in which painting projects had been previously awarded. The project manager stated the contractors would be responsible for identifying what they could install, the quality they could provide and the cost it would take to do the entire project. Some contractors, who may have been inexperienced at identifying their performance, did not participate in this process. The remaining contractors competed based on price, past performance history and their ability to identify and minimise potential risk.

The project manager did not make any decisions during the projects. He told the contractors that the minimum level of quality they should provide should be the minimum level of quality they would require if someone were painting their own homes. Once the project was awarded, the contractors began asking the project manager to check their quality of work. The project manager refused, knowing that user inspection brings risk (the contractors should be quality controlling their own work). After repeated requests, the project manager finally gave in and performed spot-checks during walk-throughs in order to assist the contractors. If there were problems, the contractor immediately responded without the assistance of the project manager. Overall, the amount of management, inspection, control and decision-making was minimised by 80%.

A couple of lessons were learned from the painting projects at the UH. The first was that the contractors requested quality control training from the union trainers. When the author was approached on the new training requirement, the author asked if they had previously received the training in their certification. The contractors had all been trained. However, when working in the low-bid award environment, the training was not utilised, and the contractors did things their own way. When the performance-based environment was implemented, the contractors now wanted to be retrained. This shows that when contractors are working in the low-bid environment, training does not increase the quality of the work. The second lesson was that the project manager asked the contractors to add value to the owner's buildings by identifying work that could be done more economically by the contractor while painting the buildings. The project managers had more funding than the projects cost. However, the contractors explained that they were used to low bidding, and that they would have to be educated in how to add value and act in the owner's best interest. The contractors had been conditioned by the low-bid environment of the UH to do work fast, cheap and according to strict directions.

Table A8 University of Hawaii PIPS painting results.

No.	Criteria	Results
1	Total number of projects awarded	11
2	Overall estimated budget	$2 310 000
3	Total award cost	$1 658 192
4	Percent over/under budget	–28%
5	Number of different contractors awarded jobs	6
6	Percent of jobs completed within budget	100%
7	Percent of jobs completed on time	100%
8	Percent of jobs completed ahead of time	90%
9	Percent of projects where contractor performed additional work at no charge	56%
10	Average post-project rating (maximum is 10)	9.8

Table A9 Overall comparison of the low-bid process and the PIPS process.

No.	Criteria	LB	PIPS
1	Ability to encourage contractors to perform high quality work	2.3	9.7
2	Overall performance of contractors	3.2	9.1
3	Overall quality of projects procured	3.2	9.1
4	Overall satisfaction	2.8	9.2

Results

The results of the UH projects were consistent with other PIPS projects (State of Hawaii PIPS Advisory Committee 2002). The projects were finished on time (90% of jobs ahead of time), within budget (with no change orders) and to very high quality. Table A8 summarises the overall results of the UH PIPS painting projects.

Table A9 illustrates the overall comparison of the low-bid process compared to the PIPS process (as identified by the UH project manager and awarded PIPS contractors). One hundred percent of the individuals stated that they were satisfied with PIPS and that they would rather use PIPS over low-bid.

Table A10 evaluates the factors relating to movement from the low-bid process to the PIPS process. The table also illustrates the percentage decrease in change orders, punch list items, specifications, design work, user inspection and user management when moving from the low-bid process to the PIPS process. The UH project manager and four contractors (who were awarded a PIPS project) evaluated the PIPS system.

Tables A11 and A12 analyse the perception of the contractors and the UH project manager towards the comfort levels of the different parties involved in construction procurement. This includes the analysis of industry training programmes, designers, procurement personnel, lawyers, engineers and university leaders with the PIPS process.

These numbers clearly indicate that UH was able to achieve its objective of procuring higher quality construction projects by minimising management, regulations, qualifications, user specifications and inspections.

Table A10 Evaluation of factors.

No.	Criteria	Contractor ratings	UH ratings
1	Percent decrease in cost-generated change orders	83%	75%
2	Percent decrease in the number of punch list items	96%	75%
3	Percent decrease in the amount of detailed specifications	91%	75%
4	Percent decrease in the amount of design work	17%	75%
5	Percent decrease in user inspections	63%	80%
6	Percent decrease in the amount of user management	13%	80%

Table A11 Comfort levels of the industry (in percentage terms) with the principles of PIPS.

No.	Criteria	UH ratings	Contractor ratings
1	Percent that felt that the PIPS process was a fair process	100%	100%
2	Percent that felt that the low bid process was a fair process	100%	33%
3	Percent of individuals who would use PIPS rather than low bid	100%	100%
4	Percent that felt that there was political pressure with the PIPS process	0%	25%
5	Percent that felt that there were legal pressures with the PIPS process	100%	25%

Table A12 Comfort levels of the industry with using PIPS.

No.	Criteria	UH ratings	Contractor ratings
1	Overall comfort level of the contractors involved with the PIPS process	9.0	9.0
2	Comfort level of the industry training programmes with the PIPS process	8.0	7.5
3	Comfort level of the designers with the PIPS process	9.0	6.7
4	Comfort level of the procurement personnel involved	3.0	7.3
5	Comfort level of the lawyers involved with the PIPS process	3.0	4.5
6	Comfort level of the facility engineers involved with the PIPS process	5.0	9.3
7	Comfort level of the university leaders with the PIPS process	5.0	8.0
8	Overall acceptability of the PIPS process	5.0	5.0

Charlie Serikawa had worked for the UH for approximately 15 years, and had previous experience working for a general contractor for another 15 years. He stated (personal communication, 2002):

> 'In all my years of construction experience (both in the private and public sectors) I have never been more impressed with a procurement process such as the process provided by PIPS. The system promotes a partnering "win–win" scenario between the owner and the contractor that requires minimum project management resulting in on time, on budget and outstanding quality construction.'

Serikawa observed that the contractors involved were extremely comfortable with the PIPS process and felt that they would rather use the PIPS process than the low-bid process in future. This indicates that high performing contractors favoured the system. He noted that one contractor was not comfortable with PIPS. This contractor had received awards under the low-bid system, but was not being awarded any PIPS projects. The contractor was not competitive when both performance and price were considered. In addition, Serikawa observed that the procurement personnel, university lawyers, facility engineers and university leaders were all uncomfortable with the PIPS process. This is not uncommon, since individuals are afraid of change (personal communication, 2003).

Conclusion

By implementing the PIPS process based on performance and price, UH procured higher quality construction projects and minimised project management requirements. The test results support the hypothesis that the low-bid award process may be the reason for poor construction performance. The high performance can be related to the minimisation of project management, inspection and specifications, which are required under the traditional low-bid environment.

The UH project manager and the majority of contractors felt very comfortable with the PIPS process and stated that PIPS resulted in a substantial increase in overall performance. However, the procurement personnel were uncomfortable with the process due to the minimisation of control and procurement functions. The UH lawyers were uncomfortable due to a change in thinking and their inability to understand the concepts of performance information.

The UH chose to return to the low-bid environment and increase project management of construction. In the past year, it has tried to come up with another performance-based process that attempts to duplicate the results of PIPS, but has not been successful. This has resulted in the retirement of the project manager who was assisting the UH in implementing PIPS or a substitute performance-based process. His final statement was, 'I am convinced that PIPS is the only way to go, especially for a public institution'.

The lessons learned from the University of Hawaii test case include the following:

- Change must be required by the owner. The procurement officer for the UH did not want to release control over the contractors. Even after seeing better results

from less effort, the procurement officer resisted the change. The owner must have enough information regarding the process to override the procurement officer's resistance to change.
- Initial education should include the owner's legal representative. After conversing with the lawyers, it was evident to the project manager and contractors that the lawyers did not understand the concepts of information and impact of minimum standards.
- There must be a requirement by the owner to increase construction performance.

The author proposes that based on the results of the University of Hawaii test, poor construction performance may be a result of the current bureaucratic construction delivery process and not solely poor performing contractors. Owners must hold their representatives accountable. Owners must be educated. More understanding and direction by the UH administration would have increased the opportunity to make the change to performance-based contracting permanent.

Reference

State of Hawaii PIPS Advisory Committee (2002) *Report for Senate Concurrent Resolution No. 39 Requesting a Review of the Performance Information Procurement System (PIPS)*. SOH.

Case study no. 4: the State of Georgia

Introduction

The State of Georgia is considered (by the author) as one of the failures of the implementation of PIPS. The failure was not PIPS, but in the user's understanding of the process. The failure was due to the lack of a mechanism that clearly defines the plan of action in case a project is over-designed and causes contractors to bid over the stated budget.

The Georgia State Financing and Investment Commission (GSFIC), tasked with the delivery of capital construction, had been analysing methods of increasing the value of construction (Butler 2002). The State of Georgia was also interested in implementing different delivery mechanisms. In 1999, it tested the PIPS process to increase the value of construction. Two tests were conducted. The first procurement of construction was a US$45 million environmental technology building, and the second was a $7.8 million occupational technology facility.

The State of Georgia was made aware of the results from both the states of Utah and Hawaii through presentations at the National Association of State Facility Administrators (NASFA). It is the perception of the author that the GSFIC personnel were looking for a process to select a high performance contractor. GSFIC would then use its traditional construction delivery methodology during the construction phase. PIPS is a process that covers the selection, procurement, construction and performance of the constructed facility. PIPS will not work if the client does not do all three phases.

The State of Georgia case studies resulted in the following conclusions:

- Designers do not design to minimise risk, even though third party cost estimators are utilised.
- High performance construction does not cost more.
- Clients and construction professionals may not understand the value of what they are procuring.
- Non-performing construction may be caused by an inefficient delivery system.
- The quadrant I environment of regulation, mistrust and owner direction causes havoc with the value of construction.
- The information environment and PIPS identifies the best performing contractor. In the two test cases, there was only one logical choice for the project.

Project 1: Environmental Science and Technology (ES&T) Building (Georgia Institute of Technology)

The scope of this project was to construct an environmental research laboratory (approximately 287 000 square feet) with facilities that included laboratories, classrooms, lecture theatres, offices and administrative areas.

The PIPS process brought high performing contractors. Three general contractors, five mechanical subcontractors and seven electrical subcontractors participated. A general contractor who withdrew from the project informed the owner that the list of mechanical and electrical contractors was one of the best prequalifications of mechanical and electrical contractors in the City of Atlanta that he had seen. When informed that they were not short-listed, but were the contractors who responded to the project, he was amazed. This is a result of PIPS. If the invitation is unrestrained, the highest performing contractors will respond.

Three general contractors submitted bid proposals for the project. The bids were $52.6, 54.9 and 56.2 million respectively. Since the original budget was $45 million, the architect and the University Board of Regents claimed the contractors had inflated their costs due to the 'best value' process, and they wanted the project to be awarded through a traditional low-bid process.

The author was very surprised that the bid results were so far off. The designer had an independent cost estimator verifying the costs. However, two factors convinced the author that something was drastically wrong:

(1) A construction professional, with knowledge of a recently completed project (similar to the facility being bid), informed the author (before bids were submitted) that the bids would come in around at $52 million.

(2) A comparison of two of the contractors' bids (the high bidder and the low bidder) with the designer's cost estimate showed that three areas (general requirements, metals and mechanical) were off by a total of $8.5 and $7.5 million respectively. The mechanical section alone was off by $6.7 and $5.4 million respectively.

The author followed up with the construction professional and found out that he had arrived at the $52 million by taking the finished cost of the facility that had recently been completed ($181 (£101) per square foot) and multiplying it by the area of the new facility (287 000 square feet).

The designer did not have an objective of risk minimisation. Instead the designer and the University Board saw the PIPS process as one that increased costs. However, the GFSIC project managers identified that they had been misled by the designer. The author encouraged them to award to the low bidder since the contractors had all been over budget. However, GFSIC felt that because the award process was not clearly defined (in the case that every bidder was over budget) it may be open to protest. They decided to redesign and award using the low-bid award process.

The process had been open to value engineering (VE) by the general contractors and subcontractors. However, the majority of the VE concepts were disapproved by the designer. After the bids came in, and the differential was noted, GFSIC and the designer became very open to the VE concepts. The VE concepts were bought from the contractors and used to reduce the project cost by $4.5 (£2.5) million from the project.

The project was rebid under the low-bid process, and the new proposals came in at $46.6, $47.0, $47.2, $47.5 and $48.0 million. Even though all the proposals were still over budget, the project was awarded to the lowest bidder (still 3% over budget at $46.6 million). Coincidentally, the general contractor who was awarded the project had the lowest past performance ratings (when compared with the generals from the first round). Out of the five rated mechanicals, the winning general contractor selected the lowest rated mechanical subcontractor. Their average performance rating was 8.2 (out of 10) in comparison with the average rating of the other mechanicals, which had 9.1. The winning general contractor did not select an electrical subcontractor that had documented past performance information.

After the conclusion of the project, the cost of using the low-bid process can be identified. The final construction cost was $48.8 million, and the project was extended over 300 days. From past experiences on PIPS implementations, the performance-based contractor had the capability to finish the project in 660 days instead of 960 days (45% earlier). If the State of Georgia had awarded to the lowest cost performance-based contractor ($52.6 million), with the $4.5 million reduction in scope, the award price would have been $48.1 million for a performance-based contractor with the best-qualified subcontractors. PIPS results have shown that performers deliver on time, on budget and do exceptional quality work. The cost of the low bid turned out to be over $1 million higher and 300 days late.

A major task in this project was commissioning, which is usually done by experts. This was made obvious by the fact that the low-bid contractor was at 90% complete and still took over a year before the user could use the facility. The quadrant I environment is managed by construction professionals. The author proposes that the construction professionals should become designers, coordinators and facilitators of construction, and not manage and direct low-bid contractors.

If PBSRG did not have a reason to track and measure the performance, it would not have been in the best interest of the University, the designer, the contractor or GSFIC to track the performance information. This conclusion matches the previous results in the states of Utah and Hawaii and later results in the Dallas Independent School

District (DISD) and the University of Hawaii. The concept that highly qualified contractors are more expensive is unproven when considering first costs and time. The author proposes that when the costs of management, control and inspection (which clients do on low-bid contractors) are added, there is no comparison. Contractors who know what they are doing get it done faster and more efficiently. They also maximise their profit.

Two sources have since contacted GSFIC to request information about PIPS and have been told that PIPS inflates prices for quality work, is not efficient and that GSFIC is not interested in PIPS. The author encourages high quality contractors and clients who are trying to efficiently deliver construction to visit PBSRG's website (www.pbsrg.com) to get the complete information on this project.

Project 2: Occupational Technology Building (Savannah Technical Institute)

A second test project was conducted by GSFIC, that of the Occupational Technology Building at Savannah Technical Institute. This project consisted of constructing two new structures on the existing Savannah campus. The first building was approximately 45 000 square feet and housed industrial laboratories, classrooms, offices, faculty offices and support spaces. The second building housed the automotive services and body repair laboratories, paint spray facility classroom and supporting spaces.

The concerns of the user were that the site was located in a rural area, which might limit the number of skilled subcontractors from Atlanta and also minimise the number of bidders. GSFIC's experience with the contractors from this area was not exceptional.

After the experience of the first test project, the designer was made aware of the importance of not overdesigning the project. However, the proposals still came in over the $7.8 million budget ($9.4, $9.4, $9.7 and $10 million respectively). Once again the designers were not minimising risk; they were trying to please the client. The State then asked the contractors to bid deductibles to see if any proposals would fall within budget. The deductibles reduced the bids by an average of $1.4 million.

Table A13 shows the performance ratings of the contractors, which includes the past performance information (of all the generals and the subcontractors) and the current capability of each company (interviews and management plans). Each rating is from 1–10, with 10 being the best.

Table A14 shows the final best value selection process. The best value was identified by using linear relationships between performance and price. Since the performance to price ratio in percentage terms was 53:47, the highest performer would get 53 points (for the performance factor) and the lowest cost contractor would get 47 points (for the cost factor). The model selected contractor #1 as the best-valued contractor. Intuitively, someone would have selected the same contractor (based on the information in Tables A13 and A14).

The following conclusions could be made: contractor #2 had the lowest rating on the management plan and interviews. This shows a relative lack of understanding of

Table A13 Contractor performance ratings.

No.	Criteria	Cont 1	Cont 2	Cont 3	Cont 4
1	Management plan	7.9	4.1	7.0	6.3
2	Project manager interview	8.2	6.3	8.1	7.4
3	Site superintendent interview	8.1	6.7	8.1	7.2
4	General contractor past performance average	9.4	8.6	8.8	9.1
5	Plumbing contractor past performance average	9.6	8.9	8.9	9.6
6	Electrical contractor past performance average	8.9	8.9	8.9	8.9
7	Mechanical contractor past performance average	9.0	9.0	9.0	8.8
8	Roofing contractor past performance average	9.1	9.1	9.6	9.1
9	Project manager past performance average	9.5	9.1	8.5	8.6
10	Site superintendent past performance average	9.6	9.1	9.0	8.1
	Overall average	**8.92**	**7.97**	**8.6**	**8.31**

Table A14 Best value selection (performance vs price).

Rank	Contractor	Total points	Performance score	Performance points	Price	Price points
1st	Contractor 1	99.6	0.05	53.0	$8 104 000	46.6
2nd	Contractor 3	56.5	0.23	11.4	$8 371 723	45.1
3rd	Contractor 4	50.9	0.39	6.8	$8 561 000	44.1
4th	Contractor 2	50.4	0.78	3.4	$8 033 645	47.0

the project by both the contractor and his key personnel. Contractor #2 is also the low bidder; contractor #1 had the best management plan rating, best project manager interview rating and second best site superintendent interview rating. Contractor #1's price was approximately 1% more than contractor #2's.

Construction management

GSFIC had instructed all the bidders that a performance-based contract would be awarded. It had instructed the contractors that it would have to set up an information system that would quickly transfer information and allow the contractor to take control of the project. However, once the construction was awarded, the State issued a standard low-bid contract to the contractor. When the contractor attempted to start the project, the designer would not allow the start unless a detailed schedule was approved. When the contractor attempted to service the client, the contractor was told that he did not have the authority to make changes in the best interest of the client. When the contractor tried to give the designer training on the information system, the designer did not show up. The contractor immediately reverted to low-bid behaviour in order to protect himself from the owner's actions.

Amazingly, without PIPS, the contractor who would have been selected was a contractor who did not understand the project. Once again, performance did not cost more. The differential between the highest and lower performers was

Table A15 User rating on PIPS and the low-bid process (Burke 2003).

No.	Criteria	Unit	Rating
1	Rate your satisfaction with the performance-based system of contracting (PIPS)	(1–10)*	5
2	Which process would you rather use: low bid or PIPS?	PIPS/LB	Low bid
3	Rate the ability (of PIPS) to encourage contractors to reduce the amount of management and inspections	(1–10)*	4
4a	Rate the ability (of PIPS) to encourage users to reduce the amount of management and inspections	(1–10)*	1
4b	What was the percentage decrease?	%	0%
5	Rate the ability (of PIPS) to encourage partnering	(1–10)*	5
6	Rate the ability (of PIPS) to reduce the number of punch list items	(1–10)*	1
6	Would you participate in another PIPS procurement?	Y/N	Y
7	Were you satisfied with PIPS?	Y/N	N
8	Rate your satisfaction with the low bid system of contracting	(1–10)*	7
9	Rate the ability (of the low bid process) to encourage contractors to perform high quality work	(1–10)*	7

* 10 = Highest score or strongly agree, 1 = lowest score or strongly disagree

insignificant. The highest performing critical subcontractors were also very cost competitive.

The client or building user was asked to rate PIPS (Tables A15 and A16). He was under the impression that the PIPS environment was being used to manage the project. He did not understand that the designer was using the low-bid environment of management and control. This is shown by his comments on the low-bid system. However, his rating of the contractor was 9 (out of 10) and his comment (Burke 2003) was:

> 'The contractor Paul Akins, Inc. made all the difference in taking a flawed system (PIPS), and making this construction experience a good one. I would strongly recommend Paul Akins, Inc. for any construction project they bid on. They exhibit a strong team-mentality combined with a professional concern for quality. I would work gladly with Paul Akins, Inc. on any future project.' (Vick Burke, Savannah Tech)

This was confirmed by the contractor's ratings (Table A17); he rated the low-bid system unsatisfactory, with no ability to motivate to encourage skilled craftspeople, training or quality work, and because the process does not assist in making a profit. The contractor gave PIPS high marks with regard to satisfaction, helping the contractor perform high quality work, making more profit, minimising management and inspection, and encouraging partnering. He would definitely participate in PIPS again, and was satisfied with the outcome. The contractor gave high marks to the inspectors, designer and user's project manager, but lower marks with regard to helping the contractor, utilising the PIPS process effectively and minimising bureaucracy. The contractor also made the following comments (Futch 2003):

Table A16 User rating on the contractor.

No.	Criteria	Unit	Rating
1	Rate the contractor's ability to use the PIPS process	(1–10)*	9
2	Rate the contractor's ability to assist you in increasing quality	(1–10)*	9
3	Rate the contractor's ability to manage the project cost (minimise change orders)	(1–10)*	8
4	Rate the contractor's ability to maintain project schedule	(1–10)*	9
5	Rate the contractor's overall quality of workmanship	(1–10)*	9
6	Rate the contractor's professionalism and ability to manage	(1–10)*	10
7	Rate the contractor's ability to close out the project (no punch list upon turnover, warranties, operating manuals, tax clearance, submitted promptly)	(1–10)*	8
8	Rate the contractor's ability to communicate, explanation of risk, and documentation	(1–10)*	9
9	Rate the contractor's ability to follow the user's rules, regulations, and requirements (housekeeping, safety, etc.)	(1–10)*	10
10	Rate your overall satisfaction and your comfort level hiring the contractor again based on performance	(1–10)*	9

* 10 = Highest score or strongly agree, 1 = lowest score or strongly disagree

Table A17 Contractor rating on PIPS, low-bid and the user.

No.	Criteria	Unit	Rating
1	Rate your satisfaction with the performance-based system of contracting (PIPS)	(1–10)*	9
2	Which process would you rather use: low bid or PIPS?	Circle	PIPS
3	Rate the ability (of PIPS) to encourage contractors to reduce the amount of management and inspections	(1–10)*	10
4a	Rate the ability (of PIPS) to allow contractors to make higher profit margins than in low bid	(1–10)*	8
4b	What was the percentage increase?	%	1.5%
5a	Rate the ability (of PIPS) to encourage users to reduce the amount of management and inspections	(1–10)*	8
5b	What was the percentage decrease?	%	5%
6	Rate the ability (of PIPS) to encourage partnering	(1–10)*	7
7	Would you participate in another PIPS procurement?	Y/N	Y
8	Were you satisfied with PIPS?	Y/N	Y
9	Rate your satisfaction with the low bid system of contracting	(1–10)*	4
10	Rate the ability (of the low bid process) to encourage contractors to maintain skilled craftspeople	(1–10)*	2
11	Rate the ability (of the low bid process) to encourage contractors to maintain training programmes	(1–10)*	2
12	Rate the ability (of the low bid process) to encourage contractors to perform high quality work	(1–10)*	2
13	Rate the ability (of the low bid process) in making larger profit margins	(1–10)*	1

* 10 = Highest score or strongly agree, 1 = lowest score or strongly disagree

Table A18 Project manager rating of PIPS, low-bid and the contractor.

No.	Criteria	Unit	Rating
1	Rate your satisfaction with the performance-based system of contracting (PIPS)	(1–10)*	6
2	Which process would you rather use: low bid or PIPS?	Circle	PIPS
3	Rate the ability (of PIPS) to encourage contractors to perform high quality work	(1–10)*	6
4a	Rate the ability (of PIPS) to encourage users to reduce the amount of management and inspections	(1–10)*	2
4b	What was the percentage decrease?	%	0%
5	Rate the ability (of PIPS) to encourage partnering	(1–10)*	7
6	Would you participate in another PIPS procurement?	Y/N	Y
7	Were you satisfied with PIPS?	Y/N	Y
8	Rate your satisfaction with the low bid system of contracting	(1–10)*	6
9	Rate the ability (of the low-bid process) to encourage contractors to perform high quality work	(1–10)*	6

* 10 = Highest score or strongly agree, 1 = lowest score or strongly disagree

- The end user wanted to use all aspects of PIPS, but the State of Georgia did not change the general conditions of the contract, so the contractor really did not have the opportunity to construct or manage the project according to the PIPS guidelines.
- The constructor was not allowed to pursue any VE items for credits, but instead was forced to do additional changes at no cost. (This is the only part of PIPS that the State heard.)
- The contractor could not make recommendations that would change the design, etc.
- The State only used PIPS to procure a quality contractor to build this project; the process (PIPS) ended at the signing of the contract.

The GSFIC project manager ratings are shown in Tables A18 and A19. He states that (Tremer 2003):

> 'PIPS would be a lot more effective if the state were committed to the process. There are no (up)coming PIPS packages for the State and the contractors know this. There is no incentive to perform under this model.'

The author concludes the following:

- The user was happy with the performing contractor.
- The contractor was happy with the process but was not allowed to service the owner.
- The GSFIC was unable to create a PIPS environment. GSFIC is not committed to the programme and does not understand that its delivery system may be the source of non-performance.

Table A19 Project manager rating of contractor's ability to use PIPS.

No.	Criteria	Unit	Rating
1	Rate the contractor's ability to use the PIPS process	(1–10)*	5
2	Rate the contractor's ability to assist you in increasing quality	(1–10)*	5
3	Rate the contractor's ability to manage the project cost (minimise change orders)	(1–10)*	6
4	Rate the contractor's ability to maintain project schedule	(1–10)*	5
5	Rate the contractor's overall quality of workmanship	(1–10)*	7
6	Rate the contractor's professionalism and ability to manage	(1–10)*	7
7	Rate the contractor's ability to close out the project (no punch list upon turnover, warranties, operating manuals, tax clearance, submitted promptly)	(1–10)*	2
8	Rate the contractor's ability to communicate, explanation of risk, and documentation	(1–10)*	6
9	Rate the contractor's ability to follow the user's rules, regulations, and requirements (housekeeping, safety, etc.)	(1–10)*	7
10	Rate your overall satisfaction and your comfort level hiring the contractor again based on performance	(1–10)*	8

* 10 = Highest score or strongly agree, 1 – lowest score or strongly disagree

Table A20 Analysis of PIPS projects and low-bid projects in Georgia (best available data).

Criteria	Savannah Tech (PIPS)	Georgia Tech	Other project averages
Original contract sum	$8 299 157	$46 595 000	$8 380 387
Adjusted cost	$8 429 198	$48 764 990	$9 085 770
Number of change orders	33	89	40
Original time (days)	425	660	485
Time extension (days)	146	338	120
Percentage increase in time	34%	45%	25%
Percentage increase in cost	1.6%	4.7%	8.4%

- Designers are overdesigning.
- Performance does not cost more.

Comparison with other low-bid projects

The results of the two PIPS projects were compared with 18 projects that are currently 90% complete (GSFIC website). Table A20 shows that even though PIPS was not properly implemented, PIPS may have the capability to increase construction performance (by substantial savings in cost increases). When compared to a 0% increase in cost and 96% on-time record for PIPS projects, there is a potential for higher performance, lower costs and maximised contractor profits if PIPS is used.

Conclusion

The State of Georgia construction projects gave a unique opportunity to:

- Compare the cost of quality with the cost of low bid
- Compare low-bid performance with PIPS performance
- Compare the perception of the user or client, the contractor and the project manager

These conclusions are preliminary, but agree with previous results of PIPS in that PIPS selects high performers and that performance does not cost more. The author concludes that the user's delivery system of management, control and low price is one of the largest sources of non-performance.

References

Burke V. (2003) *Survey Evaluation Form. User Evaluation of Contractor, Low-Bid Process, and Performance-Based Process.* Paul Akins, Inc. Savannah Tech.

Butler J. (2002) Construction quality stinks. *Engineering News Record*, **248** (10), 99.

Futch G. (2003) *Survey Evaluation Form. Contractor Evaluation of Low-Bid Process, Performance-Based Process, and User.* Paul Akins, Inc. Savannah Tech.

Tremer A. (2003) *Survey Evaluation Form. Project Manager Evaluation of Contractor, Low-Bid Process, and Performance-Based Process. Evaluation of PIPS Process.* Georgia State Financing and Investment Commission.

Index